Pseudoelasticity of Shape Memory Alloys

Pseudoelasticity of Shape Memory Alloys

Theory and Experimental Studies

Andrzej Ziółkowski
Department of Mechanics of Materials, Institute of Fundamental Technological Research, Polish Academy of Sciences, Warsaw, Poland

ELSEVIER

AMSTERDAM • BOSTON • HEIDELBERG • LONDON
NEW YORK • OXFORD • PARIS • SAN DIEGO
SAN FRANCISCO • SINGAPORE • SYDNEY • TOKYO
Butterworth-Heinemann is an imprint of Elsevier

Butterworth Heinemann is an imprint of Elsevier
The Boulevard, Langford Lane, Kidlington, Oxford OX5 1GB, UK
225 Wyman Street, Waltham, MA 02451, USA

Library of Congress Cataloging-in-Publication Data
A catalog record for this book is available from the Library of Congress

British Library Cataloguing in Publication Data
A catalogue record for this book is available from the British Library

For information on all Butterworth Heinemann publications visit our website at http://store.elsevier.com/

ISBN: 978-0-12-801697-8

Contents

To view supplementary materials online, please visit: http://booksite.elsevier.com/9780128016978/

About the author

Andrzej Ziółkowski studied engineering sciences at Warsaw Technical University, Warsaw, Poland. In 1983, he obtained a MSc degree in applied mechanics in the Department of Aviation and Power Systems, Warsaw Technical University, and in 1987 he obtained a second MSc degree in power systems and equipment in the same department at Warsaw Technical University upon directing his interests to nuclear power systems engineering. In 1988, he joined the Institute of Fundamental Technological Research, Polish Academy of Sciences (IPPT PAN) for PhD studies, where he has been employed since 1992. In 1996, he obtained a PhD degree for a dissertation devoted to constitutive and numerical modeling of shape memory alloys' behavior, which has been distinguished for excellence by the Scientific Board of IPPT PAN. In 2007, he was awarded a habilitation degree for a dissertation devoted to experimental studies and theoretical description of pseudoelasticity of shape memory alloys.

Dr. Ziółkowski's research interests are in mechanics and thermodynamics of continuous media, structural mechanics, experimental mechanics, thermodynamics of solid-solid phase transitions, and finite element code simulations, but his primary field of specialty is shape memory alloys. He has published a number of papers devoted to experimental investigation, and constitutive and numerical modeling of SMA materials (NiTi mainly). Dr. Ziółkowski was co-organizer of the Shape Memory Alloys Workshop and a member of the scientific committees of the Shape Memory Materials Symposiums of International Conferences. Much of his research work has been conducted within a framework of international collaboration that was facilitated by his short-stay visits in leading research centers such as the University of Grenoble, the University of Besancon, the Mathematical Institute in Bucharest, Metropolitan University of Tokyo, Kyoto University, Catholic University of Leuven, Cambridge University, and Metz University.

Preface

The quest for new materials with properties superior to those already attainable, more useful for a specific purpose, or exhibiting a combination of required qualities is a continuous subject of interest to researchers and engineers all over the world. A very interesting group of metallic materials make up so-called *shape memory alloys* (SMA), revealing a number of extremely useful practical applications, combinations of functional properties, and utility features not observed together in other metallic materials. For that reason, these materials frequently earn such attributes as *smart*, *intelligent*, or they are sometimes called *living composites*. Intelligence and smartness are inherent attributes of living creatures, and they can be inherited by SMA materials probably to the extent that actual users of them exhibit these virtues, but nobody should object when SMA materials are called *adaptive materials*. The nice, from an engineering point of view, features of SMA materials do not come at any cost. Their behavior is *very complex*, and their efficient use requires comprehensive experimental knowledge about their performance when submitted to various thermomechanical loadings, as well as possessing a credible theoretical model that enables SMA materials to have consistent characterization and the quantitative prediction of appearing stresses, deformations, and/or thermal effects.

The aim of this book is to provide experimental evidence of SMA materials' pseudoelastic behavior and a theoretical framework for the efficient and accurate constitutive modeling of SMA materials' pseudoelasticity. The term *pseudoelasticity* is understood here as the loading-unloading cycle involving *hysteresis*, with the underlying assumption that the state of the material remains the same/does not change after termination of such a loading cycle (no change of material properties occurs). In the materials science literature, this range of SMA materials' behavior is frequently called *superelasticity*, the term stemming from the SMA property that elastic strains involved are close to two orders of magnitude larger than in the case of classical engineering materials (e.g., structural steel). The terms *superelasticity* and *pseudoelasticity,* in the sense delineated above, can be treated as equivalent.

The above definition of SMA pseudoelasticity presumes no change of properties of the material due to the loading-unloading cycle; thus, the considerations present in this book do not embrace in a direct way the Mullins effect, which can be treated as a manifestation of pseudoelastic behavior of rubber-like materials and that is caused by progressive damage to the material due to the sequence of loadings. The general philosophy adopted in this book is that we already have an SMA material with stable properties, by whatever means it is achieved. By stable we mean stable in time, in the intended operational regime, which in one case will mean stable for one cycle of work and in another stable during 100-1000 cycles or even 1 million cycles of operation. Whatever the values of modeling parameters and functions characterizing SMA material, they are assumed to be invariable. Thus, the book does not deliberately discuss strategies on how to obtain SMA material with specific features, which the author believes is the primary task of materials science specialists. The procedures of thermomechanical treatments, training procedures of SMA, and evolution of SMA materials properties are not discussed, nor are material functional or utility properties degradation due to cyclic loadings.

The book focusses on delivering experimental evidence about the essential features of macroscopic, thermomechanical behavior of SMA materials. Upon careful analyses and discussion of the experimental results presented, the concepts and ideas are introduced useful in description of macroscopic, pseudoelastic behavior of SMA materials, within the context of *continuum mechanics* and *nonequilibrium thermodynamics*. The worked-out idealizations and simplifications serve as an introduction to and development of a family of so-called R_L models of pseudoelasticity, the first member of which was originally elaborated as a result of Polish-French-Japanese collaboration by Raniecki, Lexcellent, and Tanaka at the beginning of 1990s.

Shape memory effects (SME) of metallic materials, pseudoelasticity among them, originate on physical grounds from thermoelastic, martensitic phase transformation (p.t.). The latent heat of this phase transition is quite considerable in the case of some SMA materials, which results in strong coupling of two physical fields: mechanical and thermal. For example, in the case of the current commercially dominant SMA material (i.e., NiTi alloy), the thermomechanical coupling originating from latent heat of martensitic p.t. may lead, in adiabatic conditions, to an increase of temperature on the order of 50 °C; an effect that can hardly be disregarded.

The formalism of continuum, macroscopic, nonequilibrium thermodynamics with internal state parameters is employed in this book for phenomenological modeling of SMA materials behavior and characterization. Nevertheless, information from all levels of micro-, meso-, and macroscopic observation has been used in building an effective macroscopic theory of pseudoelasticity. Crystallographic and/or compound martensitic variants, objects that can be distinguished at a microscopic scale of observation, are grouped together in mesoscale to distinguish larger generic objects: *self-accommodating* martensite and *oriented* martensite. Information on specific properties and behavior of microlevel and mesolevel objects in their respective scale of observation enters the macroscopic, constitutive model only in indirect, tacit way.

For example, *mesomechanical* studies are executed to obtain a physical interpretation of terms present in a proposed heuristically form of SMA materials *macroscopic Gibbs free energy function*. The so-called representative volume element (RVE) of the SMA material, a material point with homogeneous properties in a macroscopic model, is treated as a two-phase (multiphase) *elastic medium with eigenstrains*, its *microstructure evolving* due to applied thermomechanical loadings. The assumption on continuity of the displacement field in the RVE done in these studies originates from *microscopic* observations of two-phase austenitic, martensitic microstructures. The incompatibility of phase eigenstrains and constraint of continuity of the displacement field required on interphase boundaries are the key sources for *coherency energy* stored in SMA material RVE. In order to determine the RVE macroscopic effective properties, it is actually treated as a composite with frozen—at some arbitrary stage of loading—and very complex mesostructure to which homogenization methodology is employed.

An effective description of any physical phenomenon requires reaching an amicable compromise between model intricacies and the accuracy required from it for specific purposes. It seems that such an amicable compromise has been reached in the family of R_L models of pseudoelasticity thanks to the introduction of only two internal state parameters. The scalar parameter is the *volume fraction of martensite* (oriented one), and the tensorial parameter is the so-called *ultimate phase eigenstrain*. The product of volume fraction of martensite and ultimate phase eigenstrain gives *macroscopic phase strain*. Evolution of these two parameters with external loading of macroscopic stress and temperature allows predicting the value of macroscopic phase strain describing deformation effects of SMA materials connected with martensitic phase formation, its presence and its mesostructural evolution—reorientation of martensitic phase.

Extensive study and analysis of experimental data on the *macroscopic* behavior of SMA materials in the pseudoelastic range of their behavior, and the belief that matter organization is subject to some higher ordering principles, led to the formulation of rules governing the evolution of internal state parameters.

The so-called rule of *optimum mesostructure rearrangement* of SMA material RVE has been noticed and worked out. It is presumed valid in this entire book. Exploitation of this law leads to the property that ultimate phase eigenstrain tensor always follows/has the direction of macroscopic stress tensor. As a consequence, R_L model theory predicts that macroscopic phase strain tensor always has the direction of macroscopic stress tensor in the pseudoelastic range of SMA materials behavior.

A primary question arises regarding what is actually the key reason for the appearance of pseudoelastic behavior—hysteresis loop in SMA materials behavior, taking into account the axiom posed earlier that upon the loading-unloading cycle the properties and the state of SMA material remains unchanged. In that respect, the fundamental conjecture is adopted that it is *phase instability* of RVE's mesostructural arrangement that causes the appearance of pseudoelastic behavior. Unstable phase equilibrium states lead to the abrupt initiation of qualitative reorganization of SMA material internal mesostructure. The conjecture is used to formulate the criteria of initiation of forward and reverse phase transitions directly, consistently, from the adopted form of Gibbs potential. No ad hoc threshold values are proposed for phase transition initiation analogous to the theory of plasticity, as is frequently done in the literature on modeling SMA materials.

Subsequent progress of forward or reverse martensitic phase transformations evolution of volume fraction of martensitic phase is governed by the kinetics rule of phase transitions. This law has been elaborated, taking advantage of the information that thermoelastic martensitic phase transition is athermal (i.e., it does not depend on time rate), and upon the analysis of experimental data. A specific mathematical form of this rule is adopted, which is responsible for the shape and size of the appearing hysteresis loops. The p.t. kinetics rule embraces a description of internal hysteresis loops. As mentioned earlier, strain effects resulting from reorientation of the already born martensitic phase are governed by optimum microstructure rearrangement principle.

In the broader perspective presented in this book, studies on the pseudoelastic behavior of SMA materials can be perceived as a theory of hysteresis loop; though, it is of a different character than the one exhibited by classical magnetic materials. The description of the pseudoelasticity of SMA materials proposed here is very attractive and opens up very promising new research directions and areas. Relatively uncomplicated adaptations allow reusing considerable parts of already developed theoretical apparatus to describe macroscopic pseudoelastic behavior encountered in magnetic SMA, light-induced shape memory materials, shape memory polymers, biological tissues, sponge and foam materials, and rubber-like materials in general materials exhibiting hysteresis loop under a mechanical load-unload cycle. Specific physical mechanisms underlying pseudoelastic behavior can be completely different in each specific case.

The adaptation process can be straightforward, thanks to the versatility of formalism of nonequilibrium thermodynamics with internal state parameters, which is presented in detail in this book. Adaptation of the present theory for the description of behavior of polymeric SMA, in addition to the modification of kinetics relations describing the progress of glass transition fixing/releasing process, will also require introduction of an additional internal state variable to relevantly describe the evolution of viscoelastic deformation effects. Adaptation of the theory to describe the Mullins effect—pseudoelasticity of rubber-like materials—will require at least replacement of the volume fraction of martensitic phase with some other internal state parameter correctly defined to reflect damage of the internal structure of the rubber material, as well as the discovery and development of its evolution rule.

This monographic book on the pseudoelasticity of SMA can serve well those who are novices in the field, as it provides extensive background information up to the present state of the art in SMA materials. Beginning with a bird's-eye view history of SMA materials development, it explains and/or precisely defines SME in thermomechanical terms, and discusses where, how, and why special behavior of SMA found numerous applications in various engineering areas. The book is valuable for students and faculty, as it explains SME at a fundamental level. It contains a survey of modeling approaches targeted at the reliable prediction of SMA materials behavior on different scales of observation: atomistic, microscopic, mesoscopic, and macroscopic.

Doctoral students and young researchers can find very detailed information on modern methodology used in the process of building constitutive models of advanced materials exhibiting complex behavior; embracing several physical fields—here thermal and mechanical—with special attention on grasping the relationships between the internal organization of materials and their thermomechanical properties. Advanced senior researchers and practicing engineers can find information on the philosophical approaches and many details used to pose theoretical modeling assumptions, leading to considerable simplifications in material model mathematical formulas and, thus, more effective computational models.

Engineering practitioners and designers may find the book to be a helpful, practical tool in carrying out estimates on the applicability of projected engineering solutions grounded on taking advantage the special features of SMA materials using such software as Mathcad and constitutive relations of the R_L models family. Complex, challenging, highly responsible devices made of or including components made of SMA materials usually require finite element code studies. In the web page accompanying the book (http://booksite.elsevier.com/9780128016978/) for demonstration purposes, detailed information is delivered on the process of implementation of R_L model relations into commercial finite element code in the case of axis symmetric problems. The code allows for running fully coupled thermomechanical analyses in which mechanical and thermal effects connected with SMA materials behavior can be reliably evaluated.

Successful exploitation of SMA materials requires extensive interdisciplinary collaboration. It is the hope of the author that this book can fill the gap between mechanicians, physicists, chemists, materials science specialists, and other experts interested in SMA materials, including medical doctors making use of NiTi implants and devices—probably the major and most important current commercial application area of SMA materials (NiTi-based alloys). This is achieved by creating a base of well-defined set of common concepts, objects, and ideas.

This book has evolved from the present author's over 20 years of research work on SMA materials, documented in scientific papers, and his doctoral and habilitation thesis, run in the Institute of Fundamental Technological Research, Polish Academy of Sciences, Warsaw, Poland.

As with any project of this size there are a number of people to thank. The first person is my mentor and afterward colleague professor Bogdan Raniecki, who introduced and inspired me into the topic of SMA. Discussions and disputes with him, looking from perspective, delivered me a lot of happiness and satisfaction. I had the privilege to experience what I believe had been experienced by Greek peripatetics in the flourishing period of ancient Greece. I wish to thank professor Lech Dietrich for his unconditional support and advice on issues connected with experimental work on SMA alloys. I express my gratitude to all colleagues who entered with me into lengthy, detailed, and sometimes energetic discussions, which I believe make the essence of academic life. Such discussions create a kind of mutual mirror for working out viable ideas, because before you can convince somebody else you must work out arguments to convince yourself.

Many of the results contained in the book have originated from studies performed over the years in cooperation with researchers from France and Japan: Professor Christian Lexcellent, Kiki Tanaka, Hisaaki Tobushi, Suichi Miyazaki. I would like to express my sincere gratitude to all of these individuals for their cooperation and support. I am grateful to Steven Merken and Jeffrey Freeland from Elsevier for their commitment and cooperation to make this project become reality. I also wish to thank Sruthi Satheesh for expertly running production process of the book.

Andrzej Ziółkowski, Warsaw, Poland, February, 2015

List of symbols

$\mathcal{A}$	*austenite* (parent phase). Phase stable at high temperatures with high symmetry crystallographic lattice
A_s^0	austenite start temperature of self-accommodating martensite (Note: superscript "0" standing at phase transition characteristic temperatures denotes that the measurement took place at zero macroscopic stress; these are simply classical DSC-determined temperatures $\left(A_s^0 = A_s\right)$.
A_s^{2-1}	austenite start temperature of oriented martensite $\left(A_s^{2-1} > A_s^0\right)$
A_f^0	austenite finish temperature of self-accommodating martensite
A_f^{2-1}	austenite finish temperature of oriented martensite $\left(A_f^{2-1} > A_f^0\right)$
$\breve{a}(\boldsymbol{x})$	any nonhomogeneous mesoscopic field in the RVE; for example, stress, strain, or elastic properties $\breve{\boldsymbol{\sigma}}(\boldsymbol{x})$, $\breve{\boldsymbol{\sigma}}(\boldsymbol{x})$, $\breve{\mathbf{L}}(\boldsymbol{x})$
$\boldsymbol{\alpha} = \alpha\mathbf{I}$	isotropic thermal expansion tensor
α	thermal expansion coefficient
$\alpha = 1$	numerical denotation of parent phase (austenite)
$\alpha = 2$	numerical denotation of oriented martensitic phase (oriented, detwinned martensitic compound)
$\alpha = 3$	numerical denotation of self-accommodating martensitic phase (self-accommodating, twinned martensitic compound)
${}^0\mathcal{B}$	initial configuration of material body at time instant $t = t_0$
${}^t\mathcal{B}$	configuration of material body at some time instant t
${}^t\mathcal{B}^{\text{in}}$	instantaneous natural configuration at time t, that is, configuration at state $\boldsymbol{\sigma} = \mathbf{0}, T = T_0, z = z_t, h_K = h_K(t)$; also called unloaded elastically configuration
$\boldsymbol{\mathcal{B}}_\alpha$	dimensionless stress concentration factor tensors
c_v	heat capacity at constant volume
c_p	heat capacity at constant pressure
D	normalized thermodynamic driving force of phase transition $D \equiv \pi^{1-2}/\pi_\text{N}^{1-2}$; π_N^{1-2} (normalizing factor)
$\boldsymbol{d}$	Eulerian strain rate tensor (stretching rate tensor)
$\boldsymbol{d}^\text{e}$	elastic stretching rate tensor
$\boldsymbol{d}^\text{in}$	inelastic stretching rate tensor
$\boldsymbol{E}(n)$	Lagrangean strain measure from Hill's family
$\boldsymbol{E}^\text{e}(1)$	elastic Green-Lagrange strain
$\boldsymbol{E}^\text{e}(0)$	elastic logarithmic strain
$\boldsymbol{e}(n)$	Eulerian strain measures
$\boldsymbol{e}^\text{e}(0)$	elastic Eulerian logarithmic strain
ε_eq	equivalent strain $\varepsilon_\text{eq} = \sqrt{\frac{2}{3}\bar{\varepsilon}_{ij}\bar{\varepsilon}_{ij}}$
$\boldsymbol{\varepsilon}$	small strains tensor
$\boldsymbol{\varepsilon}^\text{e}$	macroscopic elastic strain

$\boldsymbol{\varepsilon}^{\mathrm{pt}}$	macroscopic phase transition strain; $\boldsymbol{\varepsilon}^{pt} \equiv z\boldsymbol{\kappa}$
$\boldsymbol{\mathcal{E}}^{\mathrm{de}}$	elastic Zaremba-Jaumann of elastic logarithmic strain and elastic stretching rate connection tensor
$f(y)$	"shape" function of limit surface of pseudoelastic flow
$\boldsymbol{F}$	deformation gradient $\boldsymbol{F} \equiv \mathbf{d}\boldsymbol{x}/\mathbf{d}\boldsymbol{X}$
ϕ	macroscopic specific Helmholtz function; specific Helmholz free energy
ϕ^{tm}	thermoelastic part of free energy
ϕ^{coh}	specific coherency energy
ϕ_{it}	specific interaction energy of austenitic and martensitic phase
ϕ_2^{st}	energy stored in SMA macroelement after full completion of martensitic phase transformation
g	macroscopic, specific Gibbs function; specific Gibbs free energy; specific free enthalpy (J/kg)
g^{ult}	ultimate phase eigenstrain potential
γ	amplitude of pseudoelastic flow in pure shear
$\breve{\boldsymbol{\Gamma}}(\boldsymbol{x})$	local eigenstrain field in RVE, piece-wise continuous; strain field connected with martensitic phase transition
$\breve{\boldsymbol{\Gamma}}_{\alpha}(\boldsymbol{x})$	eigenstrain field of phase α in RVE; local eigenstrain field of austenitic phase is always zero, $\breve{\boldsymbol{\Gamma}}_1(\boldsymbol{x}) \equiv \mathbf{0}$ for $\boldsymbol{x} \in V_1$
$\boldsymbol{\Gamma}_{\boldsymbol{\alpha}}$	macroscopic value of phase α eigenstrain. Average value over volume of phase α in RVE—$\boldsymbol{\Gamma}_{\alpha} \equiv \left\langle \breve{\boldsymbol{\Gamma}}(\boldsymbol{x}) \right\rangle_{V_\alpha}$. Macroscopic eigenstrain of austenitic phase and self-accommodating martensitic phase (compound) is always zero $\boldsymbol{\Gamma}_1 = \boldsymbol{\Gamma}_3 \equiv \mathbf{0}$
h_K	a set of independent, mesoscopic, internal variables $(K = 1, 2, \ldots, N)$, describing evolving geometry of mesostructure of SMA material RVE
J	determinant of deformation gradient $(J \equiv \det(\boldsymbol{F}) = \rho^0/\rho)$
J^{e}	determinant of elastic deformation gradient $J^{\mathrm{e}} \equiv \det(\boldsymbol{F}^{\mathrm{e}})$
K	bulk modulus of elasticity $K = E/3(1-2\nu)$
$\boldsymbol{\kappa}$	macroscopic ultimate phase eigenstrain tensor of oriented martensite $(\mathrm{tr}(\boldsymbol{\kappa}) = 0)$.
l_f, $\overset{*}{l}_f$	actual latent heat of complete austenite into oriented martensite p.t., irreversible and reversible thermodynamically, respectively
l_f^{sa}, $l_f^{*\mathrm{sa}}$	actual latent heat of austenite to self-accommodating martensite p.t. irreversible and reversible thermodynamically, respectively
$\mathbf{L}$	tensor of elastic stiffness properties $(\mathbf{L} = \rho\partial^2\phi/\partial\boldsymbol{e}^{\mathrm{e}}(0)\partial\boldsymbol{e}^{\mathrm{e}}(0))$ or, in numerical terms current, instantaneous, elastic tangent $\partial\boldsymbol{\sigma}/\partial\boldsymbol{e}^{\mathrm{e}}(0) = \mathbf{L}$
$\mathbf{L}_{\alpha}$	tensor of elastic properties of specific phase
$\mathbf{L}^{(\mathrm{Je})}$	current, instantaneous, elastic Zaremba-Jaumann rate tangent $\left(\mathbf{L}^{(\mathrm{Je})} \equiv \mathbf{L}\left[\mathbf{I}^{(4)} + \boldsymbol{\mathcal{E}}^{\mathrm{de}}\right]; J\mathbf{L}^{(\mathrm{Je})} = \partial\left(\overset{\circ}{\boldsymbol{\tau}}{}^{(\mathrm{Je})}\right)/\partial\boldsymbol{d}^{\mathrm{e}}\right)$
$\boldsymbol{L}$	Eulerian velocity gradient tensor
λ_i^{e}	principal elastic stretches
$\{\boldsymbol{m}_i\}$	versors of director triad determining orientation of natural configurations ${}^t\mathcal{B}^{\mathrm{in}}$ in fixed coordinates frame
$\mathcal{M}$	martensite. Phase stable at low temperatures with lower symmetry crystallographic lattice.
$\mathcal{M}_{\mathrm{or}}$	oriented martensite; detwinned martensite $(\mathcal{M}_{\mathrm{or}} \Leftrightarrow \mathcal{M}_2)$
$\mathcal{M}_{\mathrm{sa}}$	self-accommodating martensite; twinned marensite $(\mathcal{M}_{\mathrm{sa}} \Leftrightarrow \mathcal{M}_3)$
M_{d}	limit temperature above which it is impossible to stress-induce martentic transformation

M_s^0	self-accommodating martensite start temperature; throughout the book it is accepted that it is the same as oriented martensite start temperature $\left(M_s^0 = M_s^{1-2}\right)$
M_f^0	martensite finish temperature
M_σ	temperature above which first plastic slip deformation is stress-induced and only later martensitic transformation takes place
$\mathbf{M}$	tensor of elastic compliance properties $\left(\mathbf{M} = \mathbf{L}^{-1}\right)$
μ	modulus of shear elasticity $\mu = E/2(1+\nu)$
$\boldsymbol{N}$	nominal stress tensor
$\boldsymbol{n}_i^e$	Eulerian (spatial) principal directions of elastic stretch tensors $\boldsymbol{V}^e$
$\boldsymbol{N}_i^e$	Lagrangean (material) principal directions of elastic stretch tensors $\boldsymbol{U}^e$
ν	Poisson coefficient
π^{1-2}	thermodynamic driving force of austenite-oriented martensite phase transformation $(\mathcal{A} \rightarrow \mathcal{M}_{or})$
π^{1-3}	thermodynamic driving force of austenite-self-accommodating martensite phase transformation $(\mathcal{A} \rightarrow \mathcal{M}_{sa})$
π_u^{1-2}	thermodynamic driving force at zero macroscopic stress
π_0^{1-2}	chemical driving force (t.d.f.) of phase transformation $(\mathcal{A} \rightarrow \mathcal{M}_{or})$
$\overset{(*)}{\boldsymbol{\Pi}}$	macroscopic nominal stress rate $\overset{(*)}{\boldsymbol{\Pi}} \equiv J^{-1}\boldsymbol{F}\dot{\boldsymbol{N}} = \dot{\boldsymbol{\sigma}} - \boldsymbol{L}\boldsymbol{\sigma} + \boldsymbol{\sigma}\mathrm{tr}(\boldsymbol{d})$
q	specific energy exchanged with environment in the form of heat (q has positive sign, when energy is removed from the macroelement)
$\boldsymbol{R}$	rotation tensor from polar decomposition of deformation gradient
$\boldsymbol{Q}$	proper orthogonal rotation tensor $\left(\boldsymbol{Q}\boldsymbol{Q}^T = \mathbf{I}, \det(\boldsymbol{Q}) = 1\right)$
ρ	mass density; throughout the book it is accepted that density of austenitic and martensitic phase is the same $\rho_1 = \rho_2 = \rho$
ρ_0	mass density in initial configuration
s	macroscopic specific entropy
s_1^0	entropy of austenite formation at microstress free state (m.r.s.), $\left(\breve{\boldsymbol{\sigma}}(\boldsymbol{x}) = \mathbf{0}, T = T_0\right)$
s_α^0	entropy of martensitic compound formation at microstress free state; $s_2^0 = s_3^0$
s_{12}	entropy material constant
$\Delta s^0 \equiv s_\alpha^0 - s_1^0$	entropy of formation of any object of martensite at microstress free state (m.r.s.), $(\alpha = 2,3)$
$\boldsymbol{\sigma}^{AM}$	critical stress of start of austenite to oriented martensite p.t.
$\boldsymbol{\sigma}^{MA}$	critical stress of start of reverese p.t. oriented martensite to austenite
$\boldsymbol{\sigma}$	macroscopic Cauchy stress
$\breve{\boldsymbol{\sigma}}(\boldsymbol{x})$	local mesoscopic Cauchy stress field
$\overset{o}{\boldsymbol{\sigma}}(0)$	Zaremba-Jaumann rate of Kirchoff stress referred to actual configuration $\left(\overset{o}{\boldsymbol{\sigma}}^{(0)} = J^{-1}\overset{o}{\boldsymbol{\tau}}^{(J)} = \dot{\boldsymbol{\sigma}} + \boldsymbol{\sigma}\boldsymbol{\omega} - \boldsymbol{\omega}\boldsymbol{\sigma} + \boldsymbol{\sigma}\mathrm{tr}(\boldsymbol{d})\right)$
σ_{ef}	effective stress of Cauchy stress tensor $\left(\sigma_{ef} \equiv \sqrt{3J_2} = \left[(3/2)\overline{\sigma}_{ij}\overline{\sigma}_{ij}\right]^{1/2}\right)$
$\boldsymbol{t}^{(b)}(\boldsymbol{x})$	boundary surface tractions
T	temperature
T_0	conventional reference state temperature
T_{or}^{eq}	equilibrium temperature of austenite and oriented martensite $\left(T_{or}^{eq} \equiv \Delta u_2^0/\Delta s_2^0\right)$
$\boldsymbol{T}(n)$	stress measure work conjugate to Lagrangean total strain measure $\boldsymbol{E}(n)$
$\boldsymbol{T}^e(1)$	elastic II Piola-Kirchoff stress $\left(\boldsymbol{T}^e(1) \equiv {}^{-1}\boldsymbol{F}^e\boldsymbol{\tau}^{-T}\boldsymbol{F}^e\right)$
$\boldsymbol{\tau}$	Kirchoff stress $(\boldsymbol{\tau} \equiv J\boldsymbol{\sigma})$
$\overset{o}{\boldsymbol{\tau}}^{(J)}$	Zaremba-Jaumann corotational derivative of Kirchoff stress tensor $\left(\overset{o}{\boldsymbol{\tau}}^{(J)} = \dot{\boldsymbol{\tau}} + \boldsymbol{\tau}\boldsymbol{\omega} - \boldsymbol{\omega}\boldsymbol{\tau} = (\rho^0/\rho)\left[\dot{\boldsymbol{\sigma}} + \boldsymbol{\sigma}\boldsymbol{\omega} - \boldsymbol{\omega}\boldsymbol{\sigma} + \boldsymbol{\sigma}\mathrm{tr}(\boldsymbol{d})\right]\right)$

$\overset{\circ}{\boldsymbol{\tau}}^{(\mathrm{Je})}$	elastic Zaremba-Jaumann corotational derivative of Kirchoff stress tensor $\overset{\circ}{\boldsymbol{\tau}}^{(\mathrm{Je})} = \dot{\boldsymbol{\tau}} + \boldsymbol{\tau}\boldsymbol{\omega}^e - \boldsymbol{\omega}^e\boldsymbol{\tau}$
u	specific internal energy of macroelement
u_1^0	internal energy of austenite at mesostress free referènce state (m.r.s.), $(\breve{\boldsymbol{\sigma}}(\boldsymbol{x}) = 0, T = T_0 \Rightarrow \boldsymbol{\sigma} = \mathbf{0},\ \breve{\boldsymbol{\sigma}}^f(\boldsymbol{x}) = \mathbf{0})$
u_α^0	internal energy of martensite at mesostress free reference state $(\alpha = 2,3)$; it is adopted in this book that $u_2^0 = u_3^0$
$u_2^0 - \phi_2^0$	internal energy of oriented martensite at t.r.s. $(\boldsymbol{\sigma} = \mathbf{0}, T = T_0)$
$u_3^0 - \phi_3^0$	internal energy of self-accommodated martensite at t.r.s.
$\Delta u^0 \equiv u_\alpha^0 - u_1^0$	internal energy of formation of any object of martensite at microstress free state (m.r.s.), $\left(\breve{\boldsymbol{\sigma}}(\boldsymbol{x}) = \mathbf{0}, T = T_0\right)$ $(\alpha = 2,3)$
$\breve{\boldsymbol{u}}(\boldsymbol{x})$	local displacement field
$\boldsymbol{U}^{\mathrm{e}}$	right elastic stretch tensor
$\boldsymbol{V}^{\mathrm{e}}$	left elastic stretch tensor
V_α	volume fraction of phase (α) in RVE
w	specific external mechanical work (done over macroelement) (J/kg)
w_{Disp}	specific mechanical work dissipation
w_{r}	specific mechanical work done over macroelement in equivalent thermodynamically reversible process
$\boldsymbol{\omega}$	material spin (vorticity) tensor $\left(\boldsymbol{\omega}^Q = \dot{\boldsymbol{Q}}\boldsymbol{Q}^{\mathrm{T}}\right)$
$\boldsymbol{\omega}^{\log}$	logarithmic spin tensor
$\boldsymbol{x}$	coordinates of material point in Eulerian description
$\boldsymbol{X}$	coordinates of material point in Lagrangean description
$Y_{(\alpha)}$	lag threshold value functions of delay in phase transitions start
y	Lode parameter characterizing direction of stress tensor direction $y = \cos(3\theta) = 27J_3/\left(2\sigma_{\mathrm{ef}}^3\right) = 3\sqrt{3}J_3/\left(2J_2^{3/2}\right)$
z_1	mass (volume) fraction of austenitic phase
z_2 or z	mass (volume) fraction of oriented martensite phase
z_3	mass (volume) fraction of self-accommodated martensite phase. (Note: In the pseudoelastic range of SMA behavior no self-accommodating martensite *can ever form*, for this reason throughout this book it is $z_3 = 0$, $z_1 + z_2 \equiv 1$, and subscript "2" is usually omitted (i.e., $z_2 = z$).)

Notation

$\mathbf{I}$ unit tensor	$\mathbf{I} \leftrightarrow \delta_{ij}$, $\left(\mathbf{I}^{4s}\right)_{klmn} \leftrightarrow 0.5(\delta_{km}\delta_{ln}+\delta_{kn}\delta_{lm})$
$\boldsymbol{A}^{T}$ transpose of tensor	$\boldsymbol{A}\left(\boldsymbol{A}^{\mathrm{T}}\right)_{ij}=A_{ji}$ or $\left(\boldsymbol{A}^{\mathrm{T}}\right)_{klmn}=A_{mnkl}$
A^{-1} inverse of tensor	$\boldsymbol{A}A_{kl}^{-1}A_{lm}=\delta_{km}$ or $A_{klmn}^{-1}A_{mnpr}=I_{klpr}$
$\boldsymbol{A}\otimes\boldsymbol{B}$ dyadic product	A_iB_j
$\boldsymbol{A}\boldsymbol{B}$ contraction product	$A_{kl}B_l$ or $A_{kl}B_{lm}$ or $A_{klmn}B_{mn}$ or $A_{klmn}B_{mnpr}$
$\boldsymbol{A}\cdot\boldsymbol{B}$ dot contraction product	A_iB_i or $A_{ij}B_{ij}$ $\left[\leftrightarrow \mathrm{tr}\left(\mathbf{A}\mathbf{B}^{\mathrm{T}}\right)\right]$
$\mathrm{tr}(AB)$ trace operation	$A_{ij}B_{ji}$
$(!\,\alpha)$	no summation over index α
$\boldsymbol{A}=\left\langle\breve{\boldsymbol{A}}(\boldsymbol{x})\right\rangle_V \equiv \frac{1}{V}\oint_V \breve{\boldsymbol{A}}(\boldsymbol{x})\mathrm{d}v$	total average of local field over volume of the body $\boldsymbol{x}\in V$
$\breve{\boldsymbol{A}}^{\mathrm{f}}(\boldsymbol{x})\equiv\breve{\boldsymbol{A}}(\boldsymbol{x})-\boldsymbol{A}$	fluctuation field with respect total average $\left\langle\breve{\boldsymbol{A}}^{\mathrm{f}}(\boldsymbol{x})\right\rangle_V=0$
$\boldsymbol{A}_\alpha\equiv\left\langle\breve{\boldsymbol{A}}_\alpha(\boldsymbol{x})\right\rangle_{V_\alpha}\equiv\frac{1}{V_\alpha}\int_{V_\alpha}\breve{\boldsymbol{A}}_\alpha(\boldsymbol{x})\mathrm{d}x$	phase average of local field over a region occupied by phase α $(\boldsymbol{x}\in V_\alpha)$
$\breve{\boldsymbol{A}}^{\mathrm{f}}_{(\alpha)}(\boldsymbol{x})\equiv\breve{\boldsymbol{A}}(\boldsymbol{x})-\left\langle\breve{\boldsymbol{A}}(\boldsymbol{x})\right\rangle_{V_\alpha}=\breve{\boldsymbol{A}}(\boldsymbol{x})-\boldsymbol{A}_{(\alpha)}$	phase fluctuation field with respect to its phase average $\left\langle\breve{\boldsymbol{A}}^{\mathrm{f}}_{(\alpha)}(\boldsymbol{x})\right\rangle_{V_\alpha}=0$
$\lVert\boldsymbol{a}(\boldsymbol{x})\rVert\equiv\left[\breve{\boldsymbol{a}}_2(\boldsymbol{x})-\breve{\boldsymbol{a}}_1(\boldsymbol{x})\right]\big\rvert_{\boldsymbol{x}\in\mathcal{S}}$	jump of a field on a surface $\mathcal{S}$

Abbreviations

1D one-dimensional
2D two-dimensional
3D three-dimensional
B2 cubic crystallographic structure of austenitic phase, CsCl type ordered structure
B19 orthorhombic crystallographic structure of martensitic phase
B19′ monoclinic crystallographic structure of martensitic phase
CW clockwise direction of executing multiaxial loading program
CCW counterclockwise direction of executing multiaxial loading program
CV crystallographic martensitic variant
DOF degree of freedom
DSC differential scanning calorimetry
FEM finite element method
FEA finite element analysis
HPV habit plane variants of martensite
ppm parts per million
p.t. phase transition
MD molecular dynamics method
MS melt spinning
m.r.s. mesoscopic thermodynamic reference state $(\breve{\boldsymbol{\sigma}}(\boldsymbol{x})=\mathbf{0}, T=T_0)$; fluctuation of stress field in (conceptual) martensitic phase at m.r.s. is zero $\breve{\boldsymbol{\sigma}}^{f}(\boldsymbol{x})=\mathbf{0}$
PTMC phenomenological theory of martensitic crystallography
R rhombohedral crystallographic structure of martensitic phase, also called R-phase (trigonal martensite phase)
RVE representative volume element
SD strength differential effect; a specific behavior of some materials exhibiting dependence of the value of critical flow stress on loading stress tensor direction. Colloquially, tension compression flow stress difference
SMA shape memory alloys
SME shape memory effects
TMT thermomechanical treatment
TRC twin roll casting
t.d.f. thermodynamic driving force
t.r.s. macroscopic thermodynamic reference state $(\boldsymbol{\sigma}=\mathbf{0}, T=T_0)$; it is assumed that RVE at t.r.s. is composed of pure austenitic phase $(z_1=1)$, that is, $T_0 \geq A_f^0$. Please note that fluctuation of the stress field in (conceptual) martensitic phase at t.r.s. is usually nonzero, $\breve{\boldsymbol{\sigma}}^{f}(\boldsymbol{x}) \equiv \breve{\boldsymbol{\sigma}}(\boldsymbol{x})-\boldsymbol{\sigma} \neq 0$ due to the presence of local a nonzero incompatible eigenstrain field $\breve{\boldsymbol{\Gamma}}(\boldsymbol{x}) \neq 0\ \boldsymbol{x} \in V_{2,3}$
U.B.T. uniform boundary traction conditions
VAR vacuum arc melting
VIM vacuum induction melting

Introduction

1

1.1 Shape memory effects and their technical significance

A broad stream of scientific research and numerous extremely complex projects, both scientific and industrial, are targeted at the discovery of new, advanced materials, as well as unearthing special properties of traditional engineering materials. Today, whenever information about some new material with astounding properties is reported, the presumption is that it is a man-made material. The most well known of such materials invented, designed, and manufactured by man, and already with a very strong position in engineering practice, are most likely plastics and/or composites.

An interesting and relatively new concept receiving attention from scientists and engineers all over the world is the concept of so-called smart, *adaptive or active materials*—these descriptions are treated here as more or less equivalent. The field of *smart materials*, also smart systems, has enjoyed more than two decades of extremely dynamic development and, due to its interdisciplinary or frequently multidisciplinary character, has proved to be very prolific in generating extremely interesting engineering solutions; see, for example, Schwartz (2002). The feature distinguishing smart materials from "common" ones is that the former have capability for active, purposeful adaptation or change in their behavior or properties in response to varying external environmental conditions. The active material must possess simultaneously a number of mutually coordinated functional properties in order to make such adaptation possible. These properties include the ability to sense changes in the environment, the ability to execute specific reactions in response to detected changes, and the ability to control of such reactions. Additionally, the material reaction requires providing an external or internal source of power; see Figure 1.1.

Current realizations of the concept of smart, active materials rely on use of the effects of couplings between various physical fields, for example, the phenomenon of piezoelectricity (coupling of electric and mechanical fields), the phenomenon of chemoluminescence (coupling of chemical and optical fields), and thermoelasticity (coupling of thermal and mechanical effects). Intelligent material adaptive capabilities consist of, for example, change of its properties, internal structure, functional characteristics, and execution of work in reaction to change of specific external conditions. A thermoreactive paint is a spectacular example of an adaptive material that, after exceeding a certain temperature, changes its color.

Special properties of active materials, their multifunctionality, very often allow for considerable simplification of the construction of an engineering device, which would otherwise have to be very complex if the possibility of using such materials is precluded.

Pseudoelasticity of Shape Memory Alloys. http://dx.doi.org/10.1016/B978-0-12-801697-8.00001-6

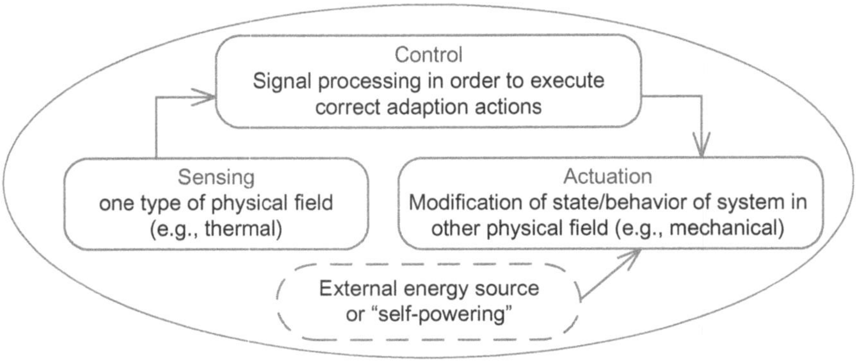

Figure 1.1 Schematic illustration of general principle of smart or adaptive material operation.

Shape memory alloys (SMA) are regarded as one of the most important adaptive materials. The special functional effects that distinguish SMA materials as a generic class of materials are pseudoelasticity, one-way memory effect, and two-way memory effect. These *shape memory effects* (SME), are illustrated schematically in Figure 1.2. It is actually the so-called *one-way shape memory* effect that earned shape memory alloys their generic name; see Figure 1.2b. This effect consists of the following behavior: when SMA material is apparently permanently deformed in a cold temperature by some mechanical loading-unloading cycle, then it reclaims its permanent deformation at no mechanical load and returns to its original shape upon heating it to a hot temperature. The complete shape recovery takes place by heating cold, deformed SMA material above a certain characteristic temperature, different for different types of shape memory alloys. Typically, in the case of NiTi alloys strains, which may be recovered, they are of the order of 8%, a value two orders larger than typical elastic strains of structural steel.

Pseudoelasticity is another extraordinary effect exhibited by shape memory alloys not encountered in the behavior of "conventional" metallic materials. It takes place at temperatures higher in comparison to those in which one-way shape memory effect can be produced for a particular SMA material. When material submitted to mechanical loading yields flows several percent upon exceeding certain critical stress to recover this strain completely following a different path upon unloading, then we speak about pseudoelastic behavior of the material; see Figure 1.2a. The prefix *pseudo-* in the name of the phenomenon emphasizes the fact that the loading-unloading cycle of SMA material, differently from for example non-linear elastic material, exhibits hysteresis loop in stress-strain space. The size of the hysteresis loop depends on many factors, for example, the specific chemical composition of SMA material and/or its thermomechanical treatment.

We speak about a *two-way shape memory* effect when material at no mechanical loading changes its shape from high-temperature shape to low-temperature shape by simple changing of the temperature; see Figure 1.2c. This happens upon lowering temperature below a certain characteristic low temperature and upon heating the material

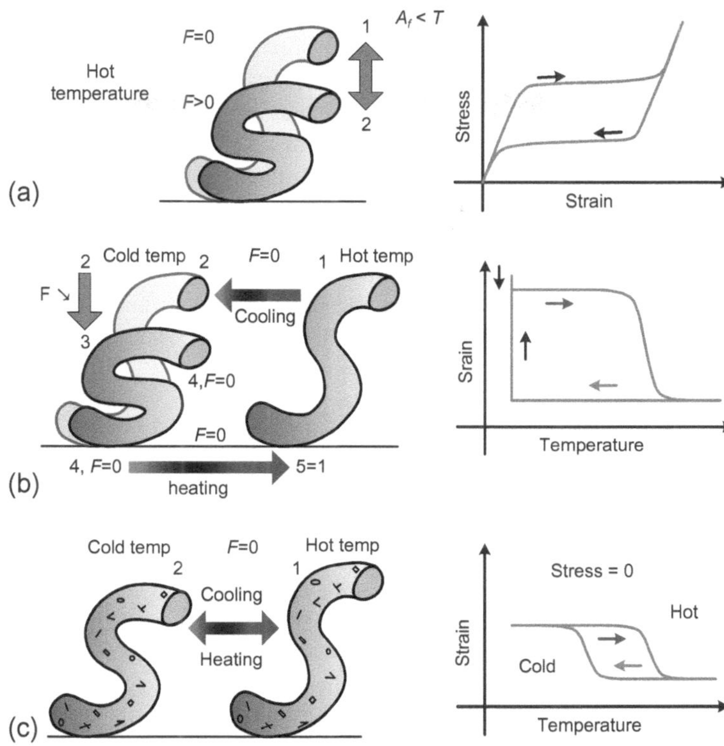

Figure 1.2 Illustrative, schematic demonstration of basic shape memory effects. (a) Pseudoelasticity; (b) One-way memory effect; (c) Two-way memory effect.

above a certain characteristic high temperature. The two-way shape memory effect can be obtained in SMA only after special thermomechanical treatment (TMT), sometimes called *training*. Range of strain difference during two-way shape memory effect strongly depends on the specific SMA alloy and at present for NiTi-based alloys it is typically at a level of 2%.

It is worth mentioning that another very interesting and useful class of materials exhibiting shape memory effects finding broad application areas in medicine, food, and/or textile industries are shape memory polymers (SMP) materials. One example that is commonly used in households is thermoshrinkable wraps belonging to this class of materials. An especially attractive feature of SMP materials is their biodegradability. The interested reader can find broad information about these materials, for example, in a book by Hu (2007). While an exhaustive discussion of SMP materials falls beyond the scope of this book due to limited space and because underlying physical mechanisms of shape memory phenomena in SMA and SMP materials are completely different, they are mentioned here because the author believes that the modeling approach presented here for SMA can be successfully applied to the description of SMP materials.

Special properties of SMA materials stimulate interest in their practical applications. In 1999, it was estimated that the number of patents in which the principal subject was the exploitation of special properties of SMA materials, exceeded 10,000 (Melton, 1999). By the year 2014, the worldwide number of patents on SMA materials is estmated to exceed 20,000 as reported by Jani et al. (2014).

Initially, SMA materials found applications in rather simple devices such as joints, clamps, pins, cellular phone antennas, or glass frames. Later, SMA applications gradually evolved toward *actuators* where special properties of the material are used for sensing conditions of a changing environment and only after fulfilling certain criteria is there an active reaction of the actuator. A suitable control function necessary for smart operation is obtained by special design (see Figure 1.1). At present the whole subassemblies fulfilling the role of mechanisms are constructed with integrated elements made of shape memory alloy, which when built from traditional engineering materials will require many more parts remaining in complex interrelations for their construction and required operation.

SMA materials find use in the most advanced and responsible applications not only due to their *functional properties* of SME but also due to the *utility properties* required from them. For example NiTi shape memory alloys find extremely widespread use in medicine because they are biocompatible. This feature enables implants, devices by their very nature requiring a long stay in body fluids, to be made of NiTi alloys. Another attractive feature is that behaviors of NiTi-based materials closely resemble behavior of other biological tissues (see Figure 1.3). Many characteristics of this behavior (e.g., height of pseudoelastic flow plateau σ^{AM} or size of hysteresis loop) can be well adapted to needs by proper alloying and/or processing of the NiTi material. NiTi alloy has a Young modulus of about 60 (GPa) a value much closer to 20 (GPa), a value of Young modulus of bone, than to 200 (GPa), a Young modulus of 316 steel. It has been shown that elastic properties of implant material as close in value as possible to natural bone facilitates and quickens the healing process of broken bones because it avoids the so-called bone shielding phenomenon. Review of enhanced properties of NiTi alloy for biomedical applications can be found in a paper by Biscarini et al. (2008).

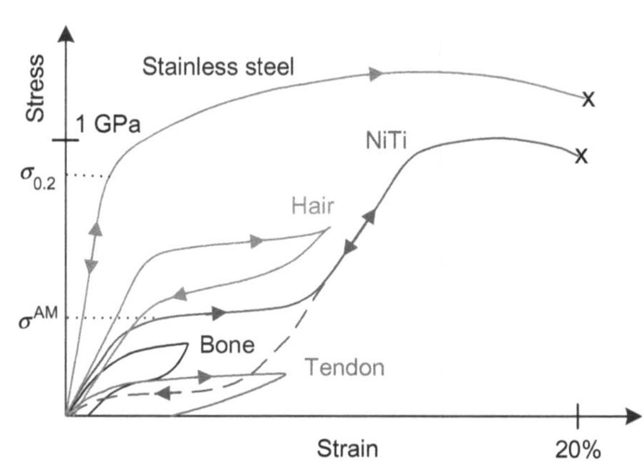

Figure 1.3 NiTi alloy behavior versus stainless steel and several biomaterials behavior.
Adapted after Figure 1 in Melzer and Stoeckel (2010).

Della Corte (2010) reported that so-called Nitinol 60, a material with 60 wt% of nickel (55 atomic percent), exhibits excellent tribological properties. It possesses excellent corrosion resistance, high hardness, shows no brittleness at cryogenic temperatures, and has a very low friction coefficient in the boundary lubrication regime. These properties, together with the fact that the material has a two to three times lower wear rate than traditionally used 440C steel, makes Nitinol 60 a prime candidate for high-performance bearings operating in spacecraft. It is also 25% less dense than steel, which additionally prolongs bearing life by limiting centrifugal forces.

The commercial market of SMA materials is dominated at present with three classes of alloys: NiTi-based alloys, copper-based alloys, and iron-based alloys.

NiTi-based SMA materials are the number one commercial SMA material, and maintain by far the most widespread, continuous, profound interest of scientists and engineers all over the world. NiTi-based shape memory alloys are available in a wide range of compositions, including third elements, and in various forms of raw and semi-finished products, such as rods, wires, plates, tubes, thin films, membranes, and springs. They find application in leading-edge industries such as medical (see, e.g., Yoneyama and Miyazaki, 2009; Yahia, 2000), aerospace and defense (see, e.g., Yamauchi et al., 2011; Lagoudas, 2008). Continuous attempts are made to use NiTi materials in bulk production in the automotive or textile and/or household industries. One very important application is microdevices made of NiTi materials; they can operate at much higher frequencies than bulk SMA devices as higher rates of operation of SMA materials is limited by heat removal capabilities. Broad discussion of opportunities and difficulties connected with the construction of SMA micro devices can be found in Kohl (2004) and Miyazaki et al. (2009).

Research attention to Cu-based SMA materials is driven by a need for high-temperature SMA materials and high electrical conductance (e.g., in household appliances). Much work has been done on Cu-Zn-Al alloys with operating temperatures up to 100 °C, and Cu-Al-Ni alloys with operating temperature up to 200 °C, but these types of alloys exhibit limited formability due to brittleness, precluding an efficient degree of cold work. Due to this, CuAlMn-type high-temperature SMA alloys have been developed (Sutou et al., 2008) and its variations, for example, CuAlMnZr (Chen, et al., 2009).

Tenacious research efforts devoted to Fe-based SMA materials are driven by prospects for opening a new broad application area in civil engineering when less expensive SMA material is found (Janke et al., 2005). Excellent pseudoelastic behavior has been obtained from Fe-28Ni-17Co-11.5Al-2.5Ta-0.05B (at.%), also known as NCATB, an Fe-based SMA material studied by a Japanese team of researchers from Tohokou University (Tanaka, et al., 2010; Ishida et al., 2012).

Frequently, this or other definite types of shape memory alloy attracts especially focused attention. For example, Ti-Nb-based alloys are investigated in the search for a nickel-free SMA material for medical implants (Miyazaki et al., 2006). The NiTiCu alloys merit attention due to their exhibition of a small hysteresis loop; an attractive feature in their application as orthodontic wires (Nespoli and Besseghini, 2011).

1.2 Historical outlook on discovery of SME in various metallic alloys

The history of research on shape memory alloys and shape memory effects is inseparably connected with the history of research devoted to *thermoelastic martensitic transformation*, a physical phenomenon responsible for shape memory effects in metallic alloys. The first observation of martensitic microstructures was done with the aid of an optical microscope by Adolf Martens around 1890. According to Christian (1982), the research on shape memory effects starts with the appearance of work by Olander (1932), in which he mentions the remarkable rubber-like elastic behavior of a AuCd alloy at room temperature. According to Delaey et al. (1974), pseudoelastic behavior and shape memory effects were implicitly studied in FeNi alloy by Scheil as early as 1932 (Scheil, 1932). According to Christian, first observations of the formation of thermoelastic martensite in a CuZn alloy were reported in 1938 by Greninger and Mooradian (1938). Results of regular studies on thermoelastic martensite were published in 1949 by Kurdjumov and Khandros (1949). The shape memory effects in metallic alloys as a new scientific research area started in 1951 with the work of Chang and Read (1951), devoted to studies on the behavior of a AuCd alloy, in which a proper understanding of pseudoelastic phenomenon mechanism was delivered. The pseudoelastic effects were soon reported in CuZn alloys by Reynolds and Bever (1952), and also in a paper by Hornbogen and Wassermann (1956). The pseudoelastic effect in an InTl alloy was found by Burkart and Read (1953) and by Basinski and Christian (1954). Its appearance in a CuAlNi alloy was discovered by Rachinger (1958). A breathrough in the field of shape memory alloys came in 1963 with the discovery of the shape memory effect in a NiTi alloy by Buehler et al. (1963). Shape memory effect in a Fe_3Pt alloy was first reported by Wayman (1971). Maki et al. (1984) reported the appearance of perfect thermoelastic martensite in an iron-based alloy (FeNiCoTi) in 1984, while Sato et al. (1982) were the first to discover the appearance of shape memory effects in a FeMnSi alloy. A large number of metallic alloys are now known to exhibit shape memory effects, but three classes receive dominant attention due to their commercial opportunities. These are NiTi-based alloys, Cu-based alloys, and Fe-based alloys. In fact, vibrant interest in shape memory alloy materials began only after the discovery of shape memory effects and thermoelastic martensitic transformation in a $Ni_{50\%at}Ti$ alloy by Buhler in 1963, which resulted in the first commercial application of a NiTi alloy used in joints of hydraulic installation of F-14 aircraft in 1971, as stated by Melton (1999).

Today, research activity in the field of shape memory alloys and martensitic transformation is extremely vivid, as three parallel in time, regular series of international conferences have been held on the subject with participation by researchers from all over the world for the last 12 years, plus the regular presence of this topic at other conferences. The three major international conferences are the Shape Memory and Superelastic Technologies Conferences (SMST), the European Symposia on Martensitic Transformations and Shape Memory Alloys (ESOMAT), and the International Conferences on Martensitic Transformations (ICOMAT). Already, 10 SMST

Conferences have taken place; the last one (SMST-2014) was held in Pacific Grove, California, on May 12-16, 2014. The ninth ESOMAT Conference was held in Saint Peterburg, Russia, on September 9-16, 2012 (ESOMAT, 2012), and ICOMAT, 2014 was held in Bilbao, Spain, on July 6-11, 2014 (ICOMAT, 2014). Typically, the proceedings from each conference are over 500 pages long (see, e.g., information on the web page SMST, 2014).

1.3 Scope and structure of the book

This book is concerned with the study of continuum, nonequilibrium thermodynamics applied to shape memory alloys. The book presents a state of the art description of pseudoelastic behavior of SMA materials with some bias toward development of numerically oriented engineering design tools. The book is organized into eight chapters. The first two chapters of the book are a general introduction into the field of shape memory alloy materials and they do not require any preliminary knowledge. All other chapters but the last contain rather advanced material, and prior knowledge is helpful, if not necessary, for a full understanding of their content. Relevant references are delivered within each chapter.

Fundamental assumptions have been made in the book during final specification of the actual SMA models: that heat capacities are at constant pressure, that heat expansion coefficients and elastic stiffness tensors are the same, and that they are constant for austenite and martensitic phases. It is also assumed that the elastic stiffness tensor is isotropic. None of these assumptions/simplifications is critical for the presentation of SMA constitutive models and any of them can be straighforwardly liberated.

Chapter 1 delivers a general introduction to the field of shape memory alloys treated as a generic instance of smart material. Special behaviors of SMA materials, so-called shape memory effects, are described. A brief history is given of the discovery of various types of SMA alloys. How special properties of SMA materials can help in the construction of more effective engineering devices is explained. It is made clear that not only shape memory effects but also their combination with the utility features of SMA, such as biocompatibility or corrosion, are key factors that have spurred wide interest in SMA materials by the research and engineering communities. Finally, the scope and structure of the book is specified.

Chapter 2 discusses the specific features of thermoelastic martensitic transformation, a phenomenon underlying the physical grounds for shape memory effects in metallic materials. Multiscale experimental evidence is provided on properties of martensitic phase and its characterization. Methodologies are described, from atomic scale to macroscopic scale of modeling material behaviors connected with martensitic transformation. Key, basic information on processing routes and methods are described by which special characteristics of SMA materials can be obtained. Groups of SMA materials of principal interest are specified and their selected properties are described. A brief review of operational modes of SMA materials in engineering devices is given, and examples of specific solutions in various application areas are provided.

In Chapter 3, a survey of experimental, thermomechanical, and pseudoelastic behavior of SMA materials submitted to uniaxial and multiaxial loadings is presented. A family of so-called thermodynamic R_L models, reference with loop, of pseudoelasticity is discussed together with their key underlying assumptions. Constitutive relations of a version of a R_L model capable of describing the so-called strength differential effect are specified in detail. A list of material model parameters and functions is presented that constitutes a full characterization of SMA material in the pseudoelastic regime of its behavior.

Chapter 4 is devoted to the theoretical study of the shape memory alloy RVE treated as a multiphase, linear thermoelastic body with eigenstrains and submitted to uniform boundary tractions conditions. Homogenization of RVE properties revealed what terms must appear in a postulated heuristically, phenomenological, macroscopic form of free energy function valid for SMA materials, in order to reliably predict, with its use, the behavior of shape memory alloys at various thermomechanical loadings. Clear physical interpretation has been obtained of individual expressions defining the value of SMA elastic energy in terms of phase averages of mesoscopic fields.

In Chapter 5, the results of a program of biaxial thermomechanical tests performed on a NiTi tubular specimen submitted to proportional loadings are presented. It is shown how to specifically process raw data to validate experimentally the admissibility of theoretical assumptions adopted in the development of the family of R_L models of pseudoelasticity. There has been indirectly, positively verified assumption on NiTi polycrystalline material isotropy. Shape of pseudoelastic flow surfaces at several temperatures has been estimated on experimental results grounds. Experimental Clusius-Clapeyron lines are determined for several directions of loading stress tensor. Their extrapolation to zero stresses delivered illuminating information that they all converge at a martensite start temperature. Estimates of the quality of predictions of two SMA models are delivered; one using only the second invariant of the stress tensor and the other using the second and third invariant of the stress tensor.

In Chapter 6, experimental data obtained for a NiTi alloy undergoing phase transition are processed and analyzed to validate the phase transition kinetics model proposed in Raniecki and Lexcellent (1994). Their analysis led to the formulation and development of a new specific model of thermoelastic martensitic phase transformation kinetics. The new phase transition (p.t.) kinetics relations model allows for better prediction of full and partial phase transformations paths (i.e., internal hysteresis loops formation).

In Chapter 7, a very comprehensive general structure of macroscopic, thermodynamic theory of pseudoelasticity of shape memory alloys is presented, taking into account geometrical effects, allowing for large rotations and large strains. Another member of the R_L models family of 3D macroscopic, phenomenological, thermodynamic, constitutive models of SMA materials' pseudoelastic behavior is formulated and presented based on multiplicative decomposition of total deformation gradient tensor. The model is formulated to be linearly elastic in terms of elastic logarithmic strain. The hypotheses on which the model is based are only partially verified experimentally at present; namely, for isothermal proportional loadings. The model enables the prediction of force, deformation, and thermomechanical coupling effects appearing during the loading of shape memory alloys. Rate equations of state have been derived to obtain exact expressions for rate tangent modules necessary to implement the model into FEM codes.

Chapter 8 contains a general summary and remarks regarding directions of near-future research work in the field of SMA materials, open scientific problems, and perspectives of SMA materials.

Shape memory effects in metallic alloys

2

2.1 Physical foundations of SME effects—thermoelastic martensitic transformation

Microscopic observations with optical, scanning, and transmission electron microscopy, differential scanning calorimetry (DSC), X-ray spectroscopy, and numerous other experimental techniques reveal that shape memory phenomena in metallic alloys appear as a result of forward and reverse thermoelastic martensitic transformation and/or martensitic variants reorientation/detwinning. Excellent materials science-orientated surveys with rich experimental evidence and profound discussions on the subject can be found, for example, in Delaey et al. (1974), Delaey (1991), Otsuka and Wayman (1999), Otsuka and Ren (2005), and Nishida et al. (2012). Extremely rich literature is devoted to microscopic observations of metallographic structures arising during the progress of martensitic phase transformation; see, for example, Olson and Owen (1992) or Christian (2002). Selected results of materials science experimental observations important from the point of view of shape memory effects (SME) exhibited by SMA materials are recalled below.

Thermoelastic martensitic transformation is diffusionless transformation of nucleation and growth type. In polycrystalline material, the first and largest martensitic plates appear constrained in size by grain boundaries, and the next and smaller plates fill up untransformed regions until all the parent phase disappears and the transformation is completed. The progress of thermoelastic martensitic transformation may be temperature-induced or stress (strain)-induced. It does not proceed when temperature and stress are kept constant. The kinetics of thermoelastic martensitic transformation is practically insensitive to rate of change of inducing it external stimuli, i.e. temperature or stress. The transformation possessing such property is called *athermal transformation*. The above qualification can lead the reader to the erroneous conclusion that thermoelastic martensitic transformation proceeds with extremely great speed. Actually, thermoelastic martensitic transformation proceeds with a speed close to the speed of sound in SMA material. Taking into account that in prevailing engineering situations stress and/or temperature change with much slower rates when compared to speed of sound allows us to neglect any time delays and describe the transition as athermal.

During *temperature-induced thermoelastic martensitic transformation,* a phase stable at higher temperatures (conventionally called *austenite*) that possesses high symmetry of crystallographic lattice transforms on cooling into a phase stable at lower temperatures (conventionally called *martensite*) that possesses lower symmetry of

Pseudoelasticity of Shape Memory Alloys. http://dx.doi.org/10.1016/B978-0-12-801697-8.00002-8

crystallographic lattice. The transformation takes place without diffusion by short-range coordinated, instantaneous movement of many atoms; due to that, it is also frequently called a *military transformation*. From the start to the end of the transformation, the relative motion of a single atom participating in the martensitic transformation does not exceed one interatomic distance. Due to that chemical composition, ordering of atomic structures, crystallographic superstructures, and the number of crystallographic lattice defects are the same in the parent and the martensitic phase. *Coherent interphase* forms between the phases, that is, resulting from the phase transition *displacement field is continuous* on the phases' interface. During *stress-induced thermoelastic martensitic transformation*, essentially the same microscopic phenomena occur as in the case of thermally induced transformation, but instead of low temperature the presence of high stress is a martensitic phase stabilization factor.

The internal structure of the martensitic phase is very complex (Figure 2.1). It exhibits multiscale organization of matter on several levels of observation. The basic building blocks of the martensitic phase, which can be distinguished on the lowest level of organization, one to two orders higher than the atomic scale of observation,

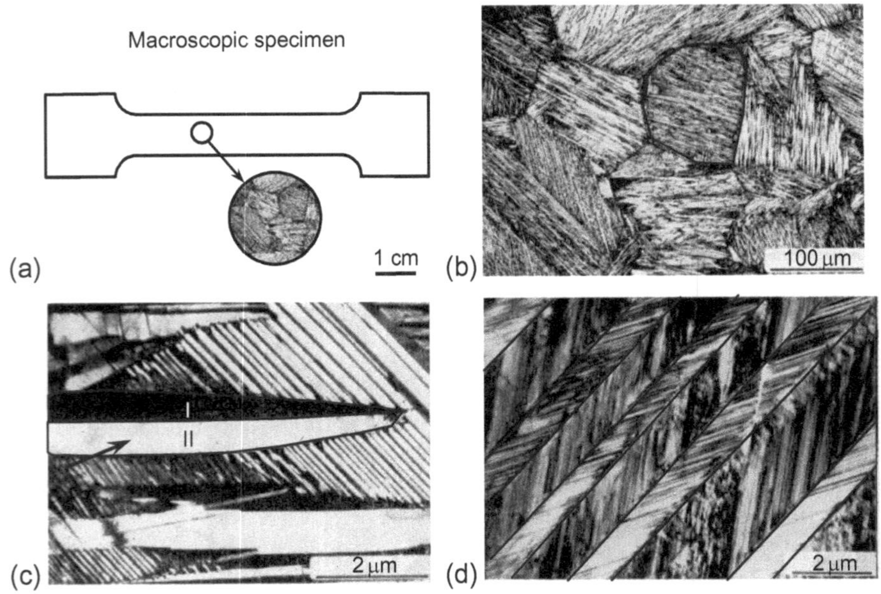

Figure 2.1 Multiscale organization of martensitic phase in polycrystalline CuAl alloy: (a) macroscopic sample in martensitic state; (b) polycrystalline structure of martensitic phase with martensitic objects—plates, confined by grain boundaries; (c) spear-like martensitic compound composed of two habit plane martensitic variants (HPV)—indicated by arrow; (d) a configuration of parallel habit plane martensitic variants (HPVs) with visible striations being a manifestation of their lower-level internal structure, composed of two martensitic lattice correspondence variants (CVs).
Photographs after Warlimont and Delaey (1974).

are the so-called martensitic *lattice correspondence variants* (CVs). They appear as striations in Figure 2.1d. They are also schematically shown in Figure 2.3. The name comes from the fact that the three orthogonal, crystallographic directions of the parent phase defining basal planes are in strict relation, orientation, with certain crystallographic directions, basal planes of the martensitic generic objects (CVs). For example, in the case of the monoclinic martensitic transition taking place in NiTi alloys, there can be distinguished 12 lattice CVs as there can be identified six, crystallographically equivalent, planes in the parent B2 lattice from which basal planes of the martensitic phase B19′ lattice originate, and on each of these planes there are possible two martensitic transformation shear directions; see, for example, Table 2 and the related text in Nishida et al. (2012). The CVs have designated numbers from 1 to 6 and the symbol prime. Two martensitic lattice CVs with the same number with and without a prime (e.g., 2 and 2′) have the same basal plane but are sheared in opposite directions; thus, they are in a mirror relationship with respect to their basal plane.

On the next level of observation, one to two orders higher than the CVs scale of organization, there can be distinguished another class of generic martensitic objects, the so-called habit plane variants (HPVs). Each HPV is composed of two lattice CVs, the so-called major CV and the minor CV remaining in strict volumetric proportion. Two CVs present in one HPV are frequently called twins. During martensitic transformations there is always observed an undistorted, unrotated plane between the parent and austenitic phase, which explains why it is called the *habit plane*. The constraint of habit plane existence led to the development of the so-called *phenomenological theory of martensite crystallography* (PTMC). Considerations of PTMC show that among all possible combinations of two different CVs coexistence there are only two distinct configurations possible, in which a specific CV is the major variant, assuring the possibility of habit plane appearance. Two possible HPVs in which a specific CV is dominant are designated with plus and minus signs, respectively (e.g., 5′(+) and 5′(−)). Returning to the example of NiTi alloys and monoclinic transition, 24 distinct HPVs are possible. Table 3 in Nishida et al. (2012) provides a complete list of these HPVs with the habit plane normals and correspondence variants combination (CVC) of each.

Formation of HPVs' martensitic region results in its deformation upon comparison to its parent phase geometrical shape, that is, the appearance of eigenstrain of martensitic transformation. This deformation leaves the habit plane undeformed; that is why it is frequently called *invariant plane strain* deformation. In general, it can have shear and dilatational components, but, in the case of thermoelastic martensitic transformation, the shear component is dominant (Figures 2.4–2.6).

The HPVs can grow to various geometrical sizes, forming plate-shaped objects, that is, objects with one dimension much smaller than the other two. When the two larger dimensions are nearly equal, they are called *martensitic plates*, a term probably most frequently encountered in the literature devoted to martensitic transformations. If the larger dimensions of martensitic objects are still very unequal, then such martensitic objects are called *martensitic lathes*; for example, a typical lath in low-carbon steel ($C < 0.4\%$) has dimensions of 0.3 x 4 x 300 μm (Delaey, 1991). In polycrystalline materials, populations of martensitic plates with considerably different geometrical sizes can be distinguished. Some researchers (e.g., Hornbogen) attempted to describe this feature of martensitic microstructures with the theory of fractals. For example, in Figure 2.1c a population of "large" martensitic plates with a lateral dimension of approximately 0.5 μm can be distinguished; each plate is composed of a single HPV. In the upper right area a second population of martensitic plates one order smaller in geometrical dimension can be distinguished, and a third population of still smaller plates can be distinguished in the lower left area. Nucleation and growth-type kinetics of martensitic transformation are the key factors in the observed pattern.

The HPVs combine together in a still higher level of organization of the martensitic phase. In particular, so-called *self-accommodating martensite* morphologies or compounds appear on a spontaneous grouping of martensitic plates (HPVs) when the martensitic phase is purely temperature-induced at zero assisting macroscopic stress. A definitional feature of a self-accommodating martensitic compound, borrowed from materials science literature, is that it has zero macroscopic strain connected with its formation—zero martensitic phase eigenstrain. For example, two plates designated by symbols I and II in Figure 2.1c taken together have zero eigenstrain and make self-accommodating morphology composed of two HPVs. The self-accommodating martensitic morphologies can be very intricate, composed of a various number of different HPVs. Recently, self-accommodating morphologies in NiTi alloy have been discovered that can be composed of two, three, four, or six HPVs; see Nishida et al. (2012) and also Figure 2.7 below. In polycrystalline alloys, martensitic internal structuring is always confined within grain boundaries; see Figure 2.1b.

Self-accommodating martensitic phase morphologies appear when martensitic transformation is temperature induced. When thermoelastic martensitic transformation is stress induced at higher stresses, so-called oriented martensite is born composed of HPVs with eigenstrains favorably oriented with respect to the direction of inducing transformation stress tensor. When already born self-accommodating martensite is submitted to mechanical loading, then, so-called reorientation and detwinning takes place, resulting in non-zero macroscopic martensitic phase eigenstrain (i.e., orientation of martensitic phase). Physically, the process of coalescence and disappearance of unfavorably oriented HPVs takes place, and HPVs favorably oriented with respect to stress tensor direction appear. So-called detwinned martensite is generated. In particular at sufficiently large stress and at the monocrystalline parent phase, the termination of coalescence in the martensitic phase may consist of only one martensitic HPV; see Figure 2.8. This process is possible because thermoelastic martensitic transformation twin boundaries preserve high mobility. When the direction of the stress tensor stabilizing martensitic phase is changed at high temperatures in pseudoelastic range of state parameters, then the martensitic phase structure evolves in such a way that phase eigenstrain direction follows the direction of loading stress.

Volumetric strain connected with classical martensitic transformation, occurring during toughening of carbon steels, is on the order of 4% of what necessarily induces plastic slip strains in the parent phase. Eigenstrains connected with thermoelastic martensitic transformation taking place in SMA materials are predominantly shear strains. They are of the order of several, and even a dozen or so percent. The volumetric strain of thermoelastic martensite is small; usually, it does not exceed the 0.2%. Such a structure of thermoelastic martensite eigenstrains can be accommodated by the parent phase in quasi-elastic way, in principle, without any plastic deformation of the parent phase. This enables reverse thermoelastic martensitic transformation to occur in a so-called *crystallographically reversible* way, contrary to the case of classical martensitic transformation occurring in carbon steels. The features of phase eigenstrains cause the appearance of essential differences in the kinetics of classical martensitic phase transformation and the thermoelastic one. *Kinetics of martensitic transformation* in carbon steels is of a *burst martensitic transformation* type. Violent, dynamic formation of martensitic plates take place, after which the plate boundaries are no longer mobile. The reverse martensite to austenite transformation in carbon steels takes place in such a way that in one plate of martensite nucleates several embryos of parent phase, which subsequently grow in size. There is no so-called *crystallographic reversibility* of the transformation (Wayman, 1992). During *thermoelastic martensitic transformation*, growth or vanishing of martensitic plates takes place in a pretty continuous manner progressively with variation of inducing it in external state parameters, that is, lowering of temperature or increasing macroscopic stress. New plates of martensite appear only then when already existing ones cannot grow in size any more due to, for example, grain boundaries constraints. Next, during reverse transformation plates of thermoelastic martensite diminish and vanish in reverse order to the sequence of their appearance during forward transformation. It is said that the transformation exhibits the property of *crystallographic reversibility*. In Figure 2.2, a comparative schematic illustration of various types phase transformation (p.t.) kinetics taking place in solid state materials is shown (i.e., *isothermal*, *burst*, *thermoelastic*).

The forward and reverse thermoelastic martensitic transformations do not run along the same paths in the space of thermodynamic parameters. Larger or smaller *hysteresis loops* are formed in a closed cycle of forward and reverse transformation. This provides evidence that thermoelastic martensitic transformation is *not thermodynamically reversible*; that is, it does not proceed through thermodynamic equilibrium states.

Martensitic phase transformation is classified as *first-order phase transition*, as it is accompanied by emission of latent heat of transformation. DSC, thermograph of $Ni_{51at\%}$-Ti polycrystalline alloy, 30% last stage cold-worked, heat-treated for 30 min at 500 °C, and water quenched is shown in Figure 2.3. The DSC cycle has been

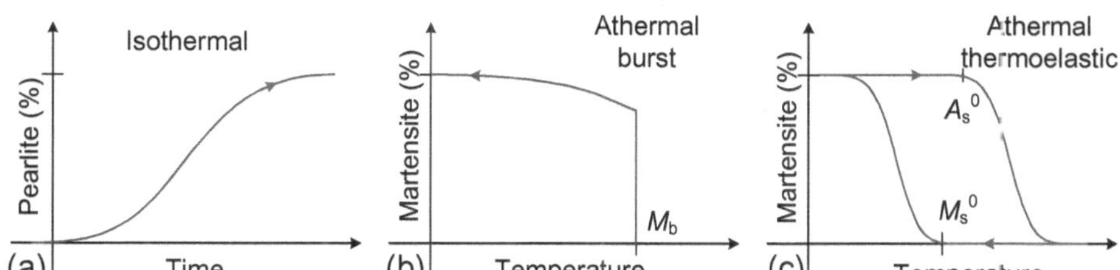

Figure 2.2 Comparative schematic illustration of various types of phase transformation (p.t.) kinetics: (a) kinetics of "classical" isothermal diffusional transformation (e.g., austenite to pearlite); (b) kinetics of athermal, burst-type martensitic transformation (e.g., in carbon steel); (c) kinetics of athermal thermoelastic martensitic transformation taking place in SMA materials.

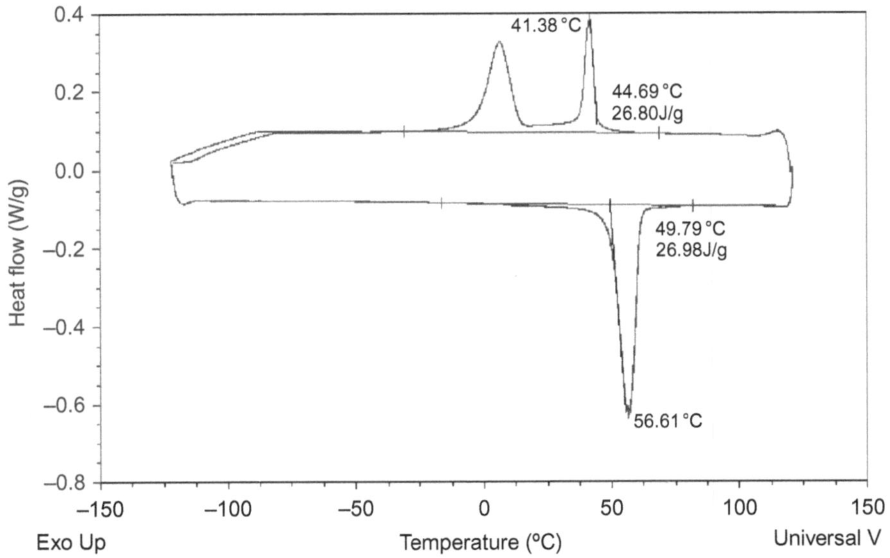

Figure 2.3 A DSC thermograph of $Ni_{51at\%}$-Ti polycrystalline alloy, 30% last stage cold-worked, heat-treated for 30 min at 500 °C, and water quenched. The sample was made from wire 1 mm in diameter. TA Instruments 2920 Apparatus.
After Ziółkowski (2012).

run with a rate of 10 K/min. Two peaks of heat emission upon the cooling part of the DSC cycle indicates that forward p.t. takes place in a two-stage manner and one peak of heat absorption on the heating part indicates that reverse phase transition takes place in a one-stage manner. Microscopic observations prove that during forward phase transition at first parent phase with B2 crystallographic structure transforms into so-called *R-phase* (rhombohedral martensitic phase), and next monoclinic B19′ martensite is formed ($\mathcal{A} \rightarrow \mathcal{R} \rightarrow \mathcal{M}$); see, for example, Otsuka and Ren (2005). Reverse phase transition takes place in one step from martensite to austenite ($\mathcal{M} \rightarrow \mathcal{A}$). The forward and reverse martensitic transformations are characterized by start and finish *characteristic temperatures*, respectively denoted by $R_s^0, R_f^0, M_s^0, M_f^0, A_s^0, A_f^0\ RA_s^0, RA_f^0$. The R_s^0, R_f^0 symbols denote start and finish temperature of $\mathcal{A} \rightarrow \mathcal{R}$ transition, and RA_s^0 and RA_f^0 are start and finish temperatures of reverse transition $\mathcal{R} \rightarrow \mathcal{A}$. The M_s^0, M_f^0 and A_s^0, A_f^0 symbols denote start and finish temperatures of $\mathcal{A} \rightarrow \mathcal{M}$ (here $\mathcal{R} \rightarrow \mathcal{M}$) transition and reverse $\mathcal{M} \rightarrow \mathcal{A}$ transition, respectively. Upper subscript "0" denotes that characteristic temperatures are measured at zero macroscopic stress assisting the transformation; standard conditions for the *DSC* measurement technique.

The characteristic temperatures of martensitic transformation and pattern of phase transitions very strongly depend on *thermomechanical treatment* (*TMT*) of specific NiTi alloy, sometimes called *training*. For example, the martensitic transformation can take place in a clearly separated two-step manner, or $\mathcal{A} \rightarrow \mathcal{R}$ and $\mathcal{A} \rightarrow \mathcal{M}$ transitions can strongly overlap, or $\mathcal{A} \rightarrow \mathcal{R}$ transition may not appear at all; see, for example, Chapter 1 of Braz Fernandez (2013).

The characteristic features of "classical" martensitic transformation taking place, for example, in carbon steels can be summarized as follows: it is a first-order phase transition (latent heat is emitted during progress of p.t.); it is displacive transformation consisting of coordinated movement of a large number of atoms with respect to the invariant (unrotated and undeformed) interphase plane called habit plane; there exists a strict crystallographic relationship between the parent phase and the martensitic phase lattices; it is diffusionless, that is, for example, the chemical composition and the degree of atomic ordering is the same in both the parent and product phase; it can be induced by temperature; it is athermal; it is coherent, that is, there is no jump of displacement field on the phases interphase; it is independent on the rate of inducing its temperature; it heterogeneous; it progresses by nucleation and growth; and martensitic eigenstrains are accommodated by the parent phase with a plastic slip.

The characteristic features distinguishing thermoelastic martensitic transformation from classical martensitic transformation can be summarized as follows: phase eigenstrains are shear strain dominant; accommodation of phase eigenstrains takes place quasi-elastically; there exists crystallographic reversibility of transformation; interphase and twin boundaries have high mobility, transformation can be induced by lowering temperature or increasing stress; and forward and reverse transition is possible.

2.2 Multiscale experimental investigation and modeling of martensitic transformation and SMA materials behavior

Experimental studies and modeling of thermoelastic martensitic transformation and shape memory alloys behavior to a large extent can be treated as investigation of the other side of the coin. In this section, experimental and theoretical research efforts devoted to understanding and describing martensite and later shape memory phenomena in metallic alloys are discussed in more detail, to some extent in chronological order.

While the first observations of martensitic structures were made by Adolf Martens around 1890, the fully developed so-called Phenomenological Theory of Martensitic Crystallography (PTMC) was formulated only 60 years later in 1953 and 1954, independently by two groups of researchers (Bowles and MacKenzie, 1954) and (Wechsler et al., 1953). The PTMC theory explains in detail the kinematical mechanism of formation of the martensitic phase, and in this way its crystallography. The input data to the theory are of atomic-scale dimensions. These are crystallographic parameters of parent-phase lattice (usually it is one value of cubic lattice) and lattice parameters of martensitic phase (here lattice can be monoclinic, tetragonal, or rhombohedral). The PTMC conjectures appearance of undistorted, unrotated plane between parent and martensitic phase, called *habit plane*. It predicts that martensitic phase is composed of two crystallographic correspondence variants (CV's), and it allows us to calculate as output quantities from the theory the components of vector normal to habit plane, shear and dilatational component of deformation connected with martensitic phase formation, and volume content of two martensitic CVs. The PTMC finds

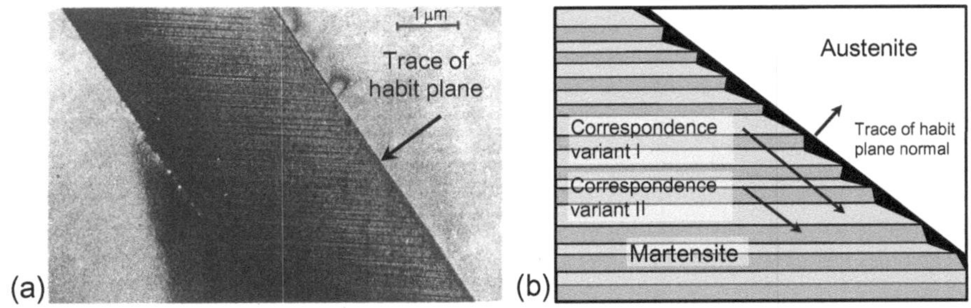

Figure 2.4 (a) Photograph of single martensitic plate (HPV), in Fe-30%Ni-0.4%C alloy with revealed twinning substructure of two martensitic crystallographic correspondence variants (CVs), obtained with transmission electron microscopy. Photo after Maki et al. (1984); (b) Schematic illustration of HPV internal structure, not to scale.

excellent experimental confirmation. Figure 2.4a shows a photo by Maki and Wayman from 1976 published in Maki (1999), where a single martensitic plate (HPV) formed in Fe-$_{30\%}$Ni-$_{0.4\%}$C with striations indicating internal substructuring of martensite is clearly visible. In Figure 2.4b the most important features of martensitic phase transition as predicted by PTMC are schematically shown.

In terms of scale of observation, taking into account characteristic geometrical dimensions of quantities entering PTMC, it should be classified as atomic scale theory. However this classification can be ambiguous, as the output of the theory (i.e., a habit plane object) can have geometrical dimensions of the order of micrometers or even millimeters. Hence, PTMC theory can be qualified equally well as mesoscopic or even macroscopic theory. In Figure 2.5a an experimental layout has been redrawn originally designed by Chang and Read in 1951, and later reported and analyzed with PTMC theory in (Lieberman et al., 1955). Chang and Read demonstrated that in a single crystal of AuCd a steep gradient of temperature induces the formation of a single plane, habit plane, with strict orientation with respect to the parent-phase lattice separating the parent and martensitic phases. The dihedral angle of habit plane observed experimentally agrees perfectly well with the values predicted from calculations of PTMC. The forward and backward motion of the phase interface plane can be temperature controlled easily by regulation of the heater power. Thus, the transformation is crystallographically perfectly reversible and can be cycled with temperature change back and forth.

Figure 2.5b shows a sequence of photos from an experiment prepared and executed by Ichinose et al. (1985). With specially designed grips they managed to induce stress formation of a single HPV of martensite in a single crystal of CuAlNi alloy. A single crystal of parent phase seen in the photo on the utmost left transforms under stress operation to a single crystal of martensite seen in the photo to the utmost right by the motion of two habit planes with a single orientation. Two habit planes are formed in the vicinity of the grips and they can be moved forward and backward or maintained in strict location under force control. Similar to previously discussed temperature-induced transformation, the dihedral angle of stress-induced habit plane agrees perfectly well with predictions of PTMC.

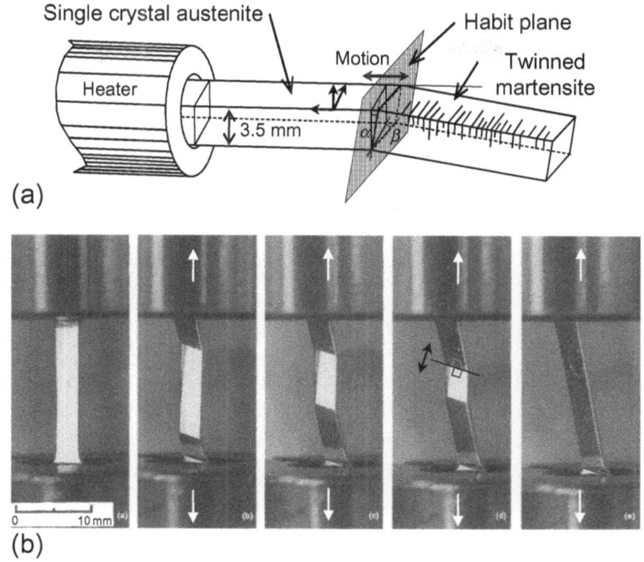

Figure 2.5 (a) Illustrative picture of temperature-induced formation of single HVP martensite in $Au_{47.5at\%}Cd$ by habit plane movement. Redrawn after the Chang and Read (1951) experiment reported in Lieberman et al. (1955); (b) Photo of stress-induced formation of single HPV martensite in CuAlNi alloy by habit plane movement. After Ichinose et al. (1985).

One of the principal results of PTMC is the finding that martensitic transformation can and does take place in ways such that the interface dividing parent and product phases during progress of p.t. (habit plane) is coherent. In other words, it means that the displacement field is continuous on such a phase interface. In the mathematical language of continuum mechanics the constraint/property of coherency translates to a requirement that deformation gradient connected with martensitic phase transition must be of a special form. Namely, it must take the form $\boldsymbol{F} \equiv \partial \boldsymbol{x}(\boldsymbol{X}, t)/\partial \boldsymbol{X} = \mathbf{1} + \boldsymbol{b}_i \otimes \boldsymbol{n}_i$, where $\boldsymbol{F}$ denotes deformation gradient, $\boldsymbol{X}$ denotes coordinates of material point in reference (initial, untransformed) configuration, $\boldsymbol{x}$ means coordinates of material point in actual (transformed) configuration, $\boldsymbol{n}_i$ is unit vector of habit plane of i-th HPV of martensite, and $\boldsymbol{b}_i$ denotes deformation vector of i-th HPV. Such form of deformation gradient is called *rank-one connection* in the mathematical literature. Parent and martensitic phase are rank-one connected. Untransformed material fiber of parent phase $\boldsymbol{n}_i$ takes the form $\boldsymbol{n}_i + \boldsymbol{b}_i$ in product phase $(\boldsymbol{F}\boldsymbol{n}_i = (\mathbf{1} + \boldsymbol{b}_i \otimes \boldsymbol{n}_i)\boldsymbol{n}_i = \boldsymbol{n}_i + \boldsymbol{b}_i)$ (Figure 2.6).

In the language of PTMC theory (Bowles-Mackenzie theory or Wechsler-Lieberman-Read theory), which elucidates in detail microscopic kinematics of martensitic transformation, martensitic phase transition "crystallographic" deformation gradient is described with the following formula: $\boldsymbol{F} = \boldsymbol{RSB}$, where $\boldsymbol{F}$ again denotes total deformation gradient of martensitic p.t., $\boldsymbol{B}$ denotes Bain strain tensor, $\boldsymbol{S}$ is lattice invariant shear strain tensor (resulting from twinning or plastic slip), and $\boldsymbol{R}$ is tensor of rigid rotation. Deformation described with such deformation

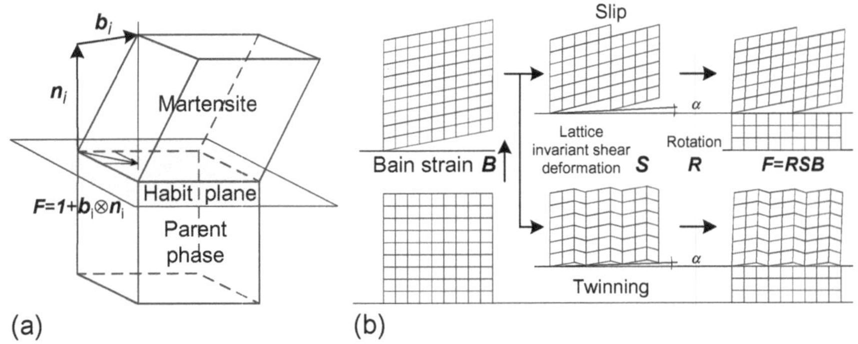

Figure 2.6 (a) Schematic illustration of eigenstrain resulting from transformation of austenite into a specific habit plane variant (HPV) of martensite, not to scale; (b) Schematic illustration of kinematics of martensitic transformation in accordance with phenomenological crystallographic theory of martensitic transformation (PTMC theory).

gradient transforms crystallographic lattice of parent phase into lattice of martensitic phase while ensuring existence of unrotated and undistorted plane (habit plane). Composition of *Bain deformation*, $\boldsymbol{B}$, and *lattice invariant shear deformation*, $\boldsymbol{S}$, gives compound deformation, with one of the principal stretches of deformation equal to one $\lambda_i = 1$. Thanks to this property after $\boldsymbol{SB}$ there exists undistorted plane; however, it is rotated with respect to its original location. Application of rigid rotation $\boldsymbol{R}$ brings it back to the original location and assures the existence of an undistorted, unrotated plane. Thus, the following equivalence between microscopic continuum mechanics and "crystallographic" description of martensitic transformation is valid $\boldsymbol{RSB} = \boldsymbol{F} = \boldsymbol{1} + \boldsymbol{b}_i \otimes \boldsymbol{n}_i$. Please note that there is no direct link between the microscopic (crystallographic) description of martensitic p.t. discussed here (total deformation gradient of HPV) and the total deformation gradient of RVE present (used) in macroscopic theory of martensitic transformation—constitutive SMA material models can be treated as such theories. The microscopic and macroscopic descriptions preserve only some properties, for example, coherency of displacement fields. One of the reasons that may be indicated for the discrepancy is the formation of martensitic superstructures; for example, the ones like self-accommodating martensitic compounds of different types as discussed in the sequel below. Macroscopic deformation gradient describes deformation of martensitic mesostructures in an averaged sense.

It is well known from continuum mechanics that the determinant of deformation gradient determines change of volume due to deformation $V(t)/V_0 = \det \boldsymbol{F}(t)$, where V_0 denotes volume before deformation and V volume after deformation.

In the case of martensitic transformation, it is $V_2/V_1 = \det \boldsymbol{F} = 1 + \boldsymbol{b}_i \cdot \boldsymbol{n}_i = 1 + \delta$, where index 2 denotes product phase and index 1 parent phase. Thus, parameter δ determines the difference in density of parent and product phase $\delta = \rho_1/\rho_2 - 1$, where ρ_1 is density of parent phase and ρ_2 density of product phase. Thermoelastic martensitic transformation is shear deformation dominant, while the dilatational part is very small and usually does not exceed 0.2% ($\delta < 0.2\%$).

Coherency constraint is the primary reason for storage of elastic energy in two-phase and martensitic states of SMA materials; so-called *coherency energy*, required to assure the said coherency. This energy very strongly influences SME, and thus is of key importance in modeling of SMA materials behavior.

Figure 2.5 shows quite exceptional situations, which can be achieved only under special and very carefully controlled experimental conditions. In "natural" conditions of operation of SMA materials when polycrystalline SMA material is thermomechanically loaded, much more complicated two-phase and martensitic-phase configurations appear. When martensitic transformation is stress induced, so-called oriented martensite is induced in SMA materials. Figure 2.7, shows a photographic sequence of oriented martensite microstructure formation generated by tensioning of a polycrystalline CuAlBe alloy sample at room temperature $\left(M_s^0 \cong 80\,°\mathrm{C}\right)$. Regions of supplementary variants formation accommodating geometrical constraints are marked with ellipses, after Chemisky et al. (2011). The *oriented martensite* microstructure

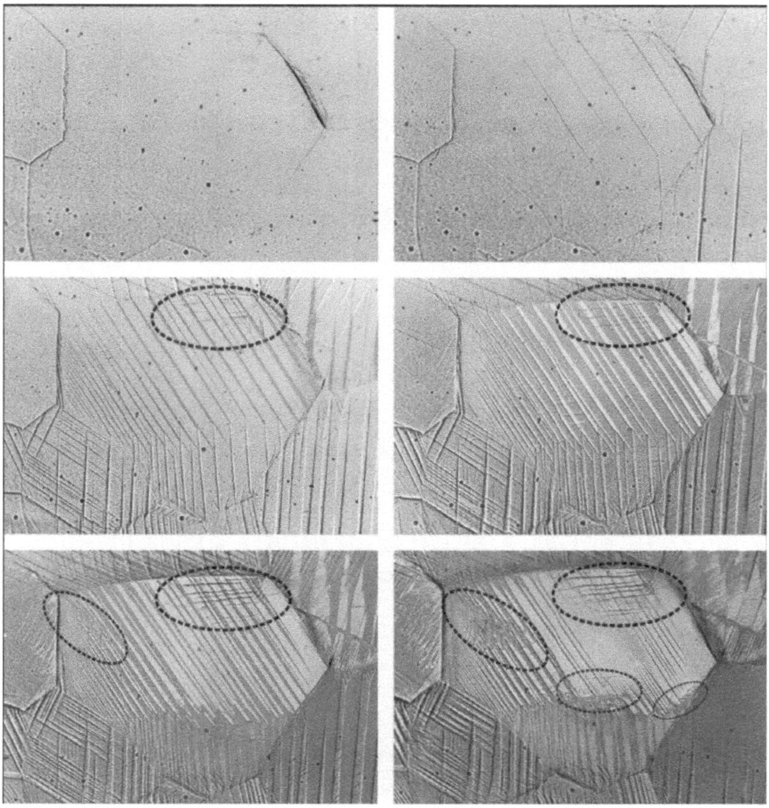

Figure 2.7 Formation of tensile stress-induced oriented martensite microstructure in CuAlBe polycrystalline alloy. Regions of supplementary variants formation accommodating geometrical constraints are marked with ellipses.
Courtesy Andre Eberhardt, LPMM, after Chemisky et al. (2011).

upon averaging its properties to a macroscopic level exhibits *nonzero macroscopic phase strain*, which is favorably oriented with respect to loading stress direction. Besides formation of favorably oriented martensitic variants other supplementary mechanisms to accommodate geometrical constraints present in polycrystalline SMA materials during martensitic phase formation are also observed (e.g., formation of so-called compound martensitic variants or plastic slip dislocations); this last mechanism is unfavorable for SME.

When martensitic transformation is temperature induced at zero assisting macroscopic stress ($\sigma = 0$) then self-accommodating martensite is induced in SMA materials. The reasons and conditions of self-accommodating morphologies formation and their internal structure are subject to intensive scientific research and attracts the attention of many researchers in the field of SMA materials. A wide variety of *self-accommodating morphologies* has been already identified. For example, in NiTi alloys (Nishida et al., 2012) identified four classes of self-accommodating clusters of HPVs: V-shaped morphology composed of two HPVs, triangular morphology composed of three HPVs, rhombic morphology composed of four HPVs, and hexangular morphology composed of six HPVs. TEM photographs of triangular morphology and rhombic morphology are shown in Figure 2.8, clearly revealing the internal structure of the compounds. Possible combinations of HPVs and CVs in self-accommodating clusters in an NiTi alloy are listed in Table 4 of Nishida et al. (2012). Depending on crystallographic conditions, for example, alloy chemical composition or type of phase transition, even self-accommodating clusters composed of the same number of HPVs can differ considerably. For example, they can create a "zig-zag" pattern as shown in Figure 2.7a or a "herringbone" pattern like that shown in Figure 31 in Otsuka and Ren (2005), both composed of four HPVs.

(a)
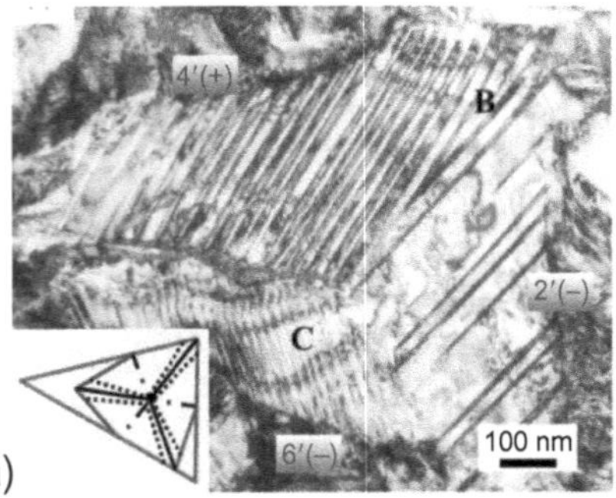

(b)
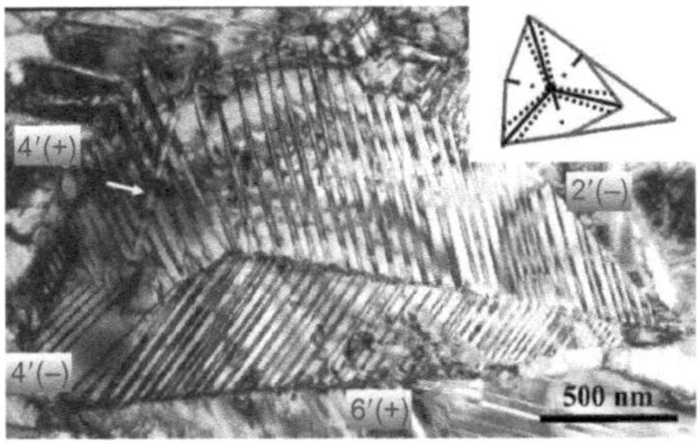

Figure 2.8 (a) TEM photo of triangular self-accommodating morphology in Ni50at%-Ti alloy composed of three HPVs 4′(+)/2′(−)/6′(−) (after Figure 9a in Nishida et al. (2012)). (b) TEM photo of rhombic self-accommodating morphology in Ti-50.0 at% Ni containing four HPVs, or looking otherwise group of clustered two V-shaped self-accommodating minimum units of HPVs 4′(+)/2′(−) and 6′(+)/4′(−) (after Figure 10b in Nishida et al. (2012)). Striations clearly show the substructure of HPVs composed of CVs pairs.
Images courtesy Taylor and Francis.

The classical phenomenological crystallographic theory of martensitic transformation (PTMC) is exclusively a kinematical one. The authors of this theory in their original work only declare that martensitic microstructures predicted by their theory minimize elastic energy of the system, but they do not demonstrate this by specifying an explicit form of elastic energy function of a two-phase system and subsequently minimizing it. The classical PTMC theory delivers an answer to the question of how, geometrically, it is possible for martensitic structures to form, but the question of why, and in what way these structures form—in terms of kinetics of p.t.—is left open. An attempt to answer the question of why microstructures of fine martensitic variants, nonuniform in close to atomic scale of observation, come into existence from an energy point of view was successfully undertaken by Ball and James (1987). These authors assumed that the elastic energy function of a martensitic phase possesses multiple minima for so-called eigenstrains of uniform martensitic variants. Next, using an invariant plane kinematical compatibility constraint, rank-one connection, of PTMC theory they showed that the elastic energy of the whole system consisting of austenitic domain and domain of martensitic HPV diminishes when variant substructure is more and more fine laminate microstructure built of crystallographic martensitic variants. Ball and James did not adopt any explicit form of elastic energy function; however, they assumed that an investigated two-phase system satisfies the boundary condition that it always remains at zero external macroscopic stress. A survey of main research areas and achievements in development of thermodynamic crystallographic theory of martensitic transformation can be found in a book by Bhattacharya (2003). Additional research has been conducted to enhance the range of applicability of classical PTMC. The invariant plane condition on the interface between the parent and product phases is supplemented by additional kinematical constraints on the interfaces between HPVs and several kinds of such interfaces have already been identified, to develop extended PTMC. The target of extended PTMC is to describe and predict crystallography of the second-tier of martensitic objects (i.e., self-accommodating morphologies); see, for example, Inamura et al. (2012).

Twin boundaries in shape memory alloys are very mobile. The application of mechanical loading to thermally induced martensite or changing the direction of the loading stress tensor in stress-induced martensite results in the evolution of martensitic-phase microstructure. This evolution finds its reflection in the macroscopic scale. Involved physical processes are still far from being well understood and are subject to intensive research. Even nomenclature is not well established. It is said that the application of stress and/or changing the stress tensor direction induces reorientation of the martensitic phase. In this book *reorientation of martensitic phase* is understood to be a change of macroscopic phase strain regardless of the underlying micro-, mesostructural processes. As a result of reorientation processes, *oriented martensite* is formed; that is, martensite with nonzero macroscopic phase strain. This can be a result of the annihilation of self-accommodating morphologies through coalescence/disappearance of unfavorably oriented martensitic HPVs and the appearance of the ones with phase eigenstrain favorably oriented with respect to the direction of the actual loading stress. The other process involves so-called *detwinning of martenstic phase*, by which is understood to be a change of proportions of major and minor

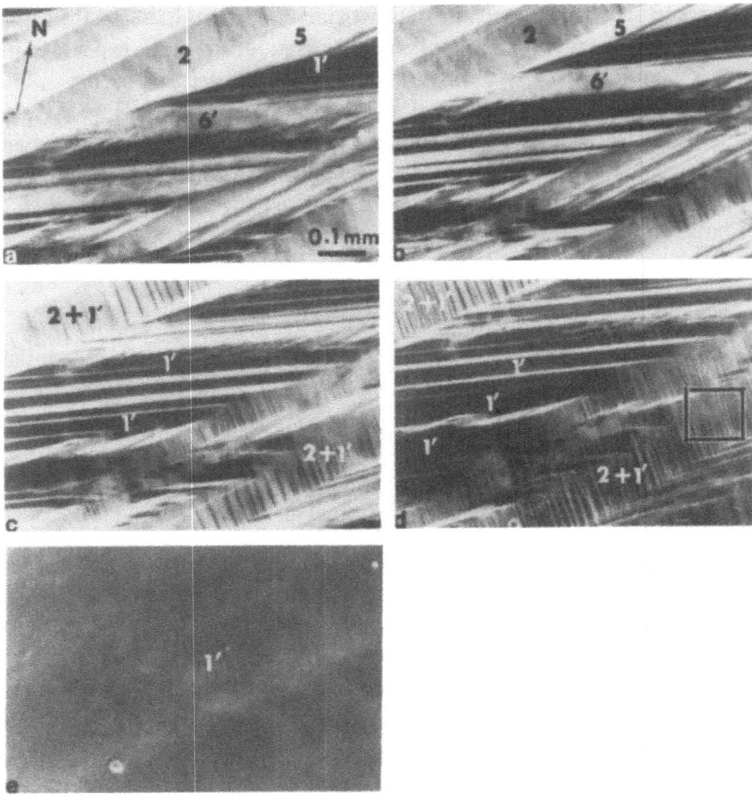

Figure 2.9 Process of stress-induced reorientation, coalescence, and detwinning of thermally induced self-accommodating morphology of HPVs in CuZnGa alloy into a single martensitic variant.
After Saburi et al. (1980).

lattice CVs in a single HPV. These mechanisms are shown in Figure 2.9. Saburi et al. (1980) cooled monocrystalline, in an austenitic state, CuZnGa alloy to obtain self-accommodating martensite. Next, they tensioned it while observing it under an optical microscope. Initial self-accommodating martensitic microstructure composed of several HPVs of martensite evolved, coalesced, into two of the most favorably oriented HPVs and then, upon detwinning, changed into single monocrystalline martensite. The coalescence and detwinning processes proceed concurrently and, more accurately, it is said that one or the other process is dominating at some stage of the deformation process.

In polycrystalline SMA materials the situation is even more complex as grain boundaries create barriers for favorable reorientation of martensite. Besides the already-mentioned mechanisms of coalescence and detwinning, new ones have been observed, for example, at more advanced stages of deformation new compound martensitic variants appear, such as in NiTi alloy $20\overline{1}$ compound twin discovered by

Nishida; see Table 4 in Otsuka and Ren (2005). Their crystallographic parameters are not predicted by the PTMC model, which means that they cannot appear at zero macroscopic stress. It can be said that this type of martensitic variant can be created, and become stable, only at a sufficiently large value of stress. Unfavorable for SME's generation of dislocations and other irreversible lattice defects can also occur. Liu et al. (2000) registered macroscopic stress-strain curves of NiTi alloy specimens tensioned, isothermally, to several strain amplitudes. Upon specimen unloading they observed the evolution of formed martensitic mesostructures with TEM. An illustrative diagram of mechanical loading with photographic inserts showing evolution of martensitic mesostructures is shown in Figure 2.10.

In this figure several stages of martensitic phase deformation under the influence of stress can be distinguished, in which certain deformation mechanisms dominate. In stage (a–b) self-accommodating martensitic morphologies are obliterated through the dominating mechanism of HPVs coalescence. In stage (b–c) further orientation of martensitic phase takes place through the prevailing mechanism of HPVs detwinning. In stage (c–d) reorientation continues through further detwinning and the formation of new "stress" variants of martensite; for example, compound twins to accommodate geometrical incompatibilities. Unfavorable irreversible processes can also take place in this stage. In stage (d–e) a plastic slip deformation mechanism dominates through lattice dislocations formation. A review of research work devoted to martensitic phase reorientation and detwinning of NiTi alloys can be found in Liu and Xie (2007).

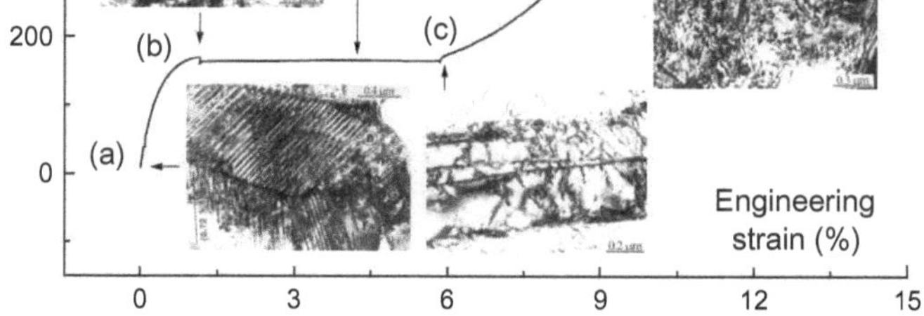

Figure 2.10 Appearing martensitic microstructures and processes taking place during simple tension loading of polycrystalline NiTi in a martensitic state; (a) Temperature-induced self-accommodating martensite, macroscopically undeformed; (b–c) reorientation of martensite, coalescence of HPVs; (c–d) Further martensite reorientation and detwinning of single HPVs, formation of compound twins; (d–e) Plastic slip becomes the dominating deformation mechanism, generation of lattice dislocations.
Adapted after Liu et al. (2000).

Advances in computer technology and the growth of computational power enabled a novel approach to modeling martensitic phase behavior, formation, and evolution of *multiscale martensitic structures*; namely, the *molecular dynamics* (MD) *method*, a kind of ab initio approach starting from atoms and their interactions. In this approach, a material volume is modeled as a large system of discrete particles interacting with each other. These interactions are described with the aid of some force potential function (e.g., Lenard-Jones potential). Presetting relevant parameters in the force potential function enables us to model the behavior of different materials. The particles motion in the system is described directly with Newton's laws. An example of the result from a simulation of martensitic transformation with the MD method performed within the scope of activities of the Network of Computational Methods of Materials Science financed by the U.S. Department of Energy is shown in Figure 2.11 (Computational Materials Science Network, 2005).

The scope of application and targets of MD models of martensitic phase transformation and the previously discussed PTMC model are very different. The MD method allows us to predict the discrete locations of individual atoms, the formation of substructures, as well as explicit energetic (force) relations, for example, stress-strain curves upon appropriate averaging. It seems that the MD method has great potential for the design of new SMA alloys. For example, at least theoretically, the effect on

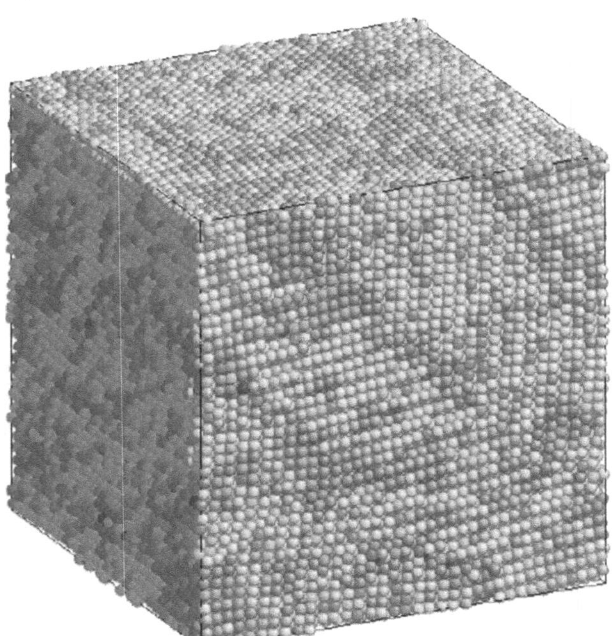

Figure 2.11 Image of microstructure of the martensitic phase, which emerged during numerical simulation of martensitic transformation by the molecular dynamics method (Computational Materials Science Network, web page).
Courtesy of University of Washington.

martensitic transformation of doping of various chemical elements can be assessed—one of the key methods of modification of properties of shape memory alloys. Uehara et al. (2009) advanced the MD methodology to simulate the influence of the grain size and orientation distribution on SMA material behavior. The disadvantage of the MD method is its frequently prohibitively high cost for modeling simulations due to a huge number particles and long computation times.

Frequently, models of martensitic transformation—models of behavior of shape memory alloys—are divided with respect to the characteristic dimension of objects described by such models: the geometrical dimension of input parameters to the model. Three levels of characteristic dimensions are most frequently distinguished: microscopic or atomic scale models—the level of observation of an electron microscope; mesoscopic scale models—the level of observation of an optical microscope; and macroscopic scale models—the level of observation of the naked eye. In this last scale no details of microstructure of the material are differentiated and the so-called representative volume element (RVE) of the material is distinguished, which is treated as homogeneous. The state of RVE internal microstructure is characterized with the aid of some number parameters in the averaged sense. The above-mentioned classification of models, taking into account the characteristic geometrical dimension, is to some extent confusing. Using the criteria indicated above, the PTMC model should be classified as atomic level theory, because input parameters for the model are dimensions of unit cell of crystallographic lattice. However, in a carefully carried out experiment martensitic plate formed can easily attain a size of several millimeters and it is as well a macroscopic object. Looked at from this perspective, the PTMC model can be classified as a macroscopic model as well. Actually, the classification of models of martensitic transformation with respect to characteristic geometrical dimension has, in fact, some other hidden rationale. The criterion is complication of martensitic microstructures, which might be described with the specific model by the very assumption of its authors. The theory on an atomic level is capable of describing the formation of one-compound martensitic plate (HPV), several plates in the form of a wedge or diamond, but it cannot be reasonably applied when one has to deal with thousands or more of martensitic plates. Then, statistical effects start to play a role, and it is necessary to perform some averaging in order to obtain any reasonable result at a prescribed realistic computational effort. In such a situation, the theories on a mesoscopic level find application, as they are assigned to deliver predictions regarding formation of martensitic microstructures consisting of thousands of martensitic plates. The distance of scale from 10^{-9} to 10^{-6} is then covered. The next level of averaging leads to macroscopic models, which are assigned to describe very complex microstructures consisting of a large number of grains and containing many martensitic variants together with other elements like superstructures (e.g., self-accommodating morphologies), and transition from scale 10^{-6} to 10^{-3} is then attained. It seems at first sight that it would be sufficient to develop only atomic scale theories and next, by proper averaging, to move to macroscopic models. However, for each kind of model some set of underlying assumptions must be adopted. Transition from a lower scale model to higher one usually requires further supplementary assumptions, which are often quite vague due to lack of information. Due to that, starting from a very low atomistic level model with the target to obtain averaged macroscopic results at present does not add much credibility to such obtained results. It is appropriate to conduct simultaneous work on atomic, mesoscopic, and macroscopic models depending on the targets to be reached. These targets are frequently quite divergent between communities.

Probably the first mesomechanical model of shape memory alloys was the model proposed by Patoor et al. (1988). The coherency condition $\boldsymbol{F}_i = \mathbf{1} + \boldsymbol{b}_i \otimes \boldsymbol{n}_i;\ i = 1,2,\ldots,n$, being principal results of PTMC, is employed in the model as fundamental conjecture (see Figure 2.8). Actually, the coherency condition constitutes a starting point for practically all mesomechanical models of shape memory alloys behavior. The HPVs normals $\boldsymbol{n}_i$, deformation vectors $\boldsymbol{b}_i$, and a number of different possible variants of martensite n are determined for particular SMA material using the PTMC model. In the model of Patoor et al. (1988) an explicit form of Gibbs function was proposed for two-phase macroelement of a SMA alloy treated as a RVE, initially by the assumption that RVE is composed of a monocrystalline austenitic phase. The state parameters in this potential are stress tensor, temperature, and n volume fractions of martensitic variants z_i treated as internal state parameters. With each variant of martensite a phase eigenstrain $e = 0.5(\boldsymbol{b}_i \otimes \boldsymbol{n}_i + \boldsymbol{n}_i \otimes \boldsymbol{b}_i)$ is associated, also called twinning systems in an analogy to slip systems appearing in the classical crystallographic theory of plasticity. The authors derive criteria that must be fulfilled for active formation of specific martensitic variant, taking advantage of the thermodynamics formalism. Next, they deliver evolution equations for strains connected with phase transformation requiring that initial criterion for active transformation be fulfilled during the whole period of phase transformation. In order to obtain constitutive equations for polycrystalline material, the authors apply a self-consistent averaging method to equations valid for monocrystalline phase. Obtained for polycrystalline SMA material, constitutive equations are so complex that in practice they require numerical computations in order to obtain any result of practical value. The authors performed experimental verification of their model, in very limited scope. They compared predictions of their model, obtained at arbitrary assumption of formation of only one crystallographic variant of martensite in one grain, with results of isothermal, uniaxial tension of a sample made of CuZnAl alloy. The Patoor et al. (1988) model is an important contribution from the point of view of indicating a certain methodology of building mesoscopic models of SMA materials behavior. The stream of mesomechanical modeling of SMA materials develops extremely dynamically and nearly each year a new mesomechanical model of SMA materials appears in the literature. They differ either by averaging method, by adopted criteria for active phase transformation progress, or by phase transformation kinetics relations. For example, Annand and Gurtin (2003) proposed a mesomechanical SMA material model formulated in finite deformations formalism. They numerically performed an averaging procedure, necessary to obtain macroscopic results that could be compared with actual experimental evidence, with the aid of finite element method computations. Such a mixed approach occurs more and more frequently among authors developing mesoscopic models of SMA materials. In addition to mainstream research on SME of pseudoelasticity and one-way memory, also other SMEs receive attention. For example, Cherkaoui et al. (1998) proposed a mesomechanical, 3D model to describe transformation induced plasticity (TRIP), in monocrystalline SMA materials. A phenomenon results from simultaneous operation and interaction of plastic slip and mechanical twinning, two inelastic deformation mechanisms. This model has been subsequently extended to predict the behavior of polycrystalline SMA materials (Cherkaoui et al., 2000); see also text below Figure 2.14. Exploitation of mesoscopic models for engineering design purposes seems to be impractical even at the computational power of today's computers. However, it is purposeful to run comprehensive numerical simulations using mesoscopic models for research purposes in order to discover certain patterns of behavior, or explain not well understood phenomena upon which there could be formulated reasonable conjectures for development and/or improvement of macroscopic models of SMA materials behavior.

It is worth it to revisit SME, uncovered in Figure 1.2, when already equipped with materials science knowledge on thermoelastic martensitic transformation. Deformation effects connected with SME with indicated microscopic and mescopic physical mechanisms responsible and enabling their apperance are schematically illustrated in Figure 2.12. Corresponding macroscopic stress-strain-temperature relations connected with SME with symbolically marked martensitic phase structures are shown schematically in Figure 2.13.

An SMA material loaded mechanically or thermomechanically, in the temperature range where austenite is stable at zero external stress (1 in Figures 2.12 and 2.13), behaves elastically until reaching some critical stress $\sigma^{AM}(T)$ when forward

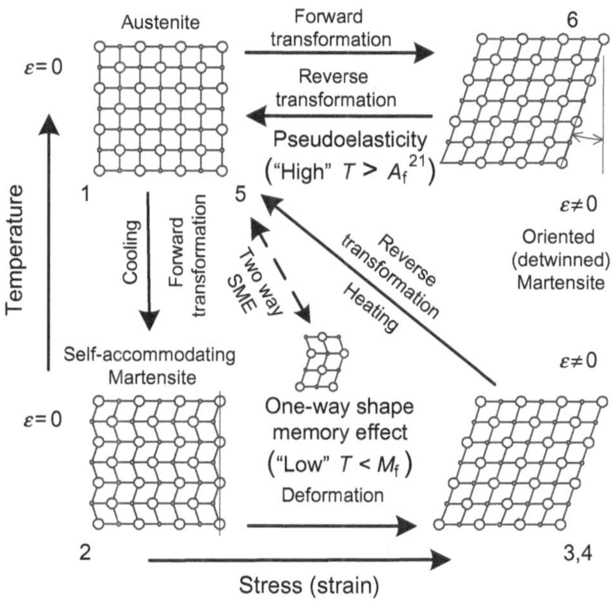

Figure 2.12 Schematic illustration of shape memory effects with underlying microscopic mechanisms.

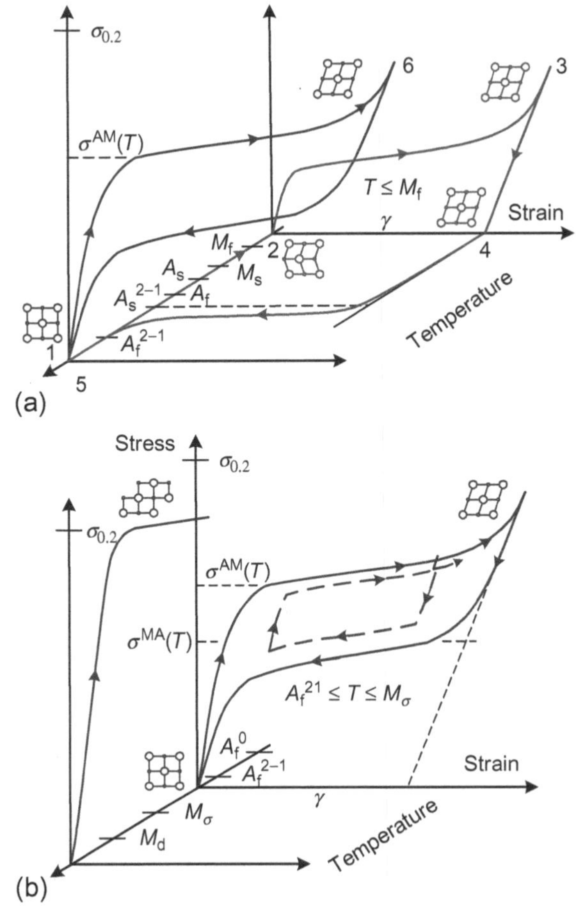

Figure 2.13 Macroscopic thermomechanical behavior of SMA materials in stress-strain-temperature coordinates. Not to scale. (a) one-way memory effect; (b) pseudoelasticity with internal hysteresis loops and plastic slip deformation.

thermoelastic martensitic transformation begins and *oriented martensite* starts to form, that is, the one exhibiting macroscopic phase transformation strain. Increasing mechanical load results in two processes proceeding simultaneously. The phase transition advances and SMA material flows pseudoelastically. At some loading stage, p.t. completes what terminates the pseudoelastic flow plateau and SMA materials again start to behave elastically (6 in Figures 2.12 and 2.13). Strain amplitude of the pseudoelastic plateau usually reaches several percent but it may reach a dozen or so percent. It is marked with the symbol γ in Figure 2.13. On removing the loading stress, SMA material starts to flow back pseudoelastically at some different critical stress $\sigma^{MA}(T)$, where reverse phase transformation begins. Upon complete loading removal, the SMA material completely recovers its initial shape (1 in Figures 2.12 and 2.13). The behavior is very similar to the behavior of nonlinearly elastic rubber materials, but a relatively large hysteresis loop is usually formed in the case of SMA materials. It is said that SMA material at "high" temperatures, with respect to characteristic temperatures of p.t., exhibits a *pseudoelastic effect*.

When SMA material in a fully austenitic state in "high" temperature (1 in Figures 2.12 and 2.13), it is cooled at zero assisting external stress then, at characteristic temperature M_s^0, *self-accommodating* martensite starts to form. If it is cooled to a sufficiently low temperature (i.e., below M_f^0), then p.t. completes and material reaches a fully single-phase martensitic state containing only self-accommodating martensite (2 in Figures 2.12 and 2.13). No change of macroscopic strain is observed during such process (1–2 in Figures 2.12 and 2.13). Subsequently, when SMA material in a self-accommodating martensitic state is loaded mechanically or thermomechanically then self-accommodating martensite, depending on temperature nearly immediately or after the short elastic behavior period, starts to flow exhibiting growth of macroscopic strain. Increasing the mechanical load results in the progress of *self-accommodating* martensite transformation into *oriented martensite*. Similar to the case of pseudoelastic behavior, at some stage of mechanical loading macroscopic reorientation strain saturates and SMA material starts to behave elastically (3 in Figures 2.12 and 2.13). This marks the completion of the transition process into an oriented martensite state. Upon removal of loading stress, the SMA material exhibits apparently permanent reorientation strain (4 in Figures 2.12 and 2.13). Experimental evidence shows that amplitude of reorientation strain, frequently called *phase strain,* can be well accepted to have the same value as pseudoelastic flow amplitude value (see Figure 2.13). The apparently permanent deformation of SMA material in "low" temperature recovers completely upon heating it above A_f (austenite finish temperature) (5 in Figures 2.12 and 2.13); that is, when the martensitic phase is completely retransformed into austenite. Such behavior of SMA materials is called *one-way memory effect*.

Specific *TMT* of SMA material, called *training,* for example consisting of cyclic straining of the SMA material in a martensitic state slightly above the martensite plastic yield limit, called *marforming*, leads to the development of special microstructure of SMA material that results in still another behavior of SMA materials. Upon cooling of "trained" SMA material in an austenitic state from high temperature at zero-assisting macroscopic stress, to low temperature partly self-accommodating partly oriented martensite is formed spontaneously. Such martensitic phase exhibits nonzero macroscopic phase strain with a value somewhere between zero and the locking value of phase strain. The actual value depends on the effectiveness of the training procedure and the type of SMA material itself. In the case of NiTi alloys, it is usually about ¼-⅓ of pseudoelastic flow amplitude (i.e., $\sim 2-3\ (\%)$). When trained SMA material is heated to high temperature, above A_f, complete reverse p.t. takes place and phase strain is fully recovered. Such behavior is called *two-way memory effect* of SMA alloys. It is indicated with a two arrow, dashed line in Figure 2.12.

Very frequently the mechanical loading process of SMA material is stopped before *complete phase transformation*, forward and/or reverse. takes place; that is, conventionally volume fraction of martensite does reach the values $z = 0.99$ or $z = 0.01$. In such situations *internal hysteresis loops* are formed, as shown in Figure 2.13b. One of the important tasks of SMA materials models is the capability to reliably describe internal hysteresis loops formation, either mechanical or thermal.

In Figure 2.13 there are marked characteristic temperatures, which are not familiar DSC characteristic temperatures. Probably the first researcher who explicitly pointed out that characterization of SMA materials with measurement of characteristic temperatures determined in classical DSC procedure only is not sufficient was Sakamoto (2002). Actually, Popov and Lagoudas (2007) were apparently the first who showed experimentally that austenite start temperatures of reverse martensitic transformation of self-accommodating and oriented martensite are considerably different. Their experimental results have been independently confirmed by the present author, as is shown in Chapter 3. The principal cause of the difference in austenite start temperature of self-accommodating and oriented martensite is the difference in their specific free energies. The austenite start temperature of self-accommodating martensite is denoted as A_s^0, where superscript "0" means that measurement takes place at zero-assisting macroscopic stress; this is simply classical DSC-determined temperature. A characteristic temperature A_s, commonly encountered in SMA literature, is equivalent with $A_s^0 = A_s$. The austenite start temperature of oriented martensite is denoted here A_s^{2-1}, also measured at zero-assisting stress. Experimental evidence shows that specific free energy of self-accommodating martensite is higher than oriented martensite, due to the fact that A_s^{2-1} is greater than A_s^0 (see Figure 2.13a). Because austenite start temperatures of reverse transformation are different, hence the austenite finish temperature of self-accommodating martensite A_f^0 and that of oriented martensite A_s^{2-1} are also considerably different. The fact that $A_s^0 \neq A_s^{2-1}$ is of key importance in modeling one-way shape memory effect. In Figure 2.13b there are marked temperatures M_σ and M_d. The M_σ temperature denotes characteristic temperature above which, when material is loaded mechanically, first plastic slip deformation takes place and only later does further inelastic deformation take place due to simultaneous plastic slip and martensitic phase transformation progress. The temperature M_d denotes the critical temperature above which it is impossible to bring about martensitic transformation by application of mechanical load.

Taking a set of characteristic temperatures as a criterion, the domain of external parameters of stress and temperature (σ, T), external in the sense that they can be fully controlled and thus preset at will by a researcher, can be divided into certain subdomains where specific SME take place and/or certain processes dominate in SMA material behavior, as shown in Figure 2.14.

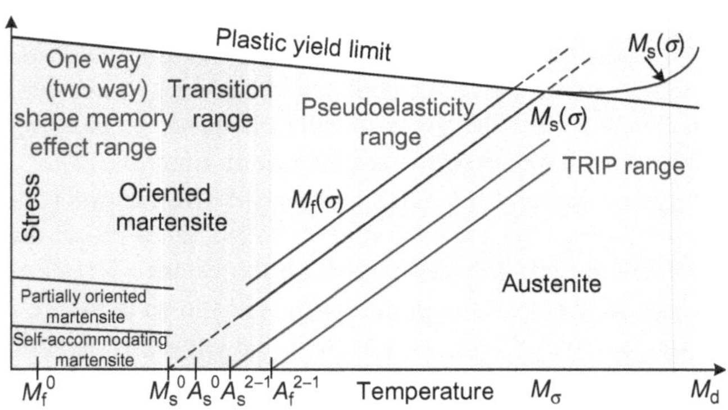

Figure 2.14 Schematic landscape picture of SMA materials behavior and SME effects in temperature and stress coordinates. Not to scale.

Previously discussed SME such as pseudoelasticity, one-way and two-way memory effects take place below M_σ temperature. Between M_σ and M_d temperatures another effect occurs—so-called *transformation induced plasticity*—that can be quite rightly qualified as another shape memory effect. In fact, the historical name of this phenomenon (i.e., TRIP) does not reflect the actual physical situation. In TRIP similarly to the case of SMEs martensitic transformation is in fact induced by stress, or combined thermomechanical load and not plastic strains, which rather facilitate nucleation and growth of the martensitic phase. Appearance of the TRIP effect in a material offers interesting prospects. Material that fractures at relatively low strain of the order of several percent without TRIP effect, a nuisance that is exhibited by many intermetallics, exhibits excellent ductility at relatively high stresses if TRIP effect is present in the material (e.g., the case of NiTi intermetallics). This utility property is present due to the mechanism of concurrent plastic slip and martensitic phase transformation.

Macroscopic constitutive models of materials behavior seem to be the most appropriate for engineering design of structures containing SMA materials elements. In the development process of such models, materials science research results on thermoelastic martensitic transformation are instrumental. They deliver leads in the adoption of the most convenient development framework, creating necessary abstract concepts and ideas, making decisions on what are state parameters and what is the minimum set of them necessary for reliable description of a specific shape memory effect. One of the most important experimental observations of SMA materials behavior is hysteresis loops formation during mechanical and/or thermal loading cycle involving forward and reverse phase transformation. This indicates that martensitic transformation runs through states of constrained thermodynamic equilibrium and hints that formalism of nonequilibrium thermodynamics is the most adequate framework for macroscopic SMA materials constitutive models development.

Let us recall in a selective manner how researchers progressed in describing different SMEs with macroscopic models of shape memory alloys behavior. The first macroscopic models of SMA materials behavior started to appear in the beginning of 1980s, nearly 20 years after the paper of Buehler et al. (1963), which might be regarded as the research that started the SMA materials era. In 1983 (Falk, 1983) proposed a one-dimensional (1D) thermodynamic model of shape memory alloys behavior, based on the Landau-Devonshire theory of phase transformations. In this model, free energy potential is accepted in the form of a sixth-order polynomial. The only state parameters used are total strain and temperature. State equations are obtained in accordance with the formalism of thermodynamics through formal differentiation of free energy potential. Falk suggests replacement of the curves ensuing from the requirement of full thermodynamic equilibrium by iso-stress lines, in the areas of mechanical instability; that is, at states where $\partial\sigma/\partial\varepsilon < 0$. Thus, he postulates an ideal flow of the material at implicitly alleged phase transformation processes, which allows for modeling observed experimentally hysteresis loops. In this model no state variable was explicitly introduced, characterizing phase transformation progress. The model of Falk is capable of describing pseudoelasticity and hysteretic behavior at low temperatures of SMA materials but only in a qualitative manner. In 1986 (Tanaka et al., 1986) proposed a 1D macroscopic model of pseudoelastic behavior of SMA materials, formally written in 3D notation. In this model, besides total strain and temperature, there also appears a new state parameter (i.e., volume fraction of martensitic phase). A big modeling advancement in the Tanaka et al. (1986) model has been explicit specification of the phase transformation kinetics rule for full forward and full reverse phase transformation. The rule is an extension of the known Koistinen-Marburger kinetics equation originally used to describe the process of steels hardening (Koistinen and Marburger, 1959). The model is capable of describing 1D pseudoelastic behavior of SMA materials in a quantitative manner with good accuracy. However, no thermodynamic criteria has been specified in this model defining when active forward or reverse transformation may take place; also, an equation for entropy evolution has not been specified. The next considerable advancement in the description of thermoelastic martensitic phase transition kinetics was made by Liang and Rogers (1990). In formulating their 1D model of SMA materials behavior, they proposed a "cosine" phase transformation kinetics rule capable of describing the formation of internal hysteresis loops. The advantage of this approach is simplicity of mathematical expressions; the drawback is that it is not rooted in an actual physical mechanism of the phase transition. In the above-mentioned models, the weight is put on modeling reconstruction of experimental curves but little attention is paid to the root causes of SME appearance. A contribution of key importance in this last aspect has been the work of Müller and Xu (1991), in which they develop a one-dimensional, macroscopic model of pseudoelasticity. In this work, the authors proposed explicit expression describing evolution of "coherence energy", that is, energy ensuring continuity of displacements field during phase transition in the form $\phi^{coh}(z) = Az(1-z)$, where A denotes material constant and z denotes volume fraction of martensitic phase. The authors justify this specific form as their conclusion from statistical studies of experimental test results performed on samples made of monocrystalline CuZnAl alloy submitted to isothermal, uniaxial tension. They also demonstrated in a mathematically rigorous way that the specific form of free energy proposed by them leads to the conclusion that two-phase states of full thermodynamic equilibrium of SMA material are unstable. Based on this finding, they formulate physically justified criterion enabling the description of internal hysteresis loops formation. The appropriateness of this criterion was confirmed experimentally for polycrystalline CuZnAl alloy. A landmark step in macroscopic modeling of SMA materials behavior was taken by Brinson (1993). Brinson, motivated by materials science observations, distinguished two martensitic phases in proposed by her 1D SMA macroscopic model. The model is capable of predicting pseudoelastic and one-way memory effect behavior of SMA materials. The weakness of the model is a lack of explicitly specified free energy function and quite vague formulation of phase transition criteria, which are effectively based on maps to be constructed for each SMA material on the base of experimental results. The Brinson 1D model has been extended to 3D situations by Panico and Brinson (2007). In the 3D model, the free energy function is already explicitly specified but the term of coherency energy in the form of parabolic function of volume fraction of oriented martensite is unrealistic, as is indicated in the mesomechanical studies presented in Chapter 4 of this book. The other deficiency of the 3D model is the rather involved way the description of deformations are connected with the martensitic phase. The evolution of volume fractions of martensitic phases and evolution of tensorial variables describing deformation effects are not clearly separated; instead, they are quite ambiguously combined together. Apart from efforts to develop models of pseudoelasticity and one-way memory effect attempts to construct macroscopic models of two-way shape memory effect also received some attention; see, for example, Zhang et al. (1992).

Genuine 3D macroscopic models of SMA materials behavior start to appear in the beginning of the 1990s. In this research stream a three-dimensional macroscopic model of pseudoelasticity proposed by Raniecki et al. (1992) seems to place a certain reference basis, in the sense that the model is complete. Complete because it contains the minimum set of objects and rules enabling reliable description of the most important observable characteristics of pseudoelasticity. These embrace an explicit form of the free energy function, including coherence energy, expressed in terms of external state parameters of stress and temperature and internal state parameters of volume fraction of martensite and ultimate phase eigenstrain tensor. The last two parameters are called *internal* because they can be only indirectly controlled by variation of external ones. Introduction of two internal parameters is necessary because volume fraction of martensitic phase is not straightforwardly linked with deformation effects as elucidated above. Raniecki et al. (1992) formulate in explicit mathematical form criteria defining when active forward or reverse phase transformation can take place, specify equations of phase transformation kinetics enabling description of internal hysteresis loops formation, and propose an evolution rule of phase transformation strain tensor enabling description of deformation effects connected with formation and reorientation of martensitic phase. An equation of entropy evolution when specified explicitly enables modeling prediction of heat effects. Actually, two versions of the model have been developed: one assuming full thermodynamic equilibrium during the p.t. processes (called R model (Reference model)) and the second version capable of description of hysteresis loops formation (called R_L model (Reference with loop model)). The constitutive relations of the model are also specified in incremental form, convenient for numerical codes implementation. The R_L model is continuously extended and fine tuned, so that today one can speak about *family of R_L models* stemming from the root work of Raniecki et al. (1992). Raniecki and Lexcellent (1994) enriched the R_L model capabilities to enable description of SMA materials for which it is necessary to take into account considerably different apparent stiffness of austenitic and martensitic phase. Leclercq and Lexcellent (1996), following Brinson, introduced two martensitic phases into the framework of the R_L model to enable description with single-model pseudoelasticity and one-way memory effect. Raniecki and Lexcellent (1998) proposed a new phase transition criteria enabling description of behavior of SMA materials exhibiting *strength differential effect* (SD); that is, SMA materials exhibiting strong dependence of critical pseudoelastic flow stress on direction of stress tensor. Ziółkowski (2006) extended the R_L model to the finite deformations regime using multiplicative decomposition of deformation gradient. The first researchers, as it seems, to propose not formal but genuine macroscopic constitutive model of SMA materials formulated in finite deformations formalism were Auricchio and Taylor (1997). Unfortunately, generic concepts of the model are adopted from classical theory of plasticity and do not take advantage of the thermodynamic framework. Due to that reason, the model is not capable of reliable description of p.t. processes; neither prediction of thermomechanical couplings effects resulting from emission nor absorption of latent heat of phase transition and dissipation of mechanical work. A momentous step forward in macroscopic modeling of SMA materials has been taken by Popov and Lagoudas (2007), who were the first to distinguish in their 3D model proposal targeted at description of pseudoelasticity and one-way memory effect the experimentally observed feature of two austenite start characteristic temperatures; that is, this of self-accommodating martensite and this of oriented martensite. The weakness of the model lies in a poor proposition regarding expression for coherency energy and a pretty obscure description of martensitic phase evolution processes. The feature that does not add value to the modeling proposal is thermodynamic formalism clumsiness. For example, the authors' Gibbs function specification explicitly contains rate terms, dependence on some processes, which is unacceptable because in accordance with thermodynamic formalism Gibbs function describes the thermodynamic state of the material.

The models and approaches noted above under no circumstances are an exhaustive list of all SMA macroscopic models, but are rather the author's selection of the marking contributions and promising research directions in the field of SMA materials behavior modeling. This area of research experiences vivacious blossoming, as each year several models of SMA materials behavior appear in the literature. Besides continuous fine-tuning and improving the accuracy of predictions of models covering a rather narrow range of SMA materials behavior, the main trend in research work concentrates rather on the extension of their predictive power to a broader range of parameters of stress and temperature and more complex thermomechanical loading paths. However, at present no commonly accepted, unified, quantitative theory describing behavior of SMA alloys in full scope of SME exists. Development of such a unified theory is an open scientific task. It should be pointed out that a universal weakness of all 3D SMA materials modeling propositions is that there is very limited or rather nearly nonexistent information about experimental test results from well-planned, comprehensive, parametric studies on SMA materials behavior when submitted to multiaxial thermomechanical loading programs.

2.3 SMA materials alloy systems, their characterization and properties

A large number of distinct alloying compositions exhibiting SME is known at present; see, for example, Otsuka and Wayman (1999) or Humbeeck and Stalmans (2002). The majority of them have been discovered and investigated out of academic curiosity. A number of quite exotic alloys such as AuCd, InTl, FePt, FePd, InCd, and NbRu containing noble and/or rare metals are still under research, but their widespread practical use is prohibited by high cost. The discovery of some SMA compositions has been driven by special demand. In search of the so-called *high-temperature shape-memory alloys* (HTSMA), that is, those operating in temperatures above 100 °C, it has been found that NiTiPd, CuAlMn, and NbRu alloys, for example, undergo thermoelastic martensitic transition at high temperatures. Some further subdivision of HTSMA depending on operational high temperature ranges is also applied as indicated by Jani et al. (2014) in Table 6. In the plethora of SMA materials, the most attention is paid to those with the greatest potential for commercial applications and, thus, potential demand. The *NiTi-based*, Cu-based, and Fe-based alloying systems dominate the SMA materials market at present, the first system being by far the most successful. Regarding commercial success, decisions are based not only on shape memory functional characteristics but also on the combined functional and utility features of a specific SMA material alloy system. For example, the widespread use of NiTi in medicine has been influenced not only by SME and characteristic temperatures of p.t. remaining in the vicinity of body temperature, but also by such *utility features* as excellent biocompatibility, corrosion resistance, ductility, fatigue resistance, and time stability of properties. An important factor is the cost of SMA material, which usually includes not only the cost of alloy preparation but also the cost of *TMT*, necessary in the case of practically all SMA materials to obtain good SME properties. The cost factor strongly depends on gained knowledge on the alloy. At the beginning of the SMA materials era, the cost of production of 1 kg of NiTi alloy was prohibitively high, on the level of $1000 USD/kg to drop below $150 USD/kg today (Alam et al., 2007). While NiTi-based alloys offer excellent SME properties, they are relatively expensive when compared to Cu-based SMA materials by a factor of about 5-10 (see Table 2.1). *Cu-based SMA materials* find application in all those situations where, besides SME effects, good electrical or thermal conductivity are of value, unit cost is an important factor, and excellent corrosion resistance, for example, is not crucial. Such conditions prevail in, for example, the home appliances market. The driving force for search and development of *Fe-based SMA materials* is cost effectiveness, as their primary target is civil engineering infrastructure applications. Recent progress by Japanese researchers (Ishida et al., 2012) reporting in a patent disclosure, the availability of manufacturing technology for Fe-based material with excellent SME properties with 13% pseudoelastic flow amplitude seems to open up excellent opportunities in this area; especially now that manufacturing costs of Fe-based SMA materials already can be two times lower than Cu-based SMA materials.

Characterizing of SMA material properties is a rather challenging task requiring good knowledge of their thermomechanical behavior. Hindrances that may be encountered in the characterization process, and selected experimental techniques to overcome them, are instructively described in a series of five papers published in *Experimental Techniques* journal under the commonly shared title "Tips and tricks for characterizing shape memory alloy wire." Only the first paper in the series is explicitly referenced here (Shaw et al., 2008). Precise technical characterization of SMA materials requires the existence of a commonly accepted constitutive model of SMA materials behavior. Unfortunately, such a commonly accepted model does not exist at present. A commonly agreed theoretical base can be identified only in a modeling description of pseudoelasticity. For example, a set of numerical values of model parameters listed at the end of Chapter 3 of this book can be understood

Table 2.1 Selected physical and thermomechanical properties of SMA alloys

Commercial SMA materials systems	**NiTi-based**	**Cu-based**	**Fe-based**
Specific examples of alloys	NiTi, NiTiCu, NiTiNb, NiTiAg	CuAlMn, CuAlMnZr, CuAlNi, CuZnAl	FeNiCoAlTaB, FeMnAlNi, FeMnSi
SME characteristics			
Transformation range (°C)	−200 to +200	−200 to +200	−200 to +150
Hysteresis (°C)	2-50	5-40	
$d\sigma^{AM}/dT$ (MPa/°C)	6.0-8.0	1.5-3.5	0.5-3.0
Specific entropy (J/kg)	12-32	7-9	
Maximum overheating temperature (°C)	400	300	
Maximum one way memory (%)	6-10	2-8.5	4-13
Typical two-way memory (%)	2.5-3	1-2	
Maximum recovery stress (MPa)	600-900	400-700	400
Typical operating stress (MPa)	100-130	40-70	
Typical number of thermal cycles	$>10^5$	$>10^4$	
SMA functional fatigue, strain (%)	6-8	4-6	
$N<10^2$	6-8	4	
$N<10^5$	2-4		
$N<10^7$	0.5		
Damping capacity (SDC %)	15-20	30-85	
Physical properties			
Density (kg/m^3)	6400-6500	7100-8000	7200-7500
Melting temperature (°C)	1250-1300	950-1050	1350
Specific heat (J/kg C)	450-520	390-440	550
Latent heat p.t. (KJ/kg)	12-28	4-7	
Thermal conductivity (W/m C)	18 (A) 8.6-10 (M)	30-120 (A)	8.5 (A)
Thermal expansion coefficient ($10^{-6}\cdot 1/C$)	6.6-11	17	17
Electrical resistivity (Ohm m$\cdot 10^{-6}$)	0.5-1.1	0.07-0.14	1.0-1.3

Continued

Table 2.1 Selected physical and thermomechanical properties of SMA alloys—Cont'd

Commercial SMA materials systems	NiTi-based	Cu-based	Fe-based
Mechanical properties			
Young modulus (GPa) austenite	65-75	70-100	140-200
Effective stiffness (GPa) martensite	30-90		
Poisson coefficient	0.34	0.34	0.36
UTS (fully annealed) (MPa)	900	400	
UTS (cold worked) (MPa)	1900	500-900	700-1000
Conventional yield $Y_{0.2}$ austenite (MPa)	100-800	150-400	200-400
Elongation fracture (fully annealed) (%)	20-60	-	-
Elongation fracture (cold worked) (%)	15-25	8-15	16-30
Structural fatigue $N=10^6$ (MPa)	350	270-350	
Grain size ($m \times 10^{-6}$)	1-100	25-150	
Other			
Corrosion resistance	Excellent	Fair-good	
Biological compatibility	Excellent	Bad	Bad
Melting, casting, and composition control	Difficult (very strict)	Fair	
Forming (rolling, extrusion)	Difficult	Easy-difficult	
Cold-working	Fair	Fair-difficult	
Machinability	Difficult	Good	Good
Cost ratio	100	1-20	8-12

The listed data can only serve for size-up purposes.
Collected from Wu (1990), Humbeeck and Stalmans (1999), Humbeeck and Stalmans (2002), Huang (2002), Janke et al. (2005), Sutou et al. (2008), Maruyama and Kubo (2011), Jani et al. (2014), and the Memory Metalle web page.

as full and precise characterization of SMA material in the pseudoelastic range of its behavior. Most often, characterization of SMA materials is fragmentary. Manufacturers of devices made of SMA materials introduce their own characterizations with specific requirements in mind that are relevant for their specific product. As NiTi SMA materials found applications in medicine as implants, government regulations for medical safety imposed a necessity to introduce standards for NiTi alloys used for manufacturing implants. These are, for example, *ASTM standards* concerning NiTi alloys specifications F2004, F2005, F2063, F2082, F2516, F2633. Unfortunately, these standards contain specifications of only the most relevant, but very limited, features of NiTi SMA materials. *TMT* of SMA materials is an essential part of the manufacturing process strongly influencing its final properties. Hence, reliable characterization of an SMA material requires determination of its specific properties after performing all technological operations in its "ready to use" state. Detailed knowledge on the history of all technological operations applied to SMA material is very helpful in preventing unexpected harmful effects.

In the engineering design process, instead of using parameters distinguished in a constitutive model it may prove more convenient to take advantage of a certain "index" of material properties in accordance with the concept introduced by Ashby (2011). For example, in practically all constitutive models of pseudoelasticity model parameters of entropy of phase transformation (Δs_2^0), amplitude of pseudoelastic flow in simple tension (γ_{tn}), and density of SMA material (ρ) are present (see also Chapter 3). While necessary for characterization of the SMA material itself, they are not very useful in the design process for evaluation of an SMA material temperature operating window; information that is very important when assessing the applicability of some SMA material to operate in open-air conditions, a typical situation in civil engineering. Knowledge of the value of temperature sensitivity of pseudoelastic flow stress, $d\sigma^{AM}/dT$, is much better for that purpose. This index allows the immediate evaluation of how stress of pseudoelastic flow plateau will move with temperature variation. The $d\sigma^{AM}/dT$ sensitivity is related to SMA material model parameters mentioned earlier with the formula $d\sigma^{AM}/dT = \rho\Delta s_2^0/\gamma_{tn}$. For example, the Fe-based alloy FeMnAlNi has $d\sigma^{AM}/dT = 0.5$ (MPa/C). So variation of temperature by 20 °C will shift pseudoelastic plateau stress by only 10 MPa. The FeNiCoAlTaB SMA alloy has $d\sigma^{AM}/dT = 3.1$ (MPa/C), which seriously limits its operating temperatures window (Omori et al., 2011). An even worse situation in that respect is an NiTi alloy for which this index has a value of 6.5 (MPa/C).

Table 2.1 provides a list of representative properties characterizing three main groups of commercial SMA materials. The data are not limited to SME functional properties but include other aspects of SMA materials important in engineering design processes; that is, the data are divided into SME functional properties, physical data, mechanical properties, and other utility characteristics. They were compiled from several sources, including Wu (1990), Humbeeck and Stalmans (1999), Humbeeck and Stalmans (2002), Huang (2002), Janke et al. (2005), Sutou et al. (2008), Maruyama and Kubo (2011), Jani et al. (2014), and the Memory Metalle web page. The data collected in Table 2.1 are intended to deliver general knowledge. They should be treated as indicators of what might be achieved by the materials supplier upon variation of chemical composition, including low additions of extra constituents and adequate technological processing. The decision on making use of a specific SMA material in a specific application must always be preceded by a detailed feasibility study. A word of caution about the frequently encountered entry of "Young modulus" in SMA literature about austenite and martensite of SMA material: a value of ~70 GPa for NiTi alloy in austenitic phase state is frequently encountered in the literature, while for martensitic phase state only about half of this value. A long-lasting controversy and a lot of confusion exists regarding this entry; see, for example, Duerig (2006). Sittner et al. (2014) added considerably to the elucidation of this problem recently as they evaluated Young's modules of austenite and stress-induced martensite of polycrystalline $Ni_{51at\%}Ti$ 0.1 mm diameter wire by three different experimental methods (tensile tests, in situ synchrotron X-ray diffraction, and dynamic mechanical analysis (DMA)) and executed some theoretical analyses. They found by all three methods that the value of Young modulus of NiTi austenite is ~70 GPa at room temperature. However, large discrepancies have been obtained for Young modulus of martensite. Tensile tests have shown 18 GPa on loading in tension, 38 GPa on unloading in tension, 51 GPa on loading in compression, and 92 GPa on unloading in compression (see Figure 10 in Sittner et al. (2014)). The DMA and neutron X-ray diffraction experimental measurements deliver still different values. The authors attribute these discrepancies to parent phase drawing texture, martensitic phase martensitic variants texture, and elastic anisotropies of single crystal of parent phase and B19′ martensite. The present author's conclusion is that the "Young modulus" entry for martensitic phase property in the SMA literature is not Young modulus in its definitional sense at all, but should instead be called and understood as "effective stiffness" of the martensitic phase. The reason is that it is very difficult, if not impossible, to separate true elastic behavior of the martensitic phase, connected with an increase of interatomic distance, from inelastic deformational effects of this phase connected with its internal rearrangement resulting from such processes as martensitic variants coalescence or detwinning when loading-unloading mechanically martensitic phase.

In Table 2.1, selected physical and *thermomechanical properties* of three main classes of SMA materials are listed.

2.4 Fabrication and processing routes of SMA materials, TMTs

General routes of *manufacturing and processing* of three commercial groups of SMA materials (i.e., NiTi-based, Cu-based, and Fe-based) have been discussed and reported in a number of publications. The conventional production technique of SMA materials is through *alloying*, melting, casting, shaping, and finishing treatment. The manufacturing process of bulk alloy starts from high-purity components or prealloys. They are melted and cast to obtain ingots with mass reaching several dozen kilograms depending on the scale of production. The ingot is homogenized, usually in a multistep procedure, and repeatedly hot forged, rolled, or extruded to obtain rod or plate of required dimensions. This is followed by multiple cold deformation; for example, cold-rolling or wire-drawing with degrees of cold work in one step reaching 40%. By the degree of cold work, it is the most often understood logarithmic degree of cross-section reduction, $\ln(A/A_0)$. As a result of these technological operations, a semi-finished product of sheet, wire, or tube is obtained. Figure 2.15a shows an illustrative scheme of the process in the case of Cu-based SMA materials (after Asanovic and Delijic (1999)).

A special processing procedure is needed to obtain the required functional properties of SMA materials, that is, good SME properties. The procedure is called TMT or, more colloquially, *training*. In fact, SMA product forming processes cannot be clearly separated from TMT procedures for obtaining good SME. They are two sides of the same

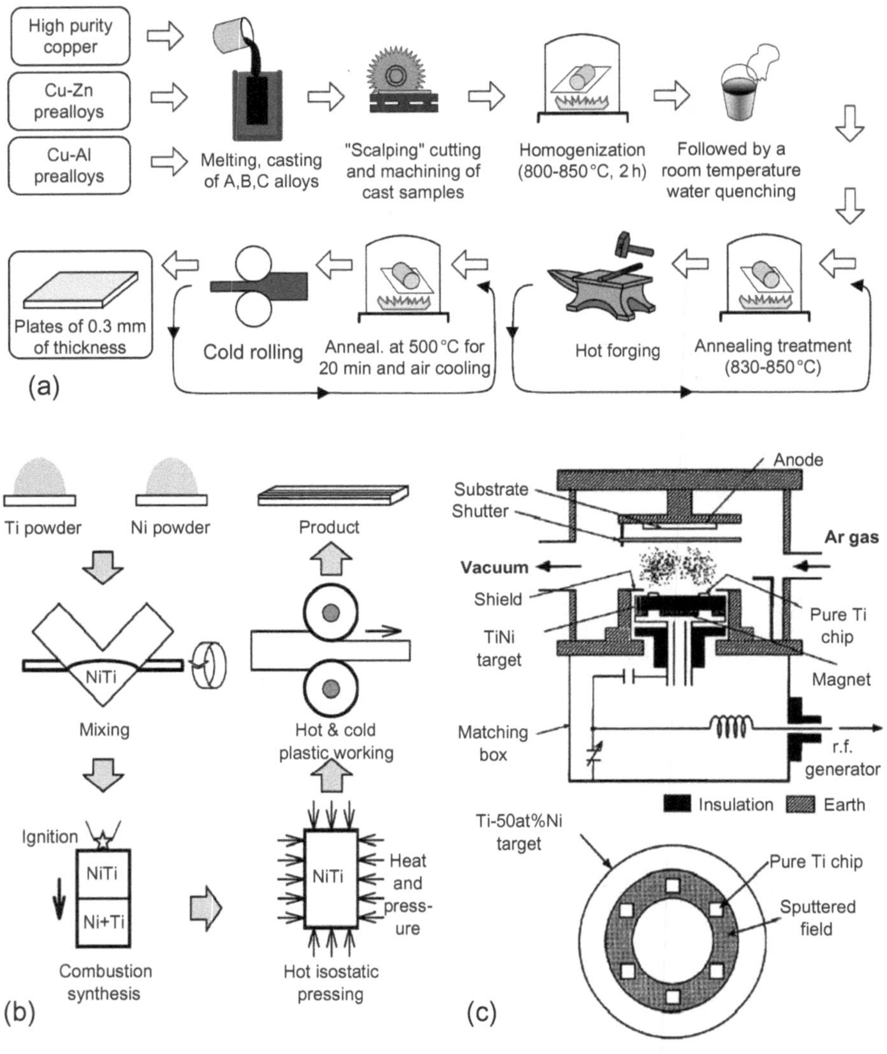

Figure 2.15 Schematic illustration of manufacturing technology: (a) conventional melting and casting of CuZnAl plate (after Asanovic and Delijic (1999)); (b) powder sintering and hot isostatic pressing of NiTi wire (after Kaieda (2003)); (c) radio frequency magnetron sputtering of NiTi thin film (after Miyazaki and Ishida (1999)).

coin and due to that they are usually highly integrated and optimized. The reason for that is that the fundamental influence on manufacturing difficulties encountered has already alloying composition itself. Thus, those processing routes are enforced, which effectively lead to obtaining the wanted shape, functional SME properties, and utility features like fatigue life or biocompatibility. The last operation in manufacturing a component made of SMA material is its high temperature *shape setting*, setting remembered shape, when the component is in an austenitic phase state. In the case of a binary NiTi alloy, the procedure in its simplest form consists of heating the SMA component fixed on a jig, in the required hot shape, to a temperature between 325 and 525 °C and keeping it at that temperature for 5-30 min (shorter times at higher temperatures). This is a one-time procedure unless a new hot shape must be preset to the SMA element.

The NiTi group of SMA materials is a most challenging area in manufacturing and TMT processing, and the most rewarding. The NiTi binary alloys with nickel content between 49 and 51 atomic percent find the most widespread use. In a conventional production procedure of the NiTi alloy melting process, it is executed in a vacuum due to the high reactivity of titanium; see, for example, Nakahata (2011). The other cause is the requirement for strict quality control of the level of oxygen, carbon, and nitrogen impurities contamination, as these impurities strongly affect transformation temperatures, hysteresis loop, and ductility of the material. A variation of chemical composition by 0.1 wt% in the Ni and Ti components leads to a 10 °C change of characteristic phase transformation temperatures.

Titanium, also in NiTi compound, has a very strong affinity to oxygen, which results in picking it up to form TiO, TiO_2, $Ti_4Ni_2O_x$, with values of x between 0 and 1, and in this way depleting the NiTi matrix of titanium. The effect has two aspects. The positive side is that TiO_2 forming due to oxidation on the free surface of NiTi, with oxygen from the atmosphere, creates a very tight protective layer that makes the NiTi alloy corrosion resistant. However, in medical applications, binding of Ti on the surface increases the concentration of relatively free Ni atoms on the NiTi surface, and this increases the risk for the appearance of effects harmful to health in patients especially sensitive to nickel atoms. The presence of oxygen in the bulk of the NiTi material, due to unavoidable contamination of raw components, is unfavorable because it creates problems with melting accuracy. Difficult to control with sufficient accuracy depletion of bulk NiTi matrix of Ti due to oxygen binding results in shifting phase transition temperatures down. It is interesting to note that $Ti_4Ni_2O_x$ particles are insoluble and unaffected by heat treatments so they are inclusions and not precipitates. Their presence can be another factor in obtaining good SME properties. The issue of the influence of oxygen-related inclusions in NiTi matrix on SME functional properties is an open scientific problem; see also Duerig et al. (2011).

Carbon is the second disadvantageous impurity in bulk NiTi that must be carefully controlled. The problem with carbon content in bulk NiTi arises only when carbon crucibles are used in the melting, casting process. Similar to the case of oxygen, carbon binds Ti atoms from NiTi matrix to form TiC inclusions, which shifts p.t. temperatures of NiTi to lower values that may result in melting inaccuracy. The other difficulty is that carbides are brittle. They initiate cracks that create manufacturing problems and can cause shortening of fatigue life.

Two melting techniques of NiTi alloys are used in industry: vacuum induced melting (VIM) and vacuum arc melting (VAR). The advantage of VIM is that it conveniently enables obtaining high homogeneity of ingot thanks to the stirring effect of eddy currents. The disadvantage is that ingots are prone to impurities due to the necessity of using crucible. The most popular are graphite crucibles due to convenience in handling and their low cost. The VAR technique does not require using crucible but requires multiple remelting to reach sufficient homogeneity. Quite frequently, NiTi producers combine both methods to reach an economic compromise in satisfying quality requirements and process economy. Technology developments allow us to keep the impurities level of oxygen and carbon on the level of 200-500 ppm, which is sufficient even for medical-grade NiTi.

Intensive research efforts are also undertaken in the development of nonconventional production techniques for manufacturing of SMA materials, predominantly NiTi-based materials, such as various variants of powder sintering metallurgy, rapid solidification methods of melt-spinning or twin roll casting, and very specialized for thin films production magnetron sputter deposition techniques.

Each manufacturing technology has some advantage that allow it to overcome specific difficulties and/or offers relative ease in reaching certain benefits in comparison to other production methods. The chief advantage of *powder sintering* metallurgy manufacturing technologies is the relative ease of use for very precise control of stoichiometry, perfect homogeneity of manufactured NiTi sinter, and the possibility to control porosity of NiTi product. This makes powder sintering technology a prime candidate to manufacture NiTi-based SMA materials with third elements additions, in which segregation is a peril. It is well known that the formation of NiTi alloy from powder Ni and Ti components is connected with with the emission of considerable heat of NiTi alloy formation, on the level of $\Delta H = 70$ KJ/mol. This creates an opportunity for self-combustion synthesis manufacturing technology of NiTi as described in Kaieda (2003). When mixed Ni and Ti powder is ignited, the sintering front self-propagates in a chain reaction until all the powder is consumed to synthesize a NiTi alloy (see Figure 2.15b). It is interesting to note that combustion synthesis does not involve macroscopic melting as the sintering process is carried out in a furnace at a temperature of about 350 °C, in a vacuum of 1×10^{-3} Pa, during several hours depending on the size of a batch. The porosity of combustion-synthesized TiNi ingots is about 60%, necessitating they receive subsequent compression to obtain the desired level of relative density. This is done in a hot isostatic pressing step, where a pressure of 100 MPa is applied and a temperature of 900 °C is maintained for several hours. In order to avoid contamination, this is done in protective capsules. Upon termination of isostatic pressing, the protective capsules are removed and the sintered "ingot" is submitted to a forming processes of hot forging, hot rolling, cold wire-drawing, and cold-rolling to obtain the required shape. Similarly as is the case of melting and casting ingots the powder metallurgy parts are submitted in final step to shape setting and TMT processing to obtain required SME functional properties. The NiTi powder metallurgy technologies arouse great interest of researchers with medical applications in view. These manufacturing techniques allow for easy control of porous material characteristics such as pore size, shape, distribution, or pores orientation and/or material final density. This, in turn, allows for alteration of mechanical properties to better match natural bone or tissue characteristics; for example, effective elastic stiffness to limit stress shielding. Optimized pore structure increases material biocompatibility, facilitates in-growth of new bone tissue, and thus quickens the healing processes. Elahinia et al. (2012) discuss advantages and drawbacks of a number of variations of powder metallurgy techniques designed for manufacturing NiTi implants.

Rapid solidification techniques prove to be economically and technically viable in a situation where it is necessary to manufacture NiTi ribbons with a thickness on the order of 30 μm. There are several variations of the technique such as melt spinning (MS,) (Anselmo et al., 2011) or twin roll casting (TRC) (Goryczka and Ochin, 2005). In principle, the technique consists of ejection of molten NiTi alloy through a small orifice, previously prepared with a conventional casting method, on a rotating water-cooled copper roller. The molten metal rapidly solidifies on the roller with very high cooling rates from 10^4 up to 10^6 K/s. Higher cooling rates are attained through an increase of tangential speed of the roller surface, which can be on the order of 50 m/s. Larger speeds result in thinner ribbon thickness and smaller grain size. Smaller grain sizes shift transformation temperatures to lower ones. In the case of some alloying composition, the cooling rates are so high that ribbon is fully amorphous after MS and it has to be submitted to crystallization processing in order to reveal SME.

There is great application potential in using NiTi alloys to manufacture SMA micropumps, microvalves, microactuators, micropositioners, or microgrippers, as discussed by Miyazaki et al. (2009) or Kohl (2004). The SMA microdevices allow overcoming one of the main obstacles in the application of SMA materials—low frequency of operation. In the case of bulk SMA materials, it is limited to a few hertz; rate of heat removal being the limiting factor. Development and efficient manufacturing of microdevices is based on the availability of NiTi thin films with thicknesses on the order of 0.5-2 μm, an order of magnitude thinner than those typically obtained by MS production technology. It appears that the manufacturing technology exhibiting the biggest advantages in production of thin film membranes has proved to be *magnetron sputter deposition*, and thus it is at present the prevailing method of choice. The technology allows overcoming such problems as nonuniformity of thin film thickness and/or composition, low deposition rate, and non-batch processing being a considerable drawback of other thin film production methods. Figure 2.14c shows schematically the principle of operation of the most commonly used radio frequency magnetron sputter deposition process (after Miyazaki and Ishida (1999)). Argon gas ions are accelerated in an electric field to high energies. Upon hitting the NiTi target they sputter Ni and Ti atoms to create vapors, which subsequently deposit on the substrate to create NiTi thin film. The thickness of the film can be easily controlled by controlling the time length of the deposition process. Additional pure Ti chip targets are used to combat problems with obtaining the required ratio of Ni/Ti. Such problems arise for many reasons; for example, due to binding of Ti atoms with oxygen. In general, sputter-deposited NiTi thin films are amorphous and must be heat-treated to crystallize. Initially, it has been perceived as a significant drawback of the technology but recently it seems that it has been turned into an advantage. Upon crystallization of the material with a laser beam in precisely controlled locations, SME are activated only in these locations while the other locations are SME unresponsive. The concept opens very interesting new efficient possibilities for manufacturing microdevices.

Depending on the specific application, the following objectives can be sought and reached by proper SMA material selection, alloying composition, and TMT processes: modify transformation temperatures and by this influence plateau stress of pseudoelastic flow, modify pseudoelastic flow amplitude, control level of recovery stress/

strain, increase or decrease hysteresis, separate R-phase and monoclinic transformation in NiTi alloys, modify amount of latent heat of phase transition, modify kinetics of phase transition, modify stiffness, increase parent phase resistance for slip, increase elongation at fracture, increase biocompatibility, increase corrosion resistance, increase creep resistance, control hardness, increase radiopacity, control brittleness, increase resistance against crack failure, and fine-tune fatigue life. Naturally, always some optimized compromise has to be reached because improvement in one property often necessitates worsening of another.

Figure 2.16 shows a schematic illustration, adapted after Treppmann and Hornbogen (1997) and Mertmann (2000), of various microstructural processes used in *manufacturing and processing of SMA materials* to modify austenite and martensite microstructure and in this way influence their structural and functional SME properties. Practically all known metallurgical techniques are used to induce this processes: cold work, hot work, work texturing, aging, annealing and quenching, thermal cycling, deformation cycling or combined cycling. The control parameters of these processes are degree of cold work, number of cycles, temperature and time of the thermal step operations and/or rate of cooling.

In the infinite possibilities and variations of manufacturing and TMT paths, simplicity and straightforwardness of actual treatment is usually very plainly encouraged and imposed by an intolerable increase in costs. For example, while third elements alloying additions to a binary NiTi alloy open up significant advantages, 90% of commercially used Nitinol is still binary NiTi at present, as evaluated by Duerig et al. (2011). Ternary alloys are used only in those specific applications where the advantages resulting from their use cannot be realized without piling up difficulties imposed by other methods. The development of optimum methods for specific SMA

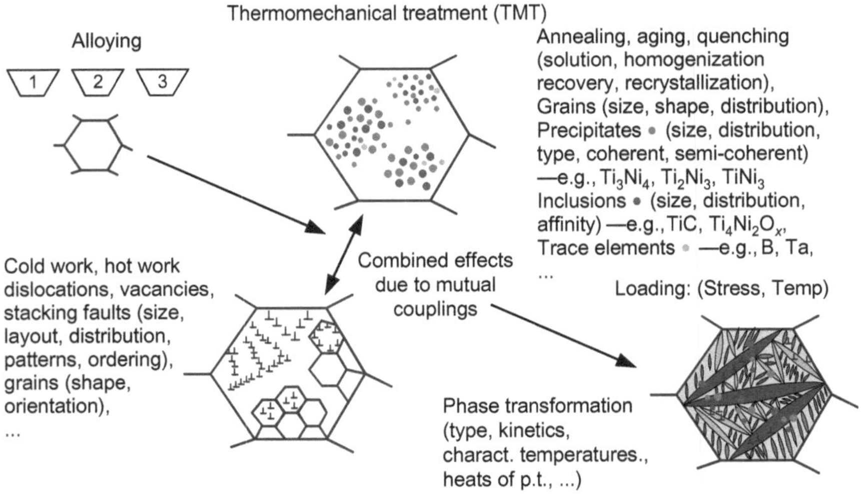

Figure 2.16 Schematic map of various thermomechanical treatments to obtain required properties of SMA alloys.
Adapted after Treppmann and Hornbogen (1997) and Mertmann (2000).

application manufacturing and TMT technology is laborious, requiring considerable time and expenditures. For that reason commercial SMA materials companies reluctantly, if at all, reveal detailed information, treating their information as a trade secret. However, information can be found about general manufacturing routes and the motivations behind them. For example, Tuissi et al. (2004) discuss manufacturing and the *TMT* process for obtaining NiTi wire with good properties to be used as an active element in a SMA actuator. The primary concern of the process has been functional fatigue and the imposed target is to manufacture a wire, which under constant, counterstress of 200 MPa, recovers 4% strain in at least 50,000 cycles. Further information on design requirements and operating conditions of such a specific application dictating specific manufacturing and TMT routes is delivered in Mertmann and Vergani (2008). The general route of processing and quality control of binary NiTi to be used in medical devices is discussed in Mertmann (2000).

Detailed, reliable information on manufacturing and TMT processing routes of SMA materials are better available in scientific papers and/or patent disclosures. For example, manufacturing and TMT processing to obtain perfect pseudoelasticity in Fe-based SMA material is revealed in Ishida et al. (2012).

Discussion devoted to SME functionality treatment, training TMT, is focused on the most important commercial group of SMA materials—NiTi-based alloys. Definite finishing TMT treatment resulting in presetting final values to different SME functional properties such as p.t. temperatures, latent heat of p.t., and pseudoelastic flow amplitude constitutes, in fact, final tuning of SMA materials properties, whose basic range is predetermined by alloying composition. The impact on SME properties of adding a relatively small amount of specific ternary elements to equiatomic $Ni_{50at\%}$-Ti alloy, in replacement of Ni or Ti, is quite well known at present; see, for example, Duerig et al. (2011) or Biscarini et al. (2008). In one example, as reported by Frenzel et al. (2008), Fe addition results in clear separation of R-phase transformation from a monoclinic one. It is well known that R-phase transition exhibits very small hysteresis loop. Separation of two transitions is advantageous when a specific SME application requires very small hysteresis loop, and due to that its operation has to be based on R-phase transition only. Replacement of nickel with a small amount of iron on the level of several percent is a method to move martensite start temperature down considerably; Fe has the effect of stabilizing austenite at higher temperatures. Similar to iron, though less pronounced, are cobalt additions. The addition of niobium as a ternary element results in nearly doubling the hysteresis loop in comparison to an equiatomic NiTi alloy. This feature is useful, for example, in NiTi pipe joints not to loose the joint tightness upon unpredicted lowering of temperature (http://www.aerofit.com). The addition of copper in place of nickel allows us to reduce the hysteresis loop from about 40 to about 10 °C, with copper content exceeding 10%. The most widespread use of NiTiCu alloys is found in actuators and orthodontic wires. Although, in this last application area, quaternary alloy NiTiCuCr is competitive because chromium increases material corrosion resistance and reduces p.t. temperatures to better match body temperature Farzin-Nia and Yoneyama (2009). Additions of Pt, Pd, Hf, or Zr to NiTi are known to increase phase transformation temperatures to above 100 °C and in this way allow obtaining so-called high-temperature SMA materials. Some ternary additions are used not for modification of SME functional properties but to improve utility properties. For example, tungsten is used to increase radioopacity of medical devices made of NiTi. A very important feature in minimal invasive surgery procedures, boron is added to control the grain size.

The TMT procedure, training, to impart final SME properties in NiTi-based alloy with specific alloying composition is usually a combination of cold work and recovery and/or precipitation heat treatment. It is commonly agreed in the materials science community that a single-phase region of binary NiTi alloy below 650 °C exists for narrow compositions in a range from 50 to 50.5 at% Ni. Precipitation treatment at temperatures between 300 and 500 °C is effective only for binary NiTi alloys with composition exceeding 50.5 at% Ni because no precipitation of Ni_xTi_y compounds takes place otherwise (Saburi, 1999). Fully annealed, for example, at a temperature of 850 ° C for 1 h and water quenched, an NiTi alloy exhibits only partial pseudoelasticity. In a fully annealed state, due to too low plastic yield stress, considerable residual plastic slip deformation is generated in the mechanical loading cycle. For alloys with near equiatomic composition, cold work and annealing heat treatment at temperatures below 600 °C —for recovery and possibly partial recrystallization—is necessary to obtain good shape memory effect. The cold work step is needed to increase resistance for slip deformation of the material. There are many variations of cold work treatment such as ausforming (plastic slip deformation of material in an austenitic state) and marforming (plastic slip deformation of material in a martensitic state); see Treppmann and Hornbogen (1997). The temperature of recrystallization depends on the degree of previous degree of cold work. For example, for 25% cold worked near-equiatomic Ti-Ni alloys it is between 500 and 600 °C. It has been found that 25% cold worked near-equiatomic Ti-Ni alloy annealed at a temperature of 600 °C or above (i.e., close to recrystallization temperature) again behaves like it is fully annealed at 850 °C, exhibiting only partial pseudoelastic strain recovery; see Saburi (1999). Thus, in order that post-cold work and annealing, *TMT* can be effective, it has to be done at temperatures limited from above by recrystallization temperature in order not to completely annihilate high dislocation-density introduced by cold work. It has been found that the degree of last step cold work before annealing treatment around 30% is favorable for fatigue life of NiTi-based alloys; see Biscarini et al. (2008).

The processing of NiTi alloys with composition exceeding 50.5 at% on one hand gives more opportunities but requires subtlety and knowledge on the other. The reason is that heat treatment operation after last cold-work step of such alloys inseparably involves two physical processes: modification of high dislocation-density structures (e.g., their partial annihilation and/or their ordering) and generation of various precipitation structures. High-degree cold-work step results in complete suppression of phase transition. Time and temperature of thermal treatment allows, through annihilation/modification of dislocation structures, for control of the extent of recovery of phase transition capability. This finds reflection in *latent heat of phase transition*, which in ready-made NiTi alloy can have a value in the range of 10-30 J/g. Cold work can also influence whether phase transition takes place in one, two, or multiple steps. The phase transformation can follow different paths: $B2 \rightarrow B19'$, $B2 \rightarrow R \rightarrow B19'$, $B2 \rightarrow B19 \rightarrow B19'$ and $B2 \rightarrow R \rightarrow B19 \rightarrow B19'$; see, for example, Braz Fernandez (2013). Similarly, very dense structure of very small precipitates also can lead to nearly complete suppression of phase transition. Precipitation is a thermally activated nucleation and growth process. Favorable conditions for nucleation or for growth can decide on the structure of precipitates and this, in turn, influences thermoelastic p.t. kinetics. Fan et al. (2004) managed to explain puzzling behavior of 50.6 at% Ni-rich NiTi alloy, which in single crystal form exhibited two-stage phase transformation and in polycrystalline form showed three-stage p.t. The alloy with 51.5 at% Ni-rich composition showed two-stage transition regardless, in single crystal and in polycrystalline form. The key factor appeared to be the large relative difference between the nucleation rate in the grain interior region I_{GI} and the boundary region I_{GB} as a function of saturation degree. For small saturation degree of 50.6 at% Ni alloy, this ratio is $I_{GB}/I_{GI}=9$ and for large saturation degree of 51.5 at% Ni alloy it is only $I_{GB}/I_{GI}=1.2$. In 50.6 at% Ni alloy supersaturation is very small and, thus, heterogeneous nucleation at the grain boundary is much favored versus homogeneous nucleation in its interior. As a result, precipitation takes place mainly at the grain boundaries and no precipitation occurs in the grain interior. In 51.5 at% Ni alloy supersaturation is high and there is a relatively small difference in the nucleation rate on the grain boundary and its interior; hence, precipitation distribution is pretty much homogenous. The difference in precipitates distribution allows us to explain the difference in behavior of two alloys; see also Section 9 in Otsuka and Ren (2005).

The grain size proved to be a key factor for success in obtaining pseudoelastic strain amplitude of CuAlMn alloy comparable to that of NiTi-based alloys. A CuAlMn alloy as potential SMA material arouses interest because of its excellent ductility in comparison to other Cu-based SMA materials. Excellent SME has been achieved in CuAlMn alloy only when relative size of grain measured as d/t, or d/D has been sufficiently increased, where d is mean grain size, t is thickness of a sheet, and D is diameter of a wire. The optimum relative size seems to be one grain per cross section of SMA component made of CuAlMn alloy; see Sutou et al. (2008).

The *TMT* of a specific shape memory alloy frequently is an artistic play to obtain the preferred flavor of final properties.

2.5 SMA materials application areas, functional and utility features

The first industrial application of SMA materials was in the military area, in 1971. This was a coupling made of NiTi alloy used to connect hydraulic tubing in F-14 aircraft. Defense, aerospace, and medical markets have been the first customers for SMA alloys, as these markets eagerly accept higher prices for superior performance. Nowadays, shape memory alloy applications slowly but surely spread out to numerous branches of industry, though the medical market attracts by far the highest interest and attention of various communities interested in shape memory alloys. Several books are available at present delivering information on successful commercial applications of SMA materials, mainly NiTi-based (see Section 1.1), besides review papers. For example, Rice (2002) lists such products as pipe couplings, muscle wires, brassiere underwires, orthodontic wires, files, clothing chivalry such as rim of hats, shoe wires to retain their shape, electrical connectors, eyeglass frames, mobile phone antennas, fasteners, connectors, medical guide wires, bendable surgical tools, implants, vascular stents, vibration dampers, structural dampers, temperature sensors, safety products such as coffeemaker switches, anti-scald valves that shut off the tap in sinks, tubs, or showers, fire sprinklers, actuators, and others. Song (2010) presents a history of the development of NiTi devices and implants used in minimally invasive surgery. He also discusses trends in SMA materials technologies and identifies future research directions and opportunities in the area. Petrini and Migliavacca (2011) prepared a report on academic research, patent activity, and perspectives related to biomedical applications of NiTi-based materials, with division to countries.

Different applications of SMA materials may require fine-tuning of their properties depending on a mode of operation of the SMA engineering element. The following *modes of operation* of SMA materials can be distinguished.

2.5.1 Free recovery

In this mode of operation, the only required functionality of an SMA element is the adoption of a prescribed shape, motion, or displacement—with no opposing force $(F=0,\ u\neq 0)$.

A good example of this mode of operation is an SMA overheating sensor rising when a prescribed temperature is surpassed and, thus, indicating the occurrence of critical conditions. Another example might be an SMA component in a safety system, which closes or opens an electrical circuit. A good example of this mode of operation is unfolding of a NiTi blood-clot filter in a vein to catch passing clots. Free recovery mode of operation requires no or minimal work execution from the SMA component.

2.5.2 Control operation

In this mode of operation, the SMA component is required to perform a control operation. It is required that the output signal (e.g., fluid flow) be controlled, possibly linearly by some input signal (e.g., temperature). For example, an SMA nozzle with its diameter changing due to temperature or a positioning mechanism working in control operation mode. The control operation mode can be treated as a variation of free recovery with a requirement for a specific quantitative relationship between the input and output signals.

2.5.3 Constrained recovery

In this mode of operation, the essential functional requirement is that a SMA component generates considerable, persistent stress (force), not relaxing in time, arising due to enforced limitation of strain recovery $(F\neq 0,\ u=0)$. Typical examples of this mode of operation are cryogenic, hydraulic, tubular couplings (Aerofit web page) and zero insertion force electronic connectors (Intrinsic Devices web page). It can also be used in sealing pegs, or anchoring pegs, which possess favorable characteristics when used in cultural heritage buildings preservation (Musolff et al., 2004).

2.5.4 Work production

In this mode of operation it is essential that an SMA component performs work against external loading. The shape recovery motion is executed against considerable opposing force ($F \neq 0$, $u \neq 0$; $W \neq 0$). This is typical operation mode for many types of actuators. For example, this is the mode of operation of orthodontic wires or an SMA spring in a water temperature regulator; see Figure 2.17.

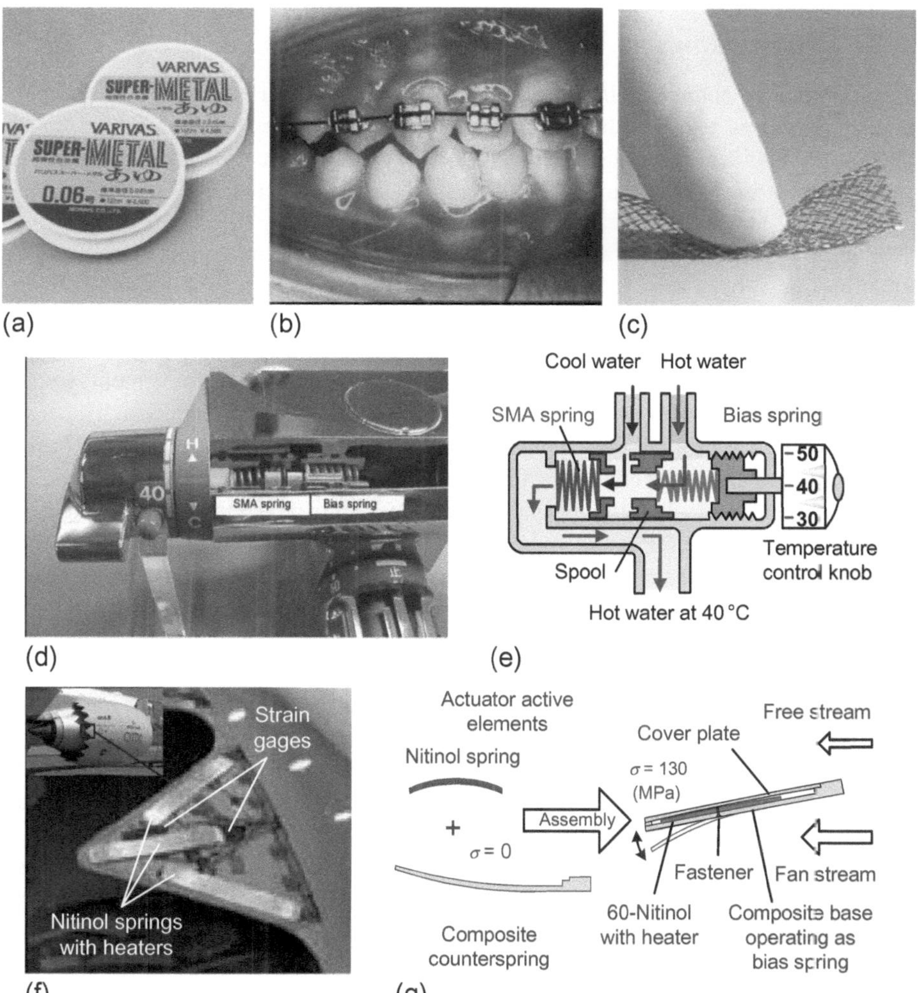

Figure 2.17 Examples of SMA alloys applications: (a) NiTi fishing line (0.045-0.085 mm) (after Morris Co Ltd. Courtesy Morris Co Ltd); (b) orthodontic wire (after Okhata (2011)); (c) NiTi stent extreme crushing resistance (Cordis SMART stent) (after Stoeckel et al. (2009)); (d and e) water temperature regulator and scheme of operation principle (after Suzuki (2011)); (f and g) variable geometry chevron for noise control of a jet engine and scheme of operation principle (after Mabe et al. (2006)).
Courtesy Boeing.

2.5.5 Pseudoelasticity (superelasticity)

In this mode of operation, the primary requirement is large flexibility together with large forces, which is impossible to attain in the case of the application of rubber materials. Typical applications are eyeglass frames and guide wires for steering catheters into a patient's body; see, for example, Melton (1999), Yamauchi et al. (2011), and Yahia (2000).

2.5.6 High damping

In this mode of operation, intrinsic damping of SMA materials is used to dissipate energy, which is released during extreme loading events of engineering structures such as car accidents, earthquakes, and blasts. Typical applications include an earthquake protection system of civil structures; see, for example, DesRoches and Delemont (2003) or Janke et al. (2005).

Figure 2.17 shows several examples of successful NiTi alloys applications. The examples have been selected to show growing engineering mastery in taking advantage of SMA functional properties. The Japanese company Morris Co. Ltd has found NiTi alloy useful in manufacturing of impact proof fishing lines with diameters of 0.045-0.085 mm (Figure 2.17a). A strong fish can, with violent movement, impart shock loadings on a fishing line, which frequently ends up with it breaking. The property of pseudoelasticity allows for mitigation of such impact loadings without breakage, as upon reaching pseudoelastic flow stress, the line starts to flow at high stress for several percent strain, and in this way can harmlessly absorb shock loading. The NiTi fishing line gives the feeling of nylon fishing line but it is much more resilient. The property of pseudoelasticity contibuted to the successful use of NiTi wires as orthodontic wires for malocclusion correction (Okhata, 2011). Brackets are cemented to the teeth to which arch wire is attached (Figure 2.17b). Adequate formation of the arch wire shape results in an imposing operation on the teeth with small constant forces, of several Newtons, which move the teeth in the gum to form the correct bite. Introduction of NiTi arch wires practically pushed traditional stainless steel arch wires out of the market. The reason is that NiTi arch wire can impose a constant force on the strain of several percent, while traditional steel or Co-Cr wires can do that on 0.5%, at most. Steel arch wires require several visits to an orthodontist to reapply stress after teeth have moved. In the case of NiTi arch wires, practically only one tensioning of arch wire is required. The orthodontist's activity can subsequently be limited to monitoring when teeth have moved to the proper positions and making a decision on the time of NiTi wire and brackets removal. This results in shorter treatment times and cost savings. Recently, NiTiCu arch wires gained popularity due to their small hysteresis loop. Figure 2.17c shows a Nitinol stent submitted to extreme deformation, crushed flat, with a completely closed cross section. The stent completely recovers its original tubular shape when the load is removed. Extreme kink resistance is a valuable feature of NiTi stents because it practically precludes the danger of permanent blocking of superficial vessels through accidental outside forces. The carotid artery is a pronounced example of such danger. Restenosis is a serious problem in angioplasty. In order to mitigate this high risk, technology has been developed for manufacturing drugs eluting NiTi stents for restenosis prevention of blood vessels; see Memry Corporation (2004). Figures 2.17d–e shows a photo of an anti-scald water temperature regulator and the scheme of its principle of operation; see Suzuki (2011). The shape memory spring works against a bias spring made of steel. When the outlet water temperature is higher than the temperature preset with the adjustment knob, the SMA spring stiffens and extends due to reverse, austenitic transformation. This results in closing the inlet valve for hot water and opening the valve for cold water, and a drop in the temperature of the mixed water. When outlet water temperature is lower than preset temperature, the SMA spring softens and contracts due to martensitic transformation to open the valve for hot water and to close the valve for cold water. This results in an increase of the outlet water temperature. Figures 2.17f–g shows a photo of a chevron actuator serving for jet engine noise attenuation and its scheme of operation; see Mabe et al. (2006). This is a very advanced application of NiTi alloy as the system can operate in fully automatic, autonomous mode or it can operate in external control mode, in which the position of chevrons can be preset through the operation of surface heaters located on SMA beams made of Nitinol 60 alloy. The Nitinol 60 alloy has been selected for this application because the required p.t. temperatures can be set by the heat treatment process only, which considerably facilitates the chevron actuator manufacturing process. Three Nitinol SMA actuator beams work against a chevron composite base acting as a kind of counterspring. The no stress shapes of actuator components are shown on the left in Figure 2.17e. They are joined together with fasteners in a "cold" temperature when the SMA beam is in a martensitic state and soft. After assembly in the "cold" state, which corresponds to plane cruise conditions, the chevron is straight and out of the fan air flow. In the "hot" state, when the plane lands or the jet engine starts, the SMA beams stiffen and tend to take a remembered, more curved, high-temperature shape. This causes the chevron to move into the fan flow and in this way attenuates the jet engine noise. It is worth mentioning that shape memory functional features of SMA materials do not exhaust their application potential. Equally important are their utility features. For example, the application domain of NiTi alloys is further extended thanks to the discovery of the excellent tribological properties of $Ni60_{wt\%}$-Ti (Nitinol 60) alloy; see Della Corte et al. (2009) and Della Corte (2010). This property makes *Nitinol 60* a favorite candidate for spacecraft bearing applications working in boundary friction conditions.

Family of thermodynamic R_L models of pseudoelasticity

3

3.1 Introduction

Reliable characterization and description of SMA materials behavior requires elaboration of a relevant phenomenological, theoretical model. Phenomenological modeling of the behavior of SMA materials has been a subject of research work since the beginning of 1980s. The modeling approach adopted strongly depends on actual targets of modeling activity, as was discussed in Section 2.2 of this book. In this chapter, attention is focused on a discussion of the development of a phenomenological model possibly best suited for SMA devices engineering design purposes. It seems that the most appropriate model for that purpose is a macroscopic model rooted in mesomechanics. Specification of constitutive relations of the SMA model in an incremental form is convenient for their implementation into finite element code, the most advanced engineering design tool at present.

Materials science research results reveal that thermoelastic martensitic transformation is responsible for the special behavior of shape memory alloy materials. Formation of hysteresis loops indicates that phase transition processes usually do not proceed through states of full thermodynamic equilibrium. This suggests formalism of continuum, nonequilibrium thermodynamics to be a suitable framework for SMA macroscopic model development. In classical, equilibrium, continuum thermodynamics, the validity of so-called *postulate of local state* is conjectured. In accordance with this postulate, the local thermodynamic state of *representative volume element* (RVE)—continuum, macroscopic material point—can be unambiguously described with the aid of a set of values relevant for the particular physical situation of so-called external state parameters (e.g., stress and temperature). The external state parameters values can be freely controlled in RVE. An RVE or macroelement of a material point of a continuous body is a material volume, which is statistically representative of the infinitesimal material neighborhood of that material point. While "infinitesimal," an RVE must be large enough to include a large number of microconstituents of macroelement, and be statistically representative of the local continuum properties; see, for example, Nemat-Naser and Hori (1993) for a detailed discussion of the concept of RVE. Effectively, RVE is treated as a homogeneous thermodynamic system, with continuum properties and state evolution results from averaged properties and the state of its microconstituents. The actual physical size of RVE depends on the subject matter of modeling works; for example, in the case of metallic materials, a macroelement with a diameter of a fraction of millimeter

Pseudoelasticity of Shape Memory Alloys. http://dx.doi.org/10.1016/B978-0-12-801697-8.00003-X

probably is the correct size, while in the case of a star the correct RVE dimension might be a kilometer or so.

In order to extend the applicability of classical thermodynamics to nonequilibrium processes, the so-called thermodynamics with *internal state variables* (ISV) theory has been developed. A fundamental theoretical assumption of ISV theory and its application for the development of inelastic constitutive relations of metal plasticity was given by Rice (1971). Horstenmeyer and Bammann (2010) deliver state of the art information on the overall concept of ISV theory and follow with a presentation of very successful applications of this apparatus to constitutive modeling of various classes of materials. ISV are introduced to enable unique definition of the thermodynamic state of the RVE undergoing irreversible processes, that is, running through states of not full but constrained thermodynamic equilibrium states. This is done by the introduction of a sufficient number of ISV, which depend on material structure, scope of the model predictions, and/or required accuracy of the constitutive model. ISV are designed, chosen in such a way as to capture macroscopically in some averaged fashion the effects connected with micro- and mesostructural evolution of RVE arrangement without a need to follow very complex mesoscopic details. Usually it is assumed that full thermodynamic equilibrium exists in RVE with respect to external state parameters but not internal ones. An important difference exists between external and internal state parameters: The latter cannot be freely controlled. Their values can be changed only indirectly by variation of external state parameters such as stress or temperature. That is why they are also called *hidden parameters*. In ISV theory, Helmholtz free energy function or Gibbs free enthalpy function is uniquely defined with an extended set of external and ISV relevant for a particular physical situation.. The *constitutive relations* resulting from the application of formalism of nonequilibrium thermodynamics with internal state parameters can be divided into two groups. The first group consists of equations of state resulting from an adopted form of thermodynamic potential. These are obtained through formal differentiation of thermodynamic potential with respect to its canonic state variables. For example, when Gibbs free enthalpy is used as thermodynamic potential for description of shape memory alloys' behavior, stress and temperature are adopted as independent external state variables then equations of state for strain and entropy are straightforwardly obtained as partial derivatives of Gibbs function. They are functions of stress, temperature, and a set of ISV. A second group of constitutive relations are kinetics relations describing the evolution of ISV. The kinetics relations make additional, independent information besides that contained in the adopted specific form of thermodynamic potential. This information is necessary to uniquely describe constitutive behavior of a material. Kinetics relations link fluxes of ISV with corresponding thermodynamic forces. Usually these relations are specified in the form of a set of mutually coupled, rate, ordinary differential equations, the right-hand sides of which are functions of external state parameters and all appearing in the modeling description of internal parameters. The evolution equations of internal state parameters must fulfill special criteria as products of respective thermodynamic fluxes, and thermodynamic forces appear in the expression for entropy production. Fulfilling these criteria assures that requirements of the second law of thermodynamics are always fulfilled during dissipative processes (i.e., Clausius-Duhem inequality).

Adoption of ISV theory formalism has considerable advantage over alternative modeling approaches. While ISV cannot be freely controlled, they are usually defined in such a way that they have a clear physical interpretation and their numerical values can be determined on relevant processing of experimental tests results. A case of such processing can be found in Chapter 5 of this book. Thus, a history of evolution of ISVs can be experimentally observed that allows for validation of posed theoretically kinetics relations of ISV evolution. The next benefit originates from capabilities offered by contemporary computing power and finite element method developments. ISV theory, with its fundamental concept of describing the present state of the material, originating from a history of loadings, only with current values of a set of state variables—external and internal—is perfectly well-suited for solving nonlinear boundary value problems with the aid of incremental numerical methods.

Due to the very complicated behavior of SMA materials, the scope of phenomenological modeling discussed here is limited to the temperature and stress range in which SMA materials exhibit pseudoelastic behavior. However, elements are introduced and discussed that may facilitate further enhancement of the model prediction range to lower temperatures and enable combined description of pseudoelasticity and one-way memory effect.

3.2 Macroscopic, thermomechanical pseudoelastic behavior of SMA

In order to construct macroscopic, thermodynamic theory of pseudoelasticity, a decision must be made about what major features of pseudoelastic behavior are to be reflected in a constitutive model of SMA material; that is, the predicted scope of the model must be established. A set of key state variables must also be identified (i.e., those having a vital influence on material behavior), allowing for proper description of the effects to be described by the phenomenological model. This is possible upon collecting an appropriate amount of experimental evidence. Extremely rich literature exists at present documenting the behavior of shape memory alloys submitted to uniaxial loadings.

The results of an experiment performed by Vacher and Lexcellent (1991) on a sample made of CuZnAl alloy deliver direct experimental proof that the deformation effect of pseudoelasticity is directly proportional to the advancement of the martensitic phase transformation, that is, the volume fraction of oriented martensite. Martensitic phase can be induced only at presence of mechanical loading in pseudoelastic range of SMA materials behavior, as it can not exist at zero stress in the range of temperatures for pseudoelasticity. The CuZnAl sample has been submitted to an isothermal tension loading-unloading cycle. Simultaneously during the test, stress $\sigma(t)$, total strain $\varepsilon(t)$, and electrical resistance $\rho_{\text{el}}(t)$ has been measured and registered. It is well known that electrical resistance varies in direct proportionality to a fraction of the martensitic phase. On the base of registered electrical resistance variation, the variation of a fraction of the martensitic phase $z(t) = k\rho_{\text{el}}(t)$ in the sample has been calculated,

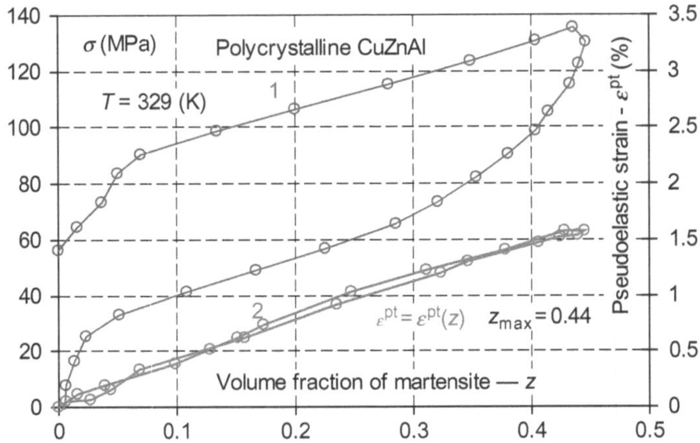

Figure 3.1 Stress-strain and martensite volume fraction↔strain curves for CuZnAl tension loading-unloading cycle. Volume fraction of martensitic phrase determined from electrical resistivity measurements.
Redrawn after Vacher and Lexcellent (1991).

where k denotes the proportionality constant. Using known experimental signals $\sigma(t)$, $\varepsilon(t)$, and $z(t)$, the quantity $\varepsilon^{pt}(z) = \varepsilon(z) - \sigma(z)/E$ has been calculated, where E denotes Young modulus and ε^{pt} is inelastic macroscopic strain connected with phase transformation. These data have been subsequently used to prepare $\sigma - z$ and $\varepsilon^{pt} - z$ graphs, shown in Figure 3.1 after Vacher and Lexcellent (1991). The graph $\sigma - z$ exhibits a hysteresis loop, while the graph $\varepsilon^{pt} - z$ does not show a hysteresis loop and can be well accepted within small experimental error as a straight line. Linking these two observations, a conclusion can be inferred that the macroscopic phase transformation strain is directly proportional to the volume fraction of martensitic phase z, when it is stress-induced. This gives a hint that the macroscopic phase transformation strain can be multiplicatively decomposed into a product of volume fraction of martensite and some p.t. eigenstrain $\varepsilon^{pt} = z \cdot \gamma_{tn}$.

Müller and his collaborators (Huo and Müller, 1993; Müller and Xu, 1991) executed a well-planned experimental program, which has been targeted to deliver experimental data for the development of a one-dimensional model of SMA materials behavior in the pseudoelastic range of their behavior; that is, in the temperature range $A_f^{2-1} < T < M_\sigma$ (also see Figure 2.14). Wire specimen made of monocrystalline CuZnAl alloy has been submitted to various thermomechanical loadings. First, isothermal tests have been performed with quite involved uniaxial tension loading-unloading cycles. The stress-strain $(\sigma_1 - \varepsilon_1)$ graph obtained for an isothermal tension test up to 9% strain performed at temperature $T = 300$ K is shown in Figure 3.1a. Initially, the sample deforms elastically until stress σ_1^{AM} is reached, when forward martensitic transformation starts $(\mathcal{A} \rightarrow \mathcal{M})$, marked with the letter A. The authors could optically observe formation of the first martensitic plate. Subsequently, the sample flows pseudoelastically at practically constant stress. Optical observation of the

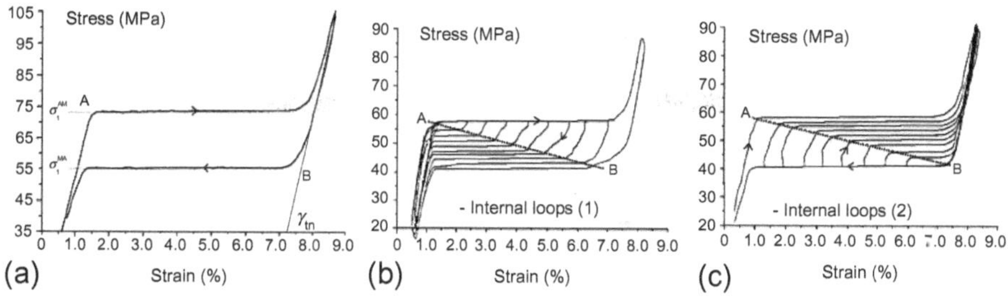

Figure 3.2 (a) Chart of uniaxial tension of monocrystalline CuZnAl sample at temperature $T=300\,°K$ (limit hysteresis loop); (b and c) Formation of internal hysteresis loops at partial phase transitions forward and reverse resulting from load, unload reversals, respectively. After Huo and Müller (1993).

sample revealed the appearance of more and more martensitic plates; evidence of phase transformation progress. This process is accompanied with heat emission. At some stage of deformation no further phase transition takes place—no more martensitic plates appear. Further loading of the sample leads to elastic deformation of the martensitic phase until the plastic yield limit of the martensitic phase is reached $\sigma_{0.2}^{M}$. When the sample loading has been stopped before reaching yield stress $\sigma_{0.2}^{M}$ and unloading has been started, then during unloading the sample behaves elastically until stress σ_1^{MA} is reached, when reverse martensitic transformation starts ($\mathcal{M} \rightarrow \mathcal{A}$), marked with the letter B. The value of stress $\sigma_1^{MA} \neq \sigma_1^{AM}$. Reverse transformation runs at practically constant stress and, after its termination, elastic behavior of the austenitic phase takes place again. When the load has been fully removed, the sample practically did not exhibit permanent strain, similar to the case of rubber-like materials behavior. During this test complete forward and complete reverse phase transition occurred. As a result, the so-called limit or *outermost hysteresis loop* has been formed.

More complex loading cycles of the sample with an interrupted loading process before full completion of forward phase transition (see Figure 3.2b), and renewed loading of the sample before full completion of reverse transformation (see Figure 3.2c), allows observing how so-called internal hysteresis loops are formed during isothermal loading conditions. These types of loadings revealed a very interesting feature; namely, when the process of pseudoelastic flow during forward transformation is interrupted with unloading before full completion of forward transformation the sample initially behaves elastically, and reverse transformation starts only in the vicinity of a diagonal line marked A-B. Similarly, when reverse transformation is stopped due to the next reloading of the sample, then the initial reaction of a two-phase material is elastic, and the next forward transformation process restarts only after reaching a state remaining in the vicinity of the diagonal line A-B. The described behavior of the SMA material indicates for a special character of states (σ, T) remaining on the diagonal line A-B, marked with letters A-B in Figure 3.2b and c. The existence of such special states delivers important experimental premises for theoretical analyses. It is shown in the sequel of the present book - cf. formula (3.14), that thermodynamic states remaining on the line A-B can be treated as states of full thermodynamic equilibrium with respect to all state variables, external and internal. At these states, the thermodynamic driving force of phase transformation takes zero value ($\pi^{1-2}=0$); see Section 3.4 for a mathematical definition of thermodynamic driving force of p.t.

The stress-strain graphs ($\sigma_1-\varepsilon_1$) of isothermal tension tests performed at several temperatures overlaid together in one plot are shown in Figure 3.3a. These graphs indicate that it can be accepted with good accuracy that the area of limit hysteresis loops are the same (i.e., they do not depend on temperature). The values of *critical stresses of pseudoelastic flow* of forward and reverse phase transition σ_1^{AM}, σ_1^{MA}, determined with the use of graphs shown in Figure 3.3a are marked in Figure 3.3b with triangle and rhomb symbols, respectively. It is clear that it can be accepted with very good accuracy that critical stresses for forward and reverse pseudoelastic flow $\sigma_1^{AM}(T)$, $\sigma_1^{MA}(T)$ are dependent linearly on temperature. Unchanging with temperature, the width of hysteresis loops results in that they are practically parallel. Extrapolation of these lines to zero stress values deliver some values of temperatures marked with a romb and triangle on x-axis in Figure 3.3b. It is a natural supposition that they represent characteristic temperatures of p.t. determined with differential scanning calorimetry (DSC). Huo and Müller (1993) have not specified DSC measurements results for investigated by them in thermomechnical tests CuZnAl alloy.

Huo and Müller (1993) also executed another type of experiment: the so-called isostress test. During such a test phase transitions are temperature induced at kept constant non-zero assisting stress. Figure 3.3c shows limit and internal hysteresis loops formation, plotted in ε_1-T coordinates, which appear as a result of stress-assisted ($\sigma_1=109$ MPa $=$ const) temperature cycling of a CuZnAl sample. Similar to the case of isothermal tests, the existence of special states located in the neighborhood of diagonal A-B can be perceived; that is, at places where partial reverse transformations start ($\mathcal{M}\rightarrow\mathcal{A}$). In Figure 3.3d the appearance of the so-called force recovery effect is shown. This effect is obtained when the SMA sample after its deformation due to tension loading-unloading cycle in cold state is constrained with fixed strain $\varepsilon_1=$ const and submitted to temperature cycling loading. The sample cannot change its shape due to imposed strain constraint, so the stress is induced in the material. The recovery force begins to increase at point B as a result of the start of reverse phase transformation ($\mathcal{M}\rightarrow\mathcal{A}$). At point A, the recovery force begins to decrease as a result of forward phase transformation ($\mathcal{A}\rightarrow\mathcal{M}$). Recalled experimental results concerning monocrystalline CuZnAl samples reveal the ideal pseudoelastic flow of the material during isothermal tests. When polycrystalline samples are tensioned, then usually some hardening is observed during pseudoelastic flow.

Tanaka et al. (1999) executed a program of tests on a rectangular specimen made of $Ni_{51.0at\%}$-Ti alloy, with cross section 1×2 mm and gauge length 30 mm, submitted to isothermal tension loading at a dozen or so temperatures. The final TMT treatment of the material consisted of annealing at 400 °C for 1 h followed by cooling in a furnace. Tanaka et al. (1999) deliver a very detailed description of the material and specimen manufacturing process. The NiTi material from the same batch has been used to manufacture tubular specimen submitted to a program of multiaxial testing discussed in Chapters 5 and 6 of this book. The results of isothermal, tension tests of Tanaka

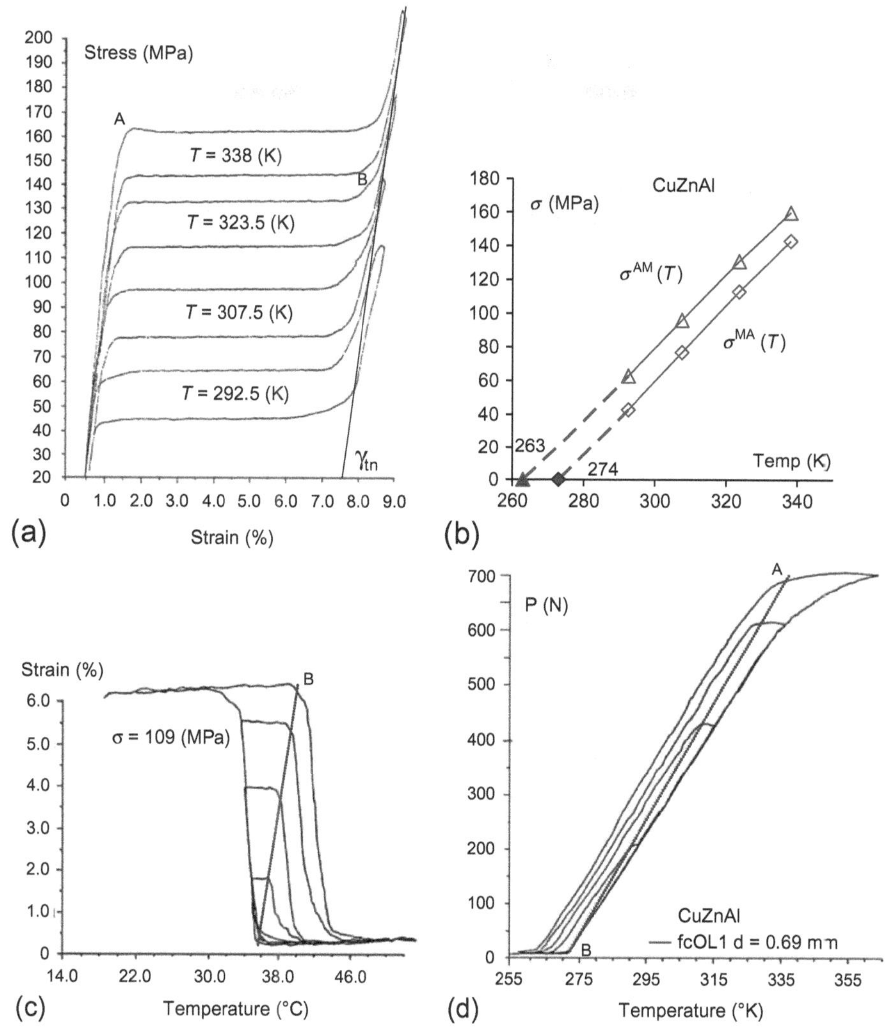

Figure 3.3 (a) Stress-strain plots of isothermal, uniaxial tension of monocrystalline CuZnAl alloy at several temperatures (limit hysteresis loops) (after Huo and Müller (1993)); (b) Generalized Clausius-Clapeyron relations of CuZnAl for forward and reverse phase transformations elaborated on data from (a); (c) Limit and internal hysteresis loops obtained in stress-assisted ($\sigma_1 = \text{const}$) temperature cycling tests of CuZnAl sample (isostatic temperature tests); (d) The effect of force recovery (drop) induced by thermal cycling of strain constrained CuZnAl sample. After Huo and Müller (1993).

et al. (1999) performed at several temperatures in the pseudoelastic range of NiTi alloy behavior have been collected and overlaid together by the present author and shown in Figure 3.4a. The stresses of martensitic phase transformation start $\sigma_1^{AM}(T)$ are marked with a filled circle symbol. Deviation from initial elastic behavior at lower temperatures ($T < 336$ K) is caused by progress of R-phase transformation taking place in

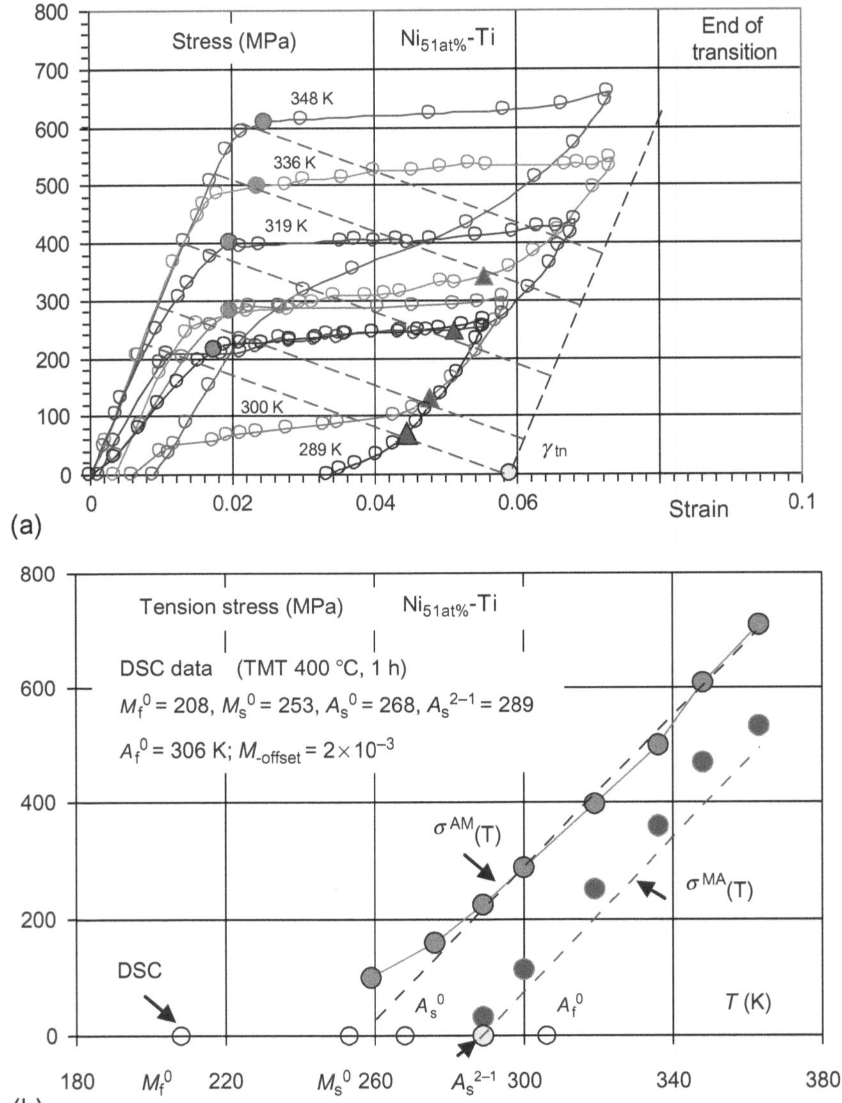

Figure 3.4 (a) Stress-strain plots of isothermal, uniaxial tension of polycrystalline $Ni_{51at\%}$-Ti alloy at several temperatures (collected after Tanaka et al. (1999)); (b) Experimental critical transformation stresses of forward and reverse transformations of $Ni_{51at\%}$-Ti alloy. Generalized Clausius-Clapeyron lines for forward and reverse p.t., and DSC measured characteristic temperatures of p.t.

investigated NiTi alloy. Unfortunately, no test has been done to determine limit hysteresis loop curves similar to the ones obtained for CuZnAl alloy and shown in Figure **3.3**a. In all the (Tanaka et al., 1999) tests, the loading of NiTi specimen has been stopped before complete phase transition takes place. The start stresses of partial reverse phase transformation of NiTi material remaining in a two-phase state delineating internal hysteresis loops, are marked with filled triangle symbol. Lack of a direct experimental data of limit hysteresis loop curve causes difficulties in determination of *pseudoelastic flow amplitude* in simple tension for investigated NiTi alloy, marked with filled circle symbol on x-axis and letter γ_{tn} in Figure 3.4a. Despite that, the value of pseudoelastic flow amplitude in simple tension for (Tanaka et al., 1999) $Ni_{51.0at\%}$-Ti alloy has been estimated to be $\gamma_{tn} \cong 0.059$. Several indirect premises have been used for that purpose: knowledge of pseudoelastic flow start stresses, thickness of outermost hysteresis loop (assessed to be about 230 MPa), and locations of twists marking start stresses of reverse phase transition, denoted with triangles in Figure 3.4a. Sittner et al. (2014) in their recent work investigate $Ni_{55.98wt\%}$-Ti ($Ni_{51att\%}$-Ti) alloy, a NiTi alloy with practically the same chemical composition as the one discussed here (i.e., $Ni_{51att\%}$-Ti). The finishing step of TMT processing of the NiTi alloy used in wire form by Sittner et al. (2014) is 41% cold work (CW) and annealing at 400 °C for 30 min, while Tanaka's finishing step for TMT annealing is 400 °C for 1 h. Sittner et al. (2014) executed a simple tension test revealing a complete pseudoelastic flow plateau and plotted the test result in their Figure 1a. Pseudoelastic flow plateau amplitude of $Ni_{51att\%}$-Ti alloy evaluated from this figure, embracing R-phase flow, amounts to $\gamma_{tn} \cong 0.0592$. More information on SMA material data identification can be found in Chapter 5. The dashed line with positive slope in Figure 3.4a originating in state $(0, \gamma_{tn})$ shows a trace of idealized elastic behavior of NiTi material in a fully martensitic state; see also Figure 3.3a. The states of full thermodynamic equilibrium of NiTi material at different isothermal tests temperatures, that is, those where the thermodynamic driving force of phase transformation has zero value—$\pi^{1-2}(T = \text{const}) = 0$, are marked with dashed lines with negative slopes in Figure 3.4a. Extrapolation of these lines and their intersection with the line of idealized elastic behavior of oriented martensite phase (i.e., the dashed line intersecting x-axis at γ_{tn}), allows for determination of oriented martensite austenite p.t. start stresses $\sigma_1^{MA}(T)$; that is, those delineating limit hysteresis loops.

The experimentally determined start stresses of forward and partial reverse p.t. in stress-temperature coordinates are shown in Figure 3.4b, marked with filled circles lying close to the lines $\Sigma^{AM}(T)$ and $\Sigma^{MA}(T)$, respectively. Critical stresses of forward p.t. determined from isothermal tests executed at temperatures $T = 276, 259$ K are also shown. The stress-strain charts for these temperatures are not shown in Figure 3.4b in order not to obscure the picture. These "low" temperature experimental points deviate from the generalized *Clausius-Claperon* line $(\sigma_1^{AM}(T), T)$; the upper dashed line, being a trend line determined on the base of pseudoelastic start stresses of forward p.t. $(T > A_s^{2-1})$. The location of extrapolated start stresses of complete, reverse p.t., delineating limit hysteresis loops, is marked with a lower dashed line in Figure 3.4b. It is reasonable to conjecture that during isothermal pseudoelastic tests only oriented martensite is generated by applying stress or equivalently strain in strain-controlled tests $(\mathcal{A} \rightarrow \mathcal{M}_{or})$. Thus, the lower dashed line shows reverse p.t. start stresses $\sigma_1^{MA}(T)$ of oriented martensite into austenite $(\mathcal{M}_{or} \rightarrow \mathcal{A})$. Observe that the start stress of oriented martensite into austenite p.t. at temperature $T = 289$ K is zero $(\sigma_1^{MA}(T = 289) = 0)$, it is marked with a filled circle located on x-axis. This characteristic temperature is also marked with symbol A_s^{2-1}. The $Ni_{51at\%}$-Ti alloy

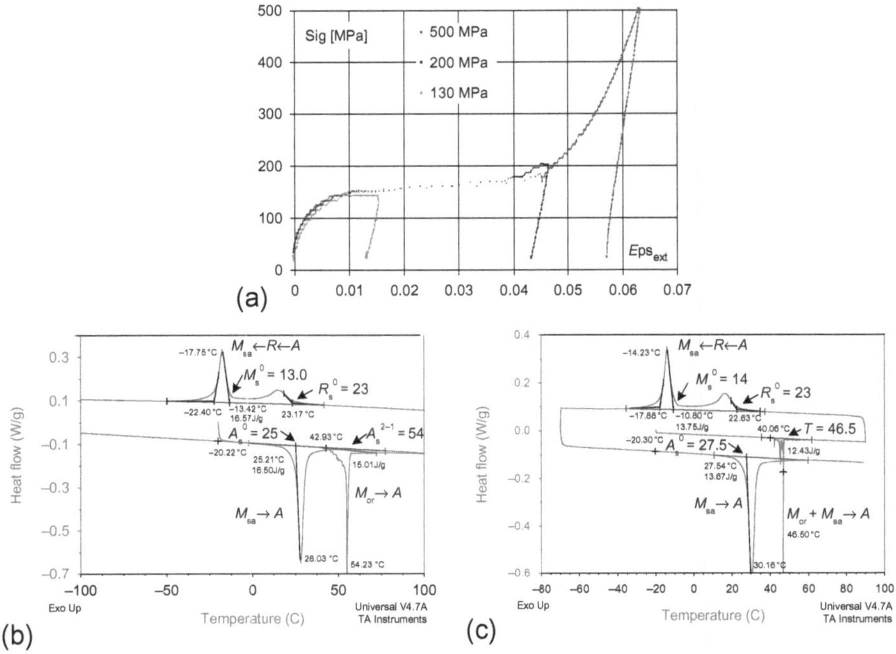

Figure 3.5 (a) Isothermal stress-strain curves of polycrystalline $Ni_{50at\%}$-Ti wire samples in martensitic state submitted to tension at temperature $65 < A_s^0 = 26$ °C; (b) DSC thermograph of specimen initially in fully oriented martensite state ($z_2 = 1$); (c) DSC thermograph of specimen initially in martensitic state with 25% oriented martensite state and 75% self-accommodating martensite ($z_2 = 0.25$, $z_3 = 0.75$).

characteristic of forward and reverse phase transition temperatures of austenite into self-accommodating martensite ($\mathcal{A} \rightarrow \mathcal{M}_{sa}$), determined from DSC, are $M_f^0 = 208, M_s^0 = 253$, $A_s^0 = 268, A_f^0 = 306$ K; see the DSC thermograph in Figure 5.1 and related text in Chapter 5. The DSC characteristic temperatures are shown with empty circles on x-axis in Figure 3.4b. The above analysis of experimental data indicates in an indirect way that the start temperature of reverse transformation of self-accommodating martensite A_s^0 (the one induced by temperature during DSC tests at zero macroscopic stress), and the start temperature of reverse transformation of oriented martensite A_s^{2-1} (the one induced by stress), differ considerably $A_s^{2-1} - A_s^0 = 289 - 268 = 21$ K. This observation allows evaluating that specific free energy of self-accommodating martensite is considerably higher than specific free energy of oriented martensite at some reference temperature and zero macroscopic stress. Popov and Lagoudas (2007) were the first to discuss the considerable difference between A_s^0 and A_s^{2-1} temperatures within the context of macroscopic modeling of SMA materials behavior. They also executed experimental work on a NiTi alloy to confirm this observation.

The present author decided to execute independent experimental work to directly confirm that A_s^0 and A_s^{2-1} temperatures are considerably different in NiTi alloys

($A_s^0 < A_s^{2-1}$). For that purpose polycrystalline $Ni_{50at\%}$-Ti material manufactured by Baoji Co Ltd was used, delivered in the form of a wire with an 0.8 mm diameter. The last manufacturing operations to obtain an "as received" state of the material consisted in cold drawing from 1 to 0.8 mm diameter (CW degree 22.3%) and straight shape setting procedure; that is, heating the material to 500 °C and keeping it at that temperature for several seconds (according to information from the manufacturer). The DSC measurement of the wire sample in an "as received" state showed that it undergoes single-step forward and single-step reverse phase transition. The DSC characteristic temperatures are $M_f^0 = 9$, $M_s^0 = 15$, $A_s^0 = 32$, and $A_f^0 = 44$ °C. Latent heat of phase transition amounted to $Q = 19.0$ J/g. In order to obtain repeatable material response in the planed tests program, each NiTi wire specimen, before actual test, has been submitted to a single cycle of tension loading-unloading to a maximum stress of 500 MPa at room temperature ($T_r = 22$ °C); TMT treatment called "TMT 500×1." After TMT, the specimens have been brought to a pure austenitic state by heating them to 100 °C and cooling them to room temperature. Plastic dislocations structure introduced by the TMT 500×1 induced change in p.t. kinetics from a single-step (B2→B19′) forward phase transition to a bimodal one (B2→R→B19′). Reverse phase transition remained as single step. DSC characteristic temperatures of $Ni_{50at\%}$-Ti alloy after TMT 500 × 1 are as follows: $M_f^0 = -16$, $M_s^0 = 1$, $R_f^0 = 4$, $R_s^0 = 21.5$, $A_s^0 = 26$, $A_f^0 = 37$ °C, and $Q = 18.0$ J/g. When $Ni_{50at\%}$-Ti alloy TMT 500×1 treated has been subsequently loaded mechanically with stresses not exceeding 500 MPa a dozen or so times, its DSC thermograph did not change considerably any more.

In order to generate oriented martensite, NiTi wire samples after TMT 500×1 were cooled in a temperature chamber to temperature $T = -65$ °C, well below $M_f^0 = -16$ °C temperature, and loaded isothermally to different maximum stresses $\sigma = 130, 200, 500$ MPa. Before mechanical loading at −65°C a sample is in a single-phase state of self-accommodating martensite generated by cooling at no macroscopic stress. After removing the mechanical load, two-phase mesostructures were obtained with different proportions of self-accommodating and oriented martensite, approximately 25%, 75%, and 100%, estimating from the value of stress-free residual strain. One of the NiTi, TMT 500×1 samples after test loading to a maximum stress of 500 MPa and unloading at −65 °C has been heated to 110 °C only to observe complete residual strain recovery. Thus, it has been demonstrated that residual strains present in specimens at −65 °C can be fully attributed to be oriented martensite phase strains and not plastic slip ones. Nominal stress (F/A_0), nominal strain charts of isothermal, mechanical loadings are shown together in Figure 3.5a, proving excellent repeatability of investigated NiTi material response. In order to avoid strain measurement artifacts connected with very high flexibility of NiTi alloy wire-executed with an extensometer, the wire sample has been preloaded with 20 MPa and at this load strain has been preset to zero value. It is interesting to note that unloading tangent Young modulus of pure oriented martensite determined from data of the curve "500 MPa" is 70 GPa; that is, it is the same value as the one commonly reported for the austenitic phase.

After mechanical test termination, each wire sample has been heated to 0 °C in a temperature chamber; well below $A_s^0 = 26$ °C temperature to preserve mixed mesostructure of the NiTi material, and about 10 mg weight of the DSC sample was cut and put into a thermos flask filled with ice and water. DSC samples so obtained, containing self-accommodating and oriented martensite in different proportions, has been submitted to DSC measurement. In the DSC apparatus the first sample was cooled to −20 °C and then heated to 90 (or 150 °C). The next two cycles of cooling-heating were executed from 90 to −70 (or 150 to −120) Celsius temperatures. During the first half-cycle of heating oriented martensite or the mixture of martensitic phases transforms into austenite. During the next two heating-cooling cycles, austenite self-accommodating martensite phase transitions take place. The DSC thermograph of the NiTi sample with a fully oriented martensite initial state is shown in Figure 3.5b. The DSC temperature rate was 10 °C/min. During the first DSC heating half-cycle, a complete reverse p.t. of oriented martensite into austenite takes place ($\mathcal{M}_{or} \rightarrow \mathcal{A}$). Some erratic stage of the transition commenced at 43 °C to be followed by a violent outburst peak of phase transition at 54 °C: $Q = 15.0$ J/g. It is this temperature that is accepted here as the characteristic p.t. temperature of oriented martensite into austenite transition: $A_s^{2-1} = 54$ °C. The violent character of oriented martensite reverse p.t. is indicated by a power of 4 W/g at the needle-shaped peak maximum of p.t. in comparison to a power of 0.65 W/g at the hill-shaped peak of self-accommodating martensite reverse p.t., at a DSC rate of 10 °C/min. During two subsequent heating-cooling cycles of DSC procedure, executed to check repeatability of material response, twice forward and reverse austenite, self-accommodating martensite phase transitions took place. The characteristic temperatures during these transitions were identical and amounted to: $R_s^0 = 23$, $M_s^0 = 13$, $A_s^0 = 25$ °C; $Q = 16.5$ J/g; see Figure 3.5b. Thus, the existence of a considerable difference between A_s^{2-1} and A_s^0 has been demonstrated by direct experimental measurement, amounting in the case of investigated $Ni_{50at\%}$-Ti alloy to $A_s^{2-1} - A_s^0 = 29$ °C.

Figure 3.5c shows a DSC thermograph of the NiTi sample with a mixture of 25% oriented martensite and 75% self-accommodating martensite. During first DSC heating half-cycle, run with a heating rate of 1.25 °C/min, a reverse p.t. of a mixture of martensites into austenite took place. Some erratic stage of the transition progress was initiated at 40 °C to be followed by a needle sharp outburst peak of phase transition at ${}_{0.25}A_s^{2-1} = 46.5$ °C; $Q = 12.4$ J/g. It is interesting to note that no separate peaks of individual p.t. of oriented and self-accommodating martensite appeared at 25 and 54 °C, which could be expected, but the mixture rather transformed like a coagulate. During two subsequent heating-cooling cycles of DSC procedure, a temperature rate of 10 °C/min, two times forward and reverse austenite, self-accommodating martensite p.t. took place. The characteristic temperatures during these transitions were identical and amounted to $R_s^0 = 23$, $M_s^0 = 14$, $A_s^0 = 27.5$ °C; $Q = 13.7$ J/g; see Figure 3.5c.

Popov and Lagoudas (2007) reported the execution of an experimental procedure similar to that shown in Figure 3.5b. The main difference between the test results reported here is that care has been taken to control the possible influence of plastic slip deformations on the $A_s^{2-1} - A_s^0$ temperature difference. This has been done by introducing possibly all plastic deformations into the material before actual discriminating tests, through the execution of the TMT 500×1 procedure. It may be conjectured that the platic slip deformations introduced during the actual testing were negligible because loading stresses did not exceed the conditioning 500 MPa stress value. Popov and Lagoudas' (2007) explanation for the appearance of $A_s^{2-1} - A_s^0$ effect is quite vague. They attribute it to the necessity for additional energy to reverse inelastic strains—work execution on inelastic mesostrains, in order to bring about oriented martensite into austenite, additional in comparison to self-accommodating martesite to austenite p.t. However, no explanation is delivered regarding what type of inelastic strains they have in mind, or what is the actual physical mechanism of the phenomenon. Theoretical investigations devoted to mesomechanical derivation of Gibbs free energy for SMA materials allow for more specific explanation. For example, it is reasonable to speculate that the process of martensitic phase reorientation leads to a decrease of fluctuation amplitude of local field of phase eigenstrains in relation to its average value ($\breve{\boldsymbol{\Gamma}}^{f}_{\alpha}(\boldsymbol{x}) \equiv \breve{\boldsymbol{\Gamma}}_{\alpha}(\boldsymbol{x}) - \tilde{\boldsymbol{\Gamma}}_{\alpha}$), where $\breve{\boldsymbol{\Gamma}}_{\alpha}(\boldsymbol{x})$ denotes local field of phase eigenstrains in the neighborhood of specific location, $\tilde{\boldsymbol{\Gamma}}_{\alpha}$ is the average value of phase eigenstrains at a specific location, and indices $\alpha = 2,3$ denote oriented and self-accommodating martensite; see also Chapter 4. A kind of smoothing of the local phase eigenstrain field takes place during the reorientation process. As a result, less elastic energy of coherence is stored in the oriented martensite than in the self-accommodating martensite in relation to a reference state with zero mesostresses. A lower energy minimum is reached by oriented martensite due to the reorientation process. In this way, specific free energy of oriented martensite is smaller than the one of self-accommodating martensite, and the temperature difference $A_s^{2-1} - A_s^0$ is a kind of macroscopic measure of this fact. Detailed mesomechanical studies justifying the above argumentation are contained in a paper prepared by Raniecki and Ziółkowski (2015). Experimental results shown in Figure 3.5c indicate that two separate physical effects induced by reorientation process, that is, growth of macroscopic phase eigenstrain and decrease of fluctuation of eigenstrain field amplitude (eigenstrain field smoothing) do not proceed at the same rate; that is, they do not proceed in linear proportionality. The issue of $A_s^{2-1} - A_s^0$ effect constitutes at present an open scientific problem requiring further careful investigation. A practical recommendation coming from the above experimental work is that whenever oriented martensite is to be exploited in an engineering application it is wise to load the SMA element to a maximum expected in specific operational conditions of stress to reach the maximum content of oriented martensite, and only later proceed with nominal SMA element exploitation. This will allow avoiding considerable, unexpected, change of the NiTi material properties originating from reaching lower specific energy levels by oriented martensite.

A survey of experimental work devoted to uniaxial tension of samples made of various shape memory alloys in the pseudoelastic range of their behavior allows preparing an overall landscape of schematic types of SMA materials behavior, shown collectively in Figure 3.6. For example, monocrystalline AuCd SMA exhibits ideal pseudoelastic flow practically without hysteresis loop. Monocrystalline CuZnAl SMA flows ideally but exhibits hysteresis loop. Polycrystalline NiTi alloy undergoing R-phase transformation (crystallographic lattice of martensite arising during R-phase transformation is rhombohedral) exhibits hardening during pseudoelastic flow but

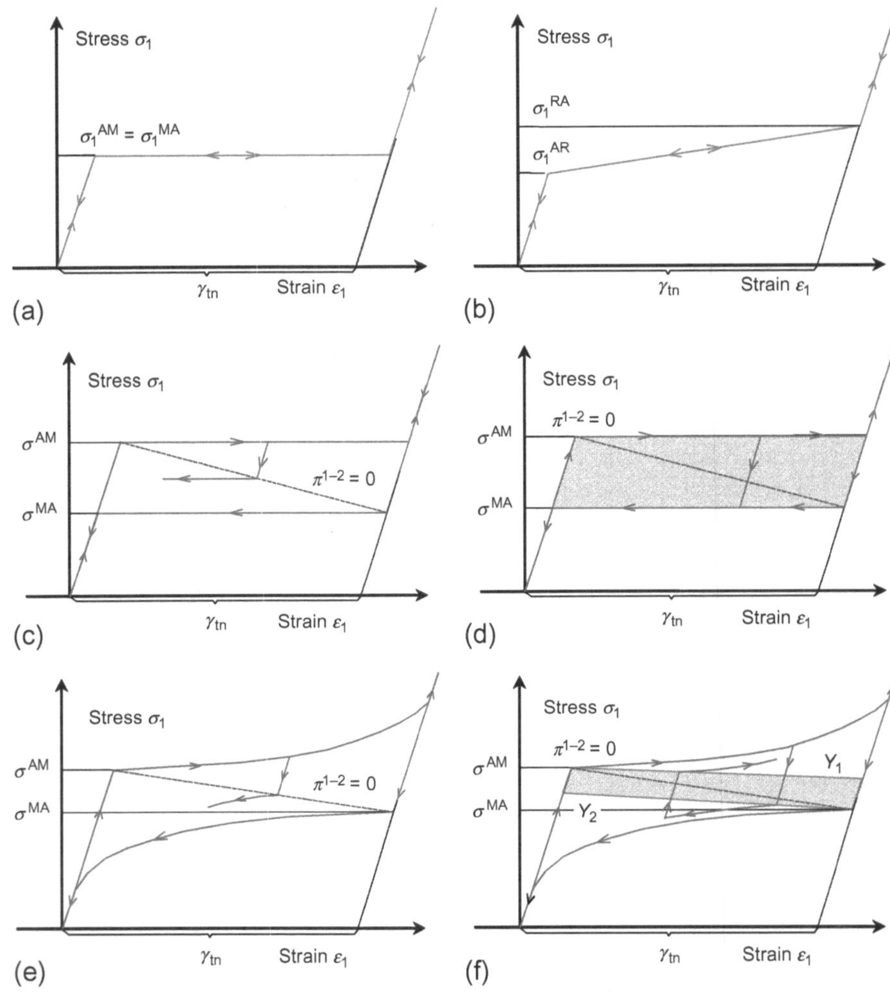

Figure 3.6 Schematic types of behavior of shape memory alloys in uniaxial tension: (a) Idealized behavior of AuCd alloy; (b) Idealized behavior of NiTi alloy undergoing R-phase transformation; (c) Idealized behavior of monocrystalline CuZnAl alloy; (d) Idealized behavior of FeNi alloy; (e) Idealized behavior of polycrystalline NiTi alloy; (f) Idealized behavior of NiTi alloy submitted to special thermomechanical treatment. The symbol γ_{tn} denotes pseudoelastic flow amplitude in simple tension.

hysteresis loop accompanying R-phase transformation is very small, practically negligible. The NiTi alloy undergoing monoclinic martensitic transformation (crystallographic lattice of martensite arising during such transformation is monoclinic) exhibits hardening and large hysteresis loop. Various ways of internal hysteresis loops formation also emerge. In the case of a CuZnAl alloy, incomplete phase transformations (forward and/or reverse) start in the neighborhood of diagonal of hysteresis loop, marked as $\pi^{1-2}=0$ in Figure 3.6c, symbol π^{1-2} denotes the thermodynamic driving force of phase transformation, and line $\pi^{1-2}=0$ determines the *states of full thermodynamic equilibrium*; see Section 3.4. In the case of a NiTi alloy, submitted to adequate TMT processing, incomplete phase transformations (forward and/or reverse) start not on the diagonal $\pi^{1-2}=0$ but only after reaching certain *lag threshold functions* values Y_α, shown in Figure 3.4f. The hatched areas in Figures 3.6d and f delineate domains of thermodynamic states, in which active phase transformation cannot take place.

The database of behavior of SMA materials submitted to multiaxial thermomechanical loadings available in the literature is very skimpy in comparison to uniaxial tests results. There are very few reports containing experimental evidence of systematic, well-planned programs of multiaxial experiments. Some selected experimental results important from the point of view of macroscopic, multiaxial, pseudoelastic modeling of SMA materials are recalled below.

Early researchers who performed multiaxial tests on SMA materials were Rogueda et al. (1996). They executed an experimental program on tubular specimens made of polycrystalline Cu-$Zn_{26wt\%}$-Al_4 alloy, manufactured by Trefimetaux company, submitted to simultaneous torsion and tension along proportional and nonproportional loading paths. In Figure 3.7a, effective stresses of forward martensitic phase transformation σ_{ef}^{AM}—$\sigma_{ef}=(\sigma^2+3\tau^2)^{1/2}$, calculated from experimental data obtained in isothermal, proportional tension-torsion tests are marked with triangles. For illustrative purposes there is also shown, marked with an "H-M" symbol, a trace of a hypothetical Huber-Mises surface that the CuZnAl alloy critical stresses σ_{ef}^{AM} would exhibit had it obey classical Huber-Mises theory. The values of critical pseudoelastic flow stress in pure torsion τ^{AM} assessed from data in the original Figures 15 and 16 of Rogueda et al.

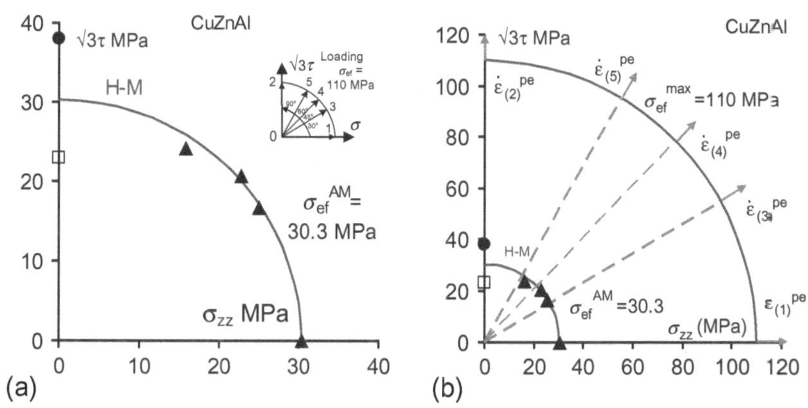

Figure 3.7 Tension-torsion, proportional, isothermal (T=303 K), strain controlled loadings: (a) Critical stresses of pseudoelastic flow of CuZnAlMn alloy—experimental data are marked with full triangles (corrected pure torsion pseudoelastic flow stress – 38 MPa, is marked with full circle). Circular line represents Huber-Mises criterion; (b) Experimental investigation of evolution of pseudoelastic strains—pseudoelastic strains directions at the value of effective stress 110 MPa are marked with arrows. Dashed lines indicate loading stress directions. Reworked after Rogueda et al. (1996).

(1996) are $\tau^{AM} \cong 22$ and $\tau^{AM} \cong 25$, respectively. This indicates that the experimental point for critical pseudoelastic flow stress in pure torsion marked in the original Figure 6 of Rogueda et al. (1996) (here Figure 3.7a), has been accepted incorrectly by the introduction of the numerical value of τ^{AM} instead of $\sqrt{3}\tau^{AM}$. The unsure data point has been marked with an empty rectangle and the value assessed from Figures 15 and 16 of about $\sqrt{3}\tau^{AM} \cong 38$ MPa is marked with a filled circle on y-axis in Figure 3.7a. This datum is consistent with other literature data on Cu-based alloys.

Rogueda et al. (1996) also investigated the evolution of phase transformation strain rates directions during proportional stress-controlled tests. In Figure 3.7b directions of phase transformation "strain rates" at maximum loading stress $\sigma_{ef}^{AM} = 110$ MPa are shown. The strain rates vectors were simply calculated from maximum values of phase transformation strains divided by the time increment of loading from zero to maximum stress, that is, $\varepsilon_{z\theta}^{pe}/\varepsilon_{zz}^{pe} = \dot{\varepsilon}_{z\theta}^{pe}/\dot{\varepsilon}_{zz}^{pe}$. Thus, in fact the strain rates directions are pseudoelastic strain directions at maximum loading stress. The authors came to the conclusion that pseudoelastic strain rate vectors are perpendicular to trace of pseudoelastic flow surface, which is a rather uncertain conclusion taking into account that this surface is sketched basing on a few experimental points. However, relying on the data shown in Figure 3.7b it can be well accepted that phase transformation strain tensor direction conforms very well to the direction of inducing its stress tensor direction at maximum loading stress; dashed lines indicate loading stress directions and arrows indicate pseudoelastic strain directions.

The proportional tests of torsion-tension-compression on samples made of NiTi alloy were performed by Orgeas and Favier (1996). These authors found considerable differences between critical effective stresses of martensitic transformation in tension and compression. The observations that these authors made regarding NiTi alloy behavior can be summarized as follows: critical stress of pseudoelastic flow in tension σ_{comp}^{AM} is considerably lower than critical stress of pseudoelastic flow in compression σ_{tens}^{AM}; critical stresses σ_{comp}^{AM}, σ_{tens}^{AM} depend linearly on temperature in the pseudoelastic range; amplitude of pseudoelastic flow in tension γ_{tens} is considerably larger than the amplitude of pseudoelastic flow in compression γ_{comp}; and energies expressed by product $\sigma_{ef}\gamma$ are approximately the same regardless of proportional loading path direction and depend only on temperature $\left(\sigma_{tens}^{AM} \cdot \gamma_{tens}\right) \cong \left(\sigma_{comp}^{AM} \cdot \gamma_{comp}\right) = C(T)$. Areas of hysteresis loops are constant at constant temperature. Speaking illustratively, Orgeas and Favier (1996) made an observation that NiTi alloy hysteresis loop is thin and long in tension while it is thick and short in compression.

A relatively rich experimental program of biaxial, *proportional tests* of tubular specimen made of $Ni_{51at\%}$-Ti alloy in the pseudoelastic range of its behavior has been executed by Raniecki et al. (1999a). Details of the material preparation, experimental testing methodology, and selected experimental and theoretical results can be found in Raniecki et al. (1999a), Raniecki et al. (1999b), and Raniecki et al. (2001). Elaboration and differently targeted theoretical analyses of raw experimental data obtained in this program are discussed in detail in Chapters 5 and 6 of this book. Figure 3.8 shows overlaid effective stress-equivalent strain curves obtained for a $Ni_{51at\%}$-Ti alloy

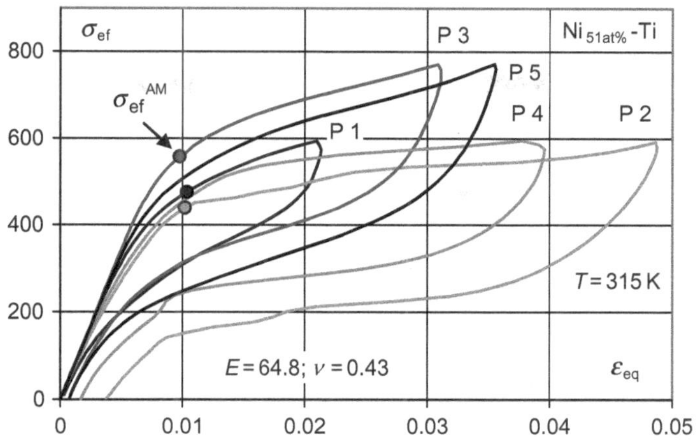

Figure 3.8 Proportional tension-torsion loadings. Stress-strain isothermal curves of $Ni_{51at\%}$-Ti alloy tubular sample submitted to stress-controlled, proportional loadings of torsion (P1), tension (P2), compression (P3), tension-torsion (P4), compression-torsion (P5), at 315 °K temperature (Ziółkowski and Raniecki, unpublished result).

tubular sample submitted to biaxial, isothermal, proportional loadings of (torsion P1; tension P2; compression P3; tension-torsion P4; compression-torsion P5; at temperature 315 K). *Effective stresses* and equivalent strains are defined in standard way as $\sigma_{ef} \equiv \left(\frac{3}{2} s_{ij} s_{ij}\right)^{1/2} = \sqrt{\sigma^2 + 3\cdot\tau^2}$, $\varepsilon_{eq} = \sqrt{\frac{2}{3}\bar{\varepsilon}_{ij}\bar{\varepsilon}_{ij}}$, where s_{ij}, $\bar{\varepsilon}_{ij}$ denote deviatoric components of stress and strain tensors, respectively. The $\varepsilon_{eq} = \sqrt{\varepsilon^2 + \frac{1}{3}(\gamma)^2}$ upon assumption of material incompressibility ($\varepsilon_{z\theta} = \frac{1}{2}\gamma$). The curves shown in Figure 3.8 represent internal hysteresis loops formation resulting from different proportional stress-controlled loadings of the same tubular NiTi specimen. Maximum stress-controlled loadings have been possibly limited in order not to destroy the NiTi specimen, which all in all endured about 250 multiaxial loading programs before damage. The data in Figure 3.8 deliver experimental evidence that NiTi material has different critical stresses pseudoelastic flow σ_{ef}^{AM} depending on the loading stress direction. It exhibits so-called strength differential effect (SD-effect).

Rogueda et al. (1996) performed several experiments on a CuZnAl alloy with non-proportional loading paths. The tests were stress-controlled with loading programs composed of segments during which either tension or shear stress was kept constant. The programs have several shapes: rectangle shape conducted in a clockwise (CW) and counterclockwise (CCW) direction and "L" shape with loading sequence of tension-shear-unshear-untension or shear-tension-unshear-untension. Maximum load of $\sigma_{eq}^{max} = 110\text{MPa}$ resulted in equivalent strains of about 1%, which is a relatively small fraction of about 7% amplitude of phase transformation eigenstrain for tested SMA alloy. Thus, small fractions of martensitic phase have been generated. In

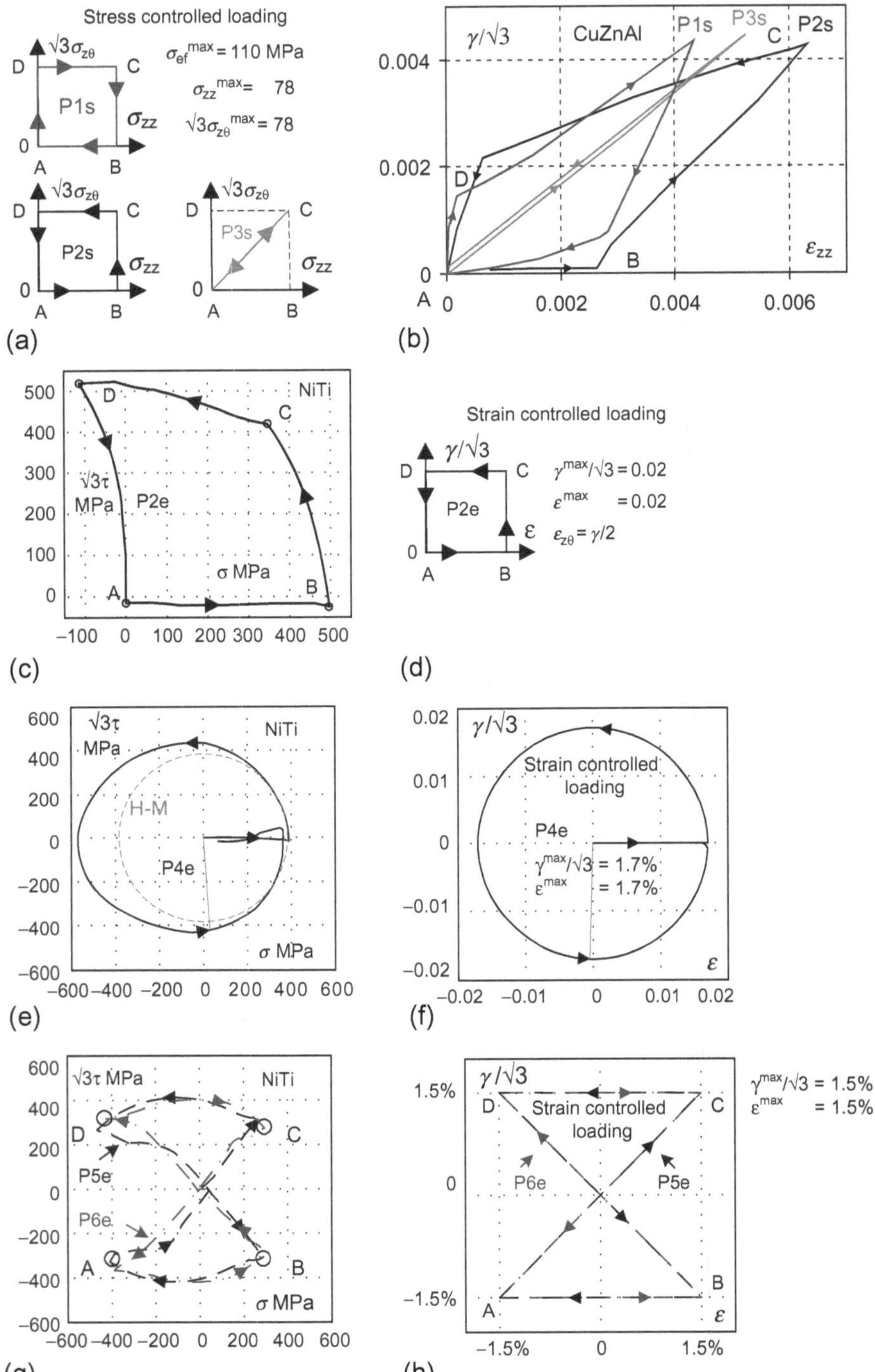

Figure 3.9 Nonproportional tension-torsion loadings: (a and b) Collective graph of first quadrant stress-controlled loading programs and polycrystalline CuZnAl alloy response (reworked after Rogueda et al. (1996)); (c and d) First quadrant, strain-controlled square loading and NiTi alloy response (after Grabe and Bruhns (2008)); (e and f) Symmetric in strain space circular loading path and NiTi alloy response; (g and h) Strain-controlled butterfly loading paths in counterclockwise (P5) and clockwise (P6) directions and NiTi alloy response (reworked after Grabe and Bruhns (2009)).

Figures 3.9a and b three nonproportional stress-controlled loading programs and overlaid together CuZnAl alloy responses are shown. The graphs have been elaborated upon with experimental data obtained by Rogueda et al. (1996). It is worth noting that different loading paths in stress space ending in the same stress state (state C in Figure 3.9a), lead to different strain states (states C in Figure 3.9b). The discrepancy is about $\pm 0.1\%$ at an equivalent strain of about 0.7%. This delineates the accuracy of the SMA material constitutive model predictions that can be expected without introducing elaborate complications into the model formulation.

A tension-torsion proportional and nonproportional testing program, with various loading paths, similar to that executed by Rogueda et al. (1996) has been executed by Sittner et al. (1995) on a CuZnAlMn alloy with qualitatively similar results. However, these last authors introduced unconventional effective stress and equivalent strain measures, making their results inconvenient to be compared quantitatively with results of other authors. They executed stress and strain-controlled tests.

Strain-controlled, *nonproportional, multiaxial tests* on $Ni_{50.7at\%}$-Ti alloy at several quasi-static strain rates in the range ($10^{-5} \div 10^{-3}$ 1/s) were carefully executed by Grabe and Bruhns (2008), who were interested in the investigation of thermal coupling effects. They experimentally demonstrated that NiTi alloy behavior is insensitive to strain rate in this range had the heat generated during the test been removed quickly enough to preserve isothermal conditions of the test. Figures 3.9c–d shows a nonproportional, square in $\gamma/\sqrt{3} - \varepsilon$ coordinates, loading program and NiTi alloy response obtained by Grabe and Bruhns (2008). The test was strain-controlled and involved maximum equivalent strain of about 2.8%, which makes about half of the investigated NiTi alloy phase transformation eigenstrain amplitude in simple tension. This means the presence of about 50% volume fraction of martensitic phase.

Experiments of Rogueda et al. (1996) show that in a stress-controlled, rectangular test (P2s) tension and subsequent shearing with assisting constant tension stress leads to a concurrent increase of shear and tension strains on segment B-C of the loading path (see Figure 3.9b). Experiments of Grabe and Bruhns (2008) show that in a strain-controlled, rectangular test (P2e) tension and subsequent shearing with keeping constant axial strain leads to an increase of shear but a decrease of axial stress on segment B-C of the loading path (see Figure 3.9c). Results of Sittner et al. (1995) who performed both stress and strain-controlled, rectangular, nonproportional tests on CuZnalMn alloy corroborate the indicated features.

Figures 3.9e and f show a CCW, circular, strain-controlled loading program ($\varepsilon_{eq}^{max} = 2.4\%$) and NiTi alloy response obtained by Grabe and Bruhns (2008). The loading program is symmetric in strain space and involves tension, torsion, and compression of the tested material. Arrows indicate mutually corresponding strain and induced-stress states. A striking feature that can be noticed is that total strain and induced-stress tensors in the investigated NiTi material can be accepted with very good approximation to have the same directions during the test execution (discrepancy is indicated with dashed lines). The stress tensor direction nicely follows total strain tensor direction though the amplitude of effective stress tensor varies when equivalent strain amplitude remains constant. This is macroscopic, quantitative manifestation of the mesoscopic mechanism of the martensitic variants reorientation process in adaptation to external loading constraints. Elastic strain tensor is always collinear with stress tensor in elastically isotropic materials. Upon acceptance that the investigated NiTi alloy is elastically isotropic, the data presented in Figure 3.9e and f allow us to accept with very good approximation a modeling conjecture that pseudoelastic, phase transformation strain also is always collinear with stress tensor. It can be noticed at the same time that at a constant value of equivalent strain, the ratio of effective stress values in compression and tension is about 1.5. Figure 3.9g and h show CW and CCW, butterfly shaped, strain-controlled loading paths ($\varepsilon_{eq}^{max} = 2.1\%$) and NiTi alloy response redrawn after experimental data obtained by Grabe and Bruhns (2009). These butterfly loading paths involve sharp changes of loading strain paths, which resulted in an equally sharp change of the response stress tensor path in stress space. It should be noted that volume fraction martensitic phase during the tests shown in Figure 3.9 did not exceed 50%. Results of these tests do not clearly reveal saturation effects, which can be properly evaluated only when complete martensitic transition is induced in the tested SMA material.

Considerable knowledge has been gained on macroscopic uniaxial behavior of SMA materials, especially NiTi alloy. It seems that the most important characteristics of their behavior in this regime of loading has been successfully identified. The weight of experimental research devoted to SMA materials behavior should be shifted in the coming years to multiaxial thermomechanical loadings, as very few reasonable experimental programs of multiaxial thermomechanical tests have been successfully executed up to now. While it is already well known that SMA materials exhibit *SD-effect*, and while it is known that pseudoelastic strain tensor direction follows that of inducing it in a stress tensor direction (for sure in rate terms), still identification of quantitative relations is necessary. Extremely limited information exists on multiaxial behavior at lower temperatures involving shape memory effect. Similarly there is very limited information on phase transformation kinetics and mutual couplings between the volume of the martensitic phase and geometrical effects.

3.3 Approaches towards modeling pseudoelasticity—hysteresis loops

In the previous section, found experimentally, characteristic features of macroscopic behavior of shape memory alloy materials in the pseudoelastic range of their behavior have been systematically discussed. The idealization of this behavior allows distinguishing two key generic features of pseudoelastic behavior of materials: the return to its original shape upon unloading (i.e., the property that at zero macroscopic stress material exhibits zero macroscopic total strain), and hysteresis loop formation on loading-unloading cycle (i.e., the property that the loading path is different from the unloading path). There arises a problem of selection of a convenient modeling framework for description of the pseudoelastic behavior. Modest engineering design purposes require the generation of modeling predictions consistent with experimental evidence (i.e., "mere" recovering of experimental curves). Scientific activity poses the demanding requirement that theoretical modeling enables deep understanding of the described phenomena and possibly enables revealing patterns and links, which cannot be immediately or easily grasped without profound analytical efforts. These two approaches are quite contradictory as the first requires modeling predictions in short time and at little cost compromising with moderate accuracy, while the second usually requires accurate and reliable predictions compromising for a long time, and a lot of effort for theoretical model development itself, and also higher qualifications of personnel generating actual modeling predictions. Fortunately, at present an appealing compromise exists.

A group of theoretical models that gained popularity in the early days of modeling SMA materials behavior has been based on the classical theory of nonlinear elasticity and the Landau-Devonshire theory developed for the description of magnetic materials behavior undergoing first-order phase transitions. The most prominent model from this group has been elaborated by Falk (1983). The fundamental assumption made in Falk's model to enable modeling thermoelastic phase transitions is nonconvexity of adopted elastic energy function expressed in terms of temperature and total strain with respect to total strain. The elastic energy potential is postulated in the form of a sixth-order polynomial with no other variables besides T, ε. Falk's model enables qualitative description of uniaxial SMA materials behavior in the full range of SME effects, reflecting generic features of pseudoelasticity, though its accuracy is very poor by contemporary standards. It can serve at present for some illustrative purposes only. The scope and predictions quality of pseudoelasticity models based on nonlinear elasticity has been considerably extended by Ogden (2001). He developed a 3D model of pseudoelasticity capable for describing the stress softening effect (also called the Mullins effect) exhibited by some rubber-like materials. It has been achieved by introducing into the elastic energy function an additional variable besides the deformation gradient variable, describing damage of elastic properties of the material. The approach adopted by Ogden (2001) can be regarded as a limited in scope approach of framework of nonequilibrium thermodynamics with hidden parameters. It is not as powerful as this last one because interests of nonlinear elasticity stream of research is by and large limited to the mechanical part of materials behavior and it does not exploit, for example, the consequences of thermodynamic couplings, to mention only one limitation.

Another stream of models developed for the description of pseudoelastic behavior of materials constitute models using so-called hysteresis operators. A typical mechanical representative of such an operator is a spring-damper unit. Probably the use of this type of model for the description of pseudoelastic behavior of SMA materials has been motivated by the use of the mathematical model of hysteretic behavior proposed by Preisach for the description of magnetic hysteresis loops in ferromagnetic materials. A correct modeling description of material behavior is attained by suitable selection of a population of hysteresis units, sometimes called *hysterons*, their proper number, and characteristics such as elastic constants, damping constants, critical stresses for switching, and others. The appropriate values of characteristic parameters is determined by best-fit procedure to experimental curves. This type of model of 3D SMA materials behavior has been elaborated by Saleeb et al. (2010). The advantage of such models is that they may be and are useful to relatively quickly and cheaply deliver modeling predictions when rather little knowledge is available on the material behavior. However, their cognitive capabilities for deep understanding of the material behavior seem to be very limited in comparison to a thermodynamics with internal parameters approach.

A popular stream of modeling efforts devoted to the pseudoelastic behavior of SMA materials make up models that are a quite straightforward adaptation of the concepts of theory of plasticity; for example, the one developed by Auricchio and Taylor (1997). The approach is pragmatic and enables obtaining quite appealing modeling predictions in the short term, but it does not rely on fundamental physical mechanisms underlying the appearance of shape memory phenomena. This leads to the introduction of modeling elements ad hoc in the form of imposing some constraints without connection with other elements of the theoretical model. For example, in SMA models stemming from the theory of plasticity, critical stress for pseudoelastic flow is preset independently, or regardless, from the specific form adopted in the model Gibbs function.

It seems that the most capable approach toward modeling macroscopic behavior of SMA materials is a framework of continuum, *nonequilibrium thermodynamics with internal state parameters*. This framework is very comprehensive, versatile, and already proved to be very successful in modeling behavior of various classes of materials as discussed, for example, by Horstenmeyer and Bammann (2010). The primary reason for the choice of this framework is knowledge that phase transition is underlying shape memory effects, and thermodynamics is a natural tool to deal with phase transition phenomena. The concept of internal state parameters proved effective to surmount the difficulty that classical equilibrium thermodynamics does not allow for the description of hysteresis loops. The thermodynamics with internal parameters methodology requires large initial investments in the form of broad background knowledge and inventiveness necessary to introduce correct and useful internal parameters. Follansbee (2014) delivers an interesting and illustrative example of the possible traps and gains in this respect. A serious benefit of the discussed methodology is the cognitive power of the developed theoretical model; for example, by taking advantage of information resulting from mutual thermodynamic couplings. It becomes possible to notice certain general regularities, which are not easily noticeable when using models basing on, for example, populations of hysteresis operators. For example, analysis of the SMA thermodynamic model reveals how elastic energy stored in martensitic mesostructures after phase transition termination influences the characteristic properties of specific SMA material. Another achievement of key importance is demonstration of the link between criteria of phase transition initiation and instability of thermodynamic states resulting from an adopted specific form of free energy function of SMA material. The framework of thermodynamics with internal parameters allows for flexible reuse of parts of the constitutive model already elaborated to enhance its predictions range. Large parts of the thermodynamic model of pseudoelasticity of shape memory metallic materials can be adapted to develop the thermodynamic model of shape memory polymers. The necessary development work in this direction would include employment of some finite deformations measure, the supplementation of the theory with rheological effects, and elaboration of new kinetics rules, to mention the most important modifications.

Macroscopic modeling of SMA materials behavior by far is not a unique methodology toward describing and understanding shape memory alloys behavior. An interesting and comprehensive survey of different approaches toward modeling SMA materials performance can be found in a two-part paper by Lagoudas et al. (2006) and Patoor et al. (2006). The remainder of this chapter presents a specific model of pseudoelastic behavior of isotropic SMA materials developed using formalism of nonequilibrium thermodynamics with internal state parameters.

3.4 Thermodynamic R_L model of pseudoelasticity with SD-effect

3.4.1 Introduction and thermodynamic equilibriums

In the years 1992-1998 Raniecki, Lexcellent, and Tanaka developed a family of macroscopic, isotropic, 3D thermodynamic models of pseudoelasticity (Raniecki et al., 1992; Raniecki and Lexcellent, 1994; Raniecki and Lexcellent, 1998). Even, the earliest model in the family is capable for the prediction of deformation, thermal effects, evolution of volume fraction of martensitic phase, and evolution of macroscopic phase transformation strain. It has been subsequently enriched to enable description of SMA materials for which it is necessary to take into account considerably different apparent stiffness of austenitic and martensitic phase. In 1998, improved phase transition criteria was proposed enabling description of SMA materials exhibiting strong dependence of critical pseudoelastic flow stress on direction of stress tensor-strength differential effect (*SD-effect*). The R_L models have been developed within the framework of small deformations theory. They are capable to reproduce all the key features of pseudoelastic behavior of SMA materials revealed by experimental results discussed in Section 3.2 and shown schematically in Figure 3.6. Due to the accepted assumption of isotropy, the R_L models are appropriate for description of polycrystalline SMA materials submitted to multiaxial thermomechanical loadings. Higher accuracy of the modeling predictions for proportional loading paths is expected. Formally, admissible values of external state parameters of stress and temperature are limited to the domain of pseudoelastic behavior of SMA materials (see Figure 2.14); that is, temperatures remaining in the range $A_f^{2-1} < T < M_\sigma$. For some loading paths models enable prediction of SMA materials behavior also in temperatures $M_s^0 < T$.

The macroscopic R_L models development is rooted in mesomechanics, where generic phases are treated as thermoelastic materials with eigenstrains. Experimental evidence discussed in Section 3.2 (see, e.g., Figure 3.5 and related text) delivers premises that martensitic phase distinguished in the atomic scale of observation on the grounds of crystallography can be further refined for macroscopic modeling purposes into self-accommodating and oriented martensite "phases" distinguished in the higher, mesoscale of observation (RVE) on the grounds of thermodynamic properties. It is reasonable to accept that, in the most general case, RVE of SMA material is composed of a conglomerate of three phases: austenite, self-accommodating martensite, and oriented martensite (see also Figure 4.1). In the range of external state parameters for which SMA material behaves pseudoelastically, no self-accommodating martensite can appear, as it can only be temperature-induced by cooling SMA material below M_s^0 temperature at zero-assisting macroscopic stress ($\sigma = 0$). It cannot be generated when temperature remains above M_s^0. Thus, RVE of SMA material in the pseudoelastic range of its behavior can be at most in a two-phase state composed of austenite ($\mathcal{A}$) and oriented martensite ($\mathcal{M}_{or}$).

Formalism of nonequilibrium thermodynamics with internal state parameters has been used for the development of the macroscopic R_L models family of SMA materials pseudoelastic behavior; see, for example, Horstenmeyer and Bammann (2010) or Rice (1971). All variants of the R_L models family rely on the following conjectures regarding equilibrium of thermodynamic states of macroelement of SMA material.

It is accepted that SMA material RVE always remains in *full equilibrium*, in the thermodynamic sense, of the following kinds:

(i) *micromechanical equilibrium*—this indicates adoption of the assumption that partial Cauchy stresses in both phases of two-phase RVE are equal to each other and equal to total macroscopic stress of RVE $\boldsymbol{\sigma}_1 = \boldsymbol{\sigma}_2 = \boldsymbol{\sigma}$. Using this property, it can be found out that then partial elastic strains in both phases are also accepted to be the same. This means that macroscopic stress of RVE can be determined from the formula $\boldsymbol{\sigma} \equiv \partial\phi/\partial\boldsymbol{\varepsilon}$, where ϕ denotes Helmholtz free energy.

(ii) *thermal equilibrium*—this indicates adoption of the assumption that entropy of RVE is determined by a value of the partial derivative of free energy potential with respect to temperature $s \equiv -\partial\phi/\partial T$.

(iii) *mesostructural equilibrium*—this indicates adoption of the assumption that RVE is in thermodynamic equilibrium with respect to all internal hidden variables h_K characterizing mesostructure of SMA material RVE. This equilibrium condition is the foundation for the acceptance of the validity of a rule of "*optimum rearrangement of mesostructure*." This conjecture leads to the acceptance of the assumption that reorientation of martensitic phase does not lead to energy dissipation in the pseudoelastic range of SMA behavior.

(iv) the SMA macroelement in general does not remain in *thermodynamic, phase equilibrium*. The consequence of this assumption is that, in general, the thermodynamic driving force of phase transition π^{1-2} ($\mathcal{A} \to \mathcal{M}_{or}$) takes the values different from zero $\pi^{1-2} \equiv -\partial\phi/\partial z \neq 0$ during active p.t... Thus, internal dissipative effects of SMA material are attributed to this nonequilibrium.

It is accepted in the R_L models family that SMA macroelement in two-phase states in general remains in *constrained thermodynamic equilibrium*.

Revised, updated, and reinterpreted in view of state of the art knowledge relations of (Raniecki and Lexcellent, 1998) variant of R_L model of pseudoelasticity are specified in the sequel below. Correspondence in notation between the work of Raniecki and Lexcellent (1998) and the present section is as follows:

$$\Delta u_2^0 \leftrightarrow \Delta u^*, \Delta s_2^0 \leftrightarrow \Delta s^*, \phi_{12} \leftrightarrow \overline{u}_0, s_{12} \leftrightarrow \overline{s}_0$$
$$\pi_0^{1-2}(T) \leftrightarrow \pi_0^{f}(T), \pi^{1-2} \leftrightarrow \pi^{f}, g^{ult} \leftrightarrow g^{ph}, \boldsymbol{\varepsilon}^{pt} \leftrightarrow \boldsymbol{\varepsilon}^{T}.$$

3.4.2 State variables

It is conjectured that the state of an SMA material element RVE, treated as an open thermodynamic system, can be completely characterized with the following set of *thermodynamic state variables*:

$\boldsymbol{\varepsilon}$—tensor of small total strains, T—temperature, $z = V^{(2)}/V$—volume fraction of oriented martensite phase; where $V^{(2)}$ is instantaneous volume of oriented martensite and V denotes total volume of macroelement (RVE). It is treated as an internal state variable, h_K—a set of independent, mesoscopic, internal variables ($K = 1, 2, \ldots N$), describing the evolving geometry of mesostructure of SMA material RVE. It is assumed that variables h_K enter the description of SMA material only indirectly through a tensor, $\boldsymbol{\kappa}(h_K)$—of macroscopic *ultimate phase eigenstrain of oriented martensite*, treated as a second tier internal state variable. The variables h_K are not specified explicitly here, as it appeared not to be necessary, but in the case of mesostructural studies they should be defined in such a way as to enable quantitative characterization of the internal martensitic pattern of the SMA macroelement. In the pseudoelastic range of SMA materials behavior it can be well accepted that tensor $\boldsymbol{\kappa}$ does not depend on temperature. For example, experimental data in Figure 3.1a show that pseudoelastic flow amplitude does not change considerably with temperature. Elaboration of a variant of the R_L model with tensor $\boldsymbol{\kappa}$ dependent on temperature $\boldsymbol{\kappa}(h_K, T)$, can be useful in the lower temperatures range. It does not present serious conceptual difficulties but leads to more complicated mathematical formulas.

The ISV $z, h_K \rightarrow \boldsymbol{\kappa}(h_K)$ require specification of additional evolution relations going beyond specification of the explicit form of Helmholtz free energy function $\phi(\boldsymbol{\varepsilon}, T, z, \boldsymbol{\kappa}(h_K))$. This additional information completes the specification of a constitutive model of shape memory alloys behavior, making it complete from a mathematical and physical point of view. The evolution relationships for volume fraction of oriented martensite z are called *phase transformation kinetics rules* in the R_L models family. The tensor $\boldsymbol{\kappa}$ is an averaged macroscopic measure describing strain effects associated with phase transformation progress and/or martensitic phase reorientation due to variation of loading state, with reference to austenitic phase. It results from a specific layout of internal pattern of martensitic habit plane variants; actual geometrical configuration of martensitic phase mesostructure. It is defined explicitly with the aid of the so-called ultimate phase strain work potential g^{ult}. *Macroscopic phase strain* connected with martenstic phase transition and macroscopic elastic strain of SMA macroelement are expressed with the formulas?

$$\boldsymbol{\varepsilon}^{\mathrm{pt}} = z\kappa, \boldsymbol{\varepsilon}^{\mathrm{e}} = \boldsymbol{\varepsilon} - \boldsymbol{\varepsilon}^{\mathrm{pt}}, \mathrm{tr}(\boldsymbol{\varepsilon}^{\mathrm{pt}}) = 0 \tag{3.1}$$

Such multiplicative decomposition of inelastic strain finds support in experimental data (e.g., as shown in Figure 3.1). As the geometry of austenitic phase is accepted to be the reference one regarding phase transition, by definition the ultimate eigenstrain tensor of austenitic phase is equal to zero ($\boldsymbol{\kappa}_1 \equiv 0$). Partial elastic strains in austenite and oriented martensite in SMA material RVE in two-phase states are the same and equal to the elastic strain of RVE. The tensor $\boldsymbol{\kappa}$ is traceless ($\mathrm{tr}(\boldsymbol{\kappa}) = 0$) in accordance with experimental results showing that the volumetric strain of thermoelastic martensite is small.

3.4.3 Free energy functions and the rule of optimum rearrangement of mesostructure

The following simplifications of physical nature are accepted. The "energy" tensors of modules of elasticity $\mathbf{L}_\alpha/\rho$—$\mathbf{L}_\alpha$ denote classical tensors of elastic properties appearing in Hooke's law relating Cauchy stress tensor with elastic strain tensor, tensor of thermal expansion coefficients $\boldsymbol{\alpha}_\alpha$, and heat capacities at constant strain c_v^α of individual phases are treated as physical constants independent of thermodynamic state and identical for both phases. The experimental observations indicate that the difference of density of thermoelastic martensite and austenite usually does not exceed 0.2%. Hence, it is also accepted that densities of austenitic and martensitic phase are always the same,

$$\boldsymbol{L}/\rho = \boldsymbol{L}_\alpha/\rho = \text{const}, \ \boldsymbol{\alpha} = \boldsymbol{\alpha}_\alpha = \text{const}, \ c_\mathrm{v} = c_\mathrm{v}^\alpha = \text{const}, \ \rho_1 = \rho_2 = \rho \tag{3.2}$$

Acceptance of the above assumptions constitutes reasonable modeling approximation, considerably simplifying mathematical formulation of the model. The case, when elastic stiffness modules are different for individual phases and dependent linearly on temperature, was investigated in the work of Raniecki and Lexcellent (1994).

The explicit form of specific *Helmholtz free energy function* valid for two-phase SMA macroelement in states of constrained equilibrium, when conditions *(i)* ÷ *(iv)* are fulfilled, is postulated as follows:

$$\begin{aligned}
\phi(\boldsymbol{\varepsilon},T,z,\boldsymbol{\kappa}(h_\mathrm{k})) &= \phi^\mathrm{th}(T,z) + \phi^\mathrm{tm}(\boldsymbol{\varepsilon},T,z,\boldsymbol{\kappa}) + \phi^\mathrm{coh}(z,T),\\
\phi^\mathrm{th}(T,z) &= u_1^0 - Ts_1^0 - [\Delta u^0 - T\Delta s^0]z + c_\mathrm{v}[(T-T_0) - T\ln(T/T_0)],\\
\phi^\mathrm{tm}(\boldsymbol{\varepsilon},T,z,\boldsymbol{\kappa}(h_\mathrm{k})) &= (\boldsymbol{\varepsilon} - z\boldsymbol{\kappa})\cdot\boldsymbol{L}(\boldsymbol{\varepsilon} - z\boldsymbol{\kappa})/(2\rho) - (T-T_0)\boldsymbol{\alpha}\cdot\boldsymbol{L}(\boldsymbol{\varepsilon} - z\boldsymbol{\kappa})/\rho,\\
\phi^\mathrm{coh}(z,T) &= z\phi_2^\mathrm{st} + (1-z)z\phi_\mathrm{it}(T), \phi_2^\mathrm{st} = \Delta u_2^\mathrm{st} - T\Delta s_2^\mathrm{st}, \phi_\mathrm{it}(T) = \phi_{12} - Ts_{12},\\
\Delta u^0 &= u_1^0 - u_2^0, \quad \Delta s^0 = s_1^0 - s_2^0, \phi_{12}, s_{12} = \text{const},
\end{aligned} \tag{3.3}$$

Justification of this form on the grounds of mesomechanical studies is delivered in Chapter 4 of this book. The terms $\phi^\mathrm{tm}(\boldsymbol{\varepsilon},T,z,\boldsymbol{\kappa})$ and $\phi^\mathrm{th}(T,z)$ denote thermoelastic and thermal part of free energy of SMA macroelement, while $\phi^\mathrm{coh}(z,T)$ is so-called energy of coherency. This energy is composed of energy of internal interactions $(1-z)z\phi_\mathrm{it}$, and *stored energy of coherency* $\phi_2^\mathrm{st}z$; ϕ_2^st denotes energy stored in SMA macroelement after full completion of martensitic phase transformation ($z=1$). The *coherency energy* is generated by a field of incompatible martensitic phase eigenstrains inducing self-equilibrating fluctuating stress field, with respect to average macroscopic stress value. Justification for this statement can be found in Chapter 4 of this book. Free energy function can be divided in many different ways; more traditionally it is divided as follows: $\phi = (1-z)\phi_1 + z\phi_2 + \phi^{\mathrm{coh}}$, where ϕ_1 is thermoelastic free energy of austenitic phase, and ϕ_2 is thermoelastic free energy of martensitic phase.

The term describing stored energy ϕ_2^{st} and the term $-\pi_0^{\mathrm{f}}$, frequently called *chemical driving force* of phase transformation in metallurgical literature, can be combined together to give

$$\pi_0^{1-2}(T) \equiv \left(\Delta u^0 - T\Delta s^0\right) - \left(\Delta u_2^{\mathrm{st}} - T\Delta s_2^{\mathrm{st}}\right) = \Delta u_2^0 - T\Delta s_2^0$$
$$\Delta u_2^0 = \Delta u^0 - \phi_2^0, \Delta s_2^0 = \Delta s^0 - \Delta s_2^{\mathrm{st}}, \Delta u_2^{\mathrm{st}} = \phi_2^0 = \mathrm{const}, \Delta s_2^{\mathrm{st}} = \mathrm{const} \tag{3.4}$$

where the material constants u_1^0, s_1^0 denote internal energy and entropy of *austenitic phase formation* in virtual *mesoscopic thermodynamic reference state* (m.r.s.); that is, at mezo-stress-free state $(\breve{\boldsymbol{\sigma}}(\boldsymbol{x}) = \mathbf{0})$ at some prescribed thermodynamic reference temperature $(T = T_0)$.

The condition $\breve{\sigma}(x) = 0$ denotes that macroscopic stress $\boldsymbol{\sigma}$ of the RVE and self-equilibrating field of internal fluctuating stresses $\breve{\sigma}^{in}(x)$ is accepted to be zero at the virtual thermodynamic reference state; see Chapter 4 for more extensive discussion. The constants u_2^0, s_2^0 denote energy and entropy of *martensitic phase formation* in mesoscopic thermodynamic reference state (m.r.s.). The values of $\Delta u_2^0, \Delta s_2^0$ are usually determined in a procedure of identification of constitutive model material constants from experimental data obtained in tests in which phase transformation is stress-induced $(A_0^{\mathrm{f}} < T)$. Hence, these material constants already include contribution connected with stored energy ϕ_2^{st}. The quantities ϕ_{12}, s_{12} are material constants characterizing free energy of austenitic and oriented martensite phases interaction.

A problem of relation describing evolution of ultimate eigenstrain of oriented martensite $\boldsymbol{\kappa}$ is not trivial. It is not reasonable to expect that pure mesomechanical studies itself themselves can furnish proper quantitative information about the macroscopic behavior of RVE of SMA. The mesostructure of RVE changes drastically in the course of complete phase transitions, and the relation between the history of thermomechanical loadings and the substructure geometry evolution is, in fact, to a large extent unknown. There are very little experimental results linking evolution of martensitic mesostructure with quantitative macroscopic effects.

The most important macroscopic observation is that martensite mesostructure evolves, adapting to changing external loading in such a way that macroscopic phase transformation strain tensor direction is collinear with stress tensor; see the experimental data shown in Figure 3.9. This delivers premises for accepting thr validity of the rule of *optimum rearrangement of mesostructure* stemming from the condition of thermodynamic mesostructural equilibrium *(iii)*.

One says that mesostructure of SMA macroelement is optimum at prescribed loading conditions when values of internal variables h_K characterizing the martensitic mesostructure are minimizing specific Helmholtz free energy function $\phi(\varepsilon, T, z, \kappa(h_K))$. Equilibrium values of the parameters h_K for which ϕ attains extremum can be determined from the equations defining necessary conditions for attaining extremum

$$\begin{aligned}(\partial\phi(\boldsymbol{\varepsilon}, T, z, \boldsymbol{\kappa}(h_K))/\partial h_K = 0) &\Rightarrow ((\boldsymbol{\sigma}/\rho)\cdot\partial\boldsymbol{\kappa}(h_K)/\partial h_K = 0)\\ &\Rightarrow (h_K = h_K^{\mathrm{eq}}(\boldsymbol{\sigma})), \quad K = 1, N\end{aligned} \tag{3.5}$$

From the above it results that equilibrium values h_K^{eq} can only depend on macroscopic stress $\boldsymbol{\sigma}$. Then, equilibrium value of macroscopic eigenstrains of martensite $\boldsymbol{\kappa}^{\mathrm{eq}}$ must be a function of stress

$$\boldsymbol{\kappa}(h_K)|_{h_K=h_K^{\mathrm{eq}}} = \boldsymbol{\kappa}^{\mathrm{eq}}(h_K^{\mathrm{eq}}(\boldsymbol{\sigma})) = \boldsymbol{\kappa}^{\mathrm{eq}}(\boldsymbol{\sigma}), \tag{3.6}$$

and tensor of eigenstrains must fulfill the condition

$$\sum_K \boldsymbol{\sigma}\cdot\frac{\partial\boldsymbol{\kappa}}{\partial h_K}\frac{\partial h_K^{\mathrm{eq}}}{\partial\boldsymbol{\sigma}} = \boldsymbol{\sigma}\cdot\frac{\partial\boldsymbol{\kappa}^{\mathrm{eq}}}{\partial\boldsymbol{\sigma}} = 0) \Leftrightarrow \left(\sigma_{ij}\frac{\partial\kappa_{ij}^{\mathrm{eq}}}{\partial\sigma_{mn}} = 0\right), \tag{3.7}$$

resulting from multiplying of N equations $(3.5)_2$ with $\partial h_K^{\mathrm{eq}}/\partial\boldsymbol{\sigma}$, respectively, and next their summation. If the second order tensor $\boldsymbol{\sigma}\cdot(\partial\boldsymbol{\kappa}^{\mathrm{eq}}/\partial\boldsymbol{\sigma})$ is contracted with stress rate tensor $\dot{\boldsymbol{\sigma}}$ then one obtains another important relation, which is always fulfilled at states of mesostructural equilibrium

$$\left(\boldsymbol{\sigma}\cdot\frac{\partial\boldsymbol{\kappa}^{\mathrm{eq}}}{\partial\boldsymbol{\sigma}}\dot{\boldsymbol{\sigma}} = \boldsymbol{\sigma}\cdot\dot{\boldsymbol{\kappa}}^{\mathrm{eq}} = 0\right) \Leftrightarrow \left(\sigma_{ij}\frac{\partial\kappa_{ij}^{\mathrm{eq}}}{\partial\sigma_{mn}}\dot{\sigma}_{mn} = 0\right) \tag{3.8}$$

In $\mathrm{R_L}$ models family it is postulated that for equilibrium values of macroscopic *ultimate phase eigenstrains tensor* $\boldsymbol{\kappa}^{\mathrm{eq}}(\boldsymbol{\sigma})$ there exists potential $g^{\mathrm{ult}}(\boldsymbol{\sigma}) \geq 0$ such that

$$\begin{aligned} &\rho g^{\mathrm{ult}}(\boldsymbol{\sigma}) = \boldsymbol{\kappa}^{\mathrm{eq}}\cdot\boldsymbol{\sigma},\ \boldsymbol{\kappa}^{\mathrm{eq}}(\boldsymbol{\sigma}) = \rho\frac{\partial g^{\mathrm{ult}}(\boldsymbol{\sigma})}{\partial\boldsymbol{\sigma}}, \\ &\sigma_{ij}\frac{\partial\kappa_{ij}^{\mathrm{eq}}}{\partial\sigma_{mn}} = \sigma_{ij}\frac{\rho\partial^2 g^{\mathrm{ult}}}{\partial\sigma_{ij}\partial\sigma_{mn}} = \sigma_{ij}\frac{\partial\kappa_{mn}^{\mathrm{eq}}}{\partial\sigma_{ij}} = 0. \end{aligned} \tag{3.9}$$

The *ultimate phase eigenstrains potential* ρg^{ult} must be a homogeneous function of order one with respect to stress $\boldsymbol{\sigma}$, in order that the condition (3.7) be fulfilled identically. Then $\boldsymbol{\kappa}^{\mathrm{eq}}(\boldsymbol{\sigma})$ is a homogeneous function of order zero with respect to stress $\boldsymbol{\sigma}$. Fulfillment of condition (3.7) identically means that as a result of microstructure reorientation (i.e., reorientation of tensor $\boldsymbol{\kappa}$), no entropy production takes place. The subscript "eq" at $\boldsymbol{\kappa}$ is omitted below in understanding that it fulfills constraints of mesostructural thermodynamic equilibrium and consequential condition (3.7). The optimum rearrangement rule indicates properties, which must possess explicit functional relation for $g^{\mathrm{ult}}(\boldsymbol{\sigma})$ defining tensor $\boldsymbol{\kappa}$ (see (3.32)).

One can obtain specific Gibbs free energy g corresponding to specific Helmholtz free energy ϕ by applying the Legendre transformation

$$g \equiv \phi(\boldsymbol{\varepsilon}, T, z) - (1/\rho)\boldsymbol{\sigma}\cdot\boldsymbol{\varepsilon} \tag{3.10}$$

After replacing expression $\boldsymbol{\kappa}\cdot\boldsymbol{\sigma}/\rho$ with $g^{\mathrm{ult}}(\boldsymbol{\sigma})$ (see $(3.9)_1$), the SMA macroelement Gibbs free energy valid in states of constrained thermodynamic equilibrium (no phase equilibrium), becomes the function of $(\boldsymbol{\sigma}, T, z)$ only and takes the explicit form

$$g(\boldsymbol{\sigma}, T, z) = u_1^0 - Ts_1^0 - z(\pi_0^f(T) - \phi_2^{st}) + c_p[(T - T_0) - T\ln(T/T_0)]$$
$$- \boldsymbol{\sigma}\cdot\mathbf{M}\boldsymbol{\sigma}/(2\rho) - (T - T_0)\boldsymbol{\alpha}\cdot\boldsymbol{\sigma}/\rho - z\underbrace{g^{ult}(\boldsymbol{\sigma})}_{=\boldsymbol{\kappa}\cdot\boldsymbol{\sigma}/\rho} + \phi_{it}(T)(1 - z)z, \tag{3.11}$$

where $\mathbf{M} = \mathbf{L}^{-1}$ is the classical tensor of elastic compliances, while c_p denotes specific heat at constant stress. In the above the following identity has been used:

$$c_p \ln(T/T_0) = c_v \ln(T/T_0) + \boldsymbol{\alpha}\cdot\mathbf{L}\boldsymbol{\alpha}(T - T_0)/\rho,$$
$$c_v/T \equiv -\partial^2\phi/\partial T^2\big|_{\boldsymbol{\varepsilon}=\text{const}}, c_p/T = -\partial^2 g/\partial T^2\big|_{\sigma=\text{const}}, \tag{3.12}$$

3.4.4 Thermal equations of state

Thermal *equations of state* of SMA macroelement are as follows:

$$\boldsymbol{\varepsilon} = -\rho\partial g/\partial\boldsymbol{\sigma} = \boldsymbol{M\sigma} + \boldsymbol{\varepsilon}^{pt} + \boldsymbol{\alpha}(T - T_0), \boldsymbol{\varepsilon}^{pt} = z\rho\partial g^{ult}/\partial\boldsymbol{\sigma} = z\boldsymbol{\kappa},$$
$$s \equiv -\partial g/\partial T = s_1^0 - z\Delta s_2^0 + c_p\ln(T/T_0) + \boldsymbol{\alpha}\cdot\boldsymbol{\sigma}/\rho + s_{12}(1 - z)z,$$
$$\pi^{1-2} \equiv -\partial g/\partial z = \pi_0^{1-2}(T) - (1 - 2z)\phi_{it}(T) + g^{ult}(\boldsymbol{\sigma}). \tag{3.13}$$

The condition $\pi^{1-2} = 0$ determines two-phase states of *full thermodynamic equilibrium* of SMA material RVE

$$\pi^{1-2} = \pi_0^{1-2}(T) - (1 - 2z)\phi_{it} + \boldsymbol{\kappa}\cdot\boldsymbol{\sigma}/\rho = 0 \tag{3.14}$$

Full thermodynamic equilibrium values of volume fraction of martensitic phase z^{eq} corresponding to $(\boldsymbol{\varepsilon}, T)$ values of external state parameters can be determined with the formula $z^{eq} = \{\boldsymbol{\kappa}\mathbf{L}[(\boldsymbol{\varepsilon} - \boldsymbol{\alpha}(T - T_0)] + \rho(\pi_0^{1-2} - \phi_{it})\}/[\boldsymbol{\kappa}\mathbf{L}\boldsymbol{\kappa} - 2\rho\phi_{it}]$ obtained from (3.14) after simple manipulations.

The states of full thermodynamic equilibrium of SMA macroelement are unstable when $\partial\pi^{1-2}/\partial z \equiv -(\partial^2 g/\partial z^2) = 2\phi_{it}(T) \geq 0$. Experimental results indicate that this condition is fulfilled for many shape memory alloys. The instability of full thermodynamic equilibrium states causes real thermodynamic processes that actually take place in the SMA material macroelement not to run through such states. The theoretical developments of the R_L models family delineated above is based on the conjecture that thermodynamic instability of phase equilibrium states is the primary cause for the formation of hysteresis loops.

A natural consequence is to adopt conditions $\pi^{1-2}(\boldsymbol{\sigma}, T, z) = 0$ as criteria determining the critical states, at which the processes of active forward and reverse phase transformation start, regardless of complete or partial (i.e., those connected) with full or internal hysteresis loops formation. The criteria originate directly from the adopted specific form of free energy function without any additional assumptions. Substitution of $(3.3)_6$ and $(3.4)_1$ into (3.14) and simple manipulations give criteria for transformation initiation of pure phases

$$F^{\mathrm{AM}} \equiv \pi^{1-2}(\boldsymbol{\sigma},T,z=0) = g^{\mathrm{ult}}(\boldsymbol{\sigma}) - \left(\Delta s_2^0 - s_{12}\right)\left[T - \frac{\left(\Delta u_2^0 - \phi_{12}\right)}{\left(\Delta s_2^0 - s_{12}\right)}\right] = 0,$$
$$F^{\mathrm{MA}} \equiv \pi^{1-2}(\boldsymbol{\sigma},T,z=1) = g^{\mathrm{ult}}(\boldsymbol{\sigma}) - \left(\Delta s_2^0 + s_{12}\right)\left[T - \frac{\left(\Delta u_2^0 + \phi_{12}\right)}{\left(\Delta s_2^0 + s_{12}\right)}\right] = 0 \tag{3.15}$$

They constitute counterparts of known, from the theory of mechanics of fluids, *Clausius-Clapeyron relation*. Graphs shown in Figures 3.3b and 3.4b show that experimentally observed Clausius-Clapeyron relations are linear in $\boldsymbol{\sigma} - T$ coordinates with respect to temperature in the pseudoelastic range of SMA materials behavior. This delivers the experimental argument for accepting linear dependence of potential g^{ult} on stress and its independence on temperature, as upon fulfilling these constraints modeling Clausius-Clapeyron relations (3.15) become linear in $\boldsymbol{\sigma} - T$ coordinates. Additionally, in order that lines $F^{\mathrm{AM}} = 0$ and $F^{\mathrm{MA}} = 0$ be parallel, a feature exhibited by many SMA materials, the constant s_{12} must be zero $(s_{12} = 0)$.

Extrapolation of Clausius-Clapeyron relations (3.15) into low temperatures to stress-free states $\boldsymbol{\sigma} = 0$ gives modeling expressions defining certain stress-free martensite start and austenite start temperatures. Experimental results obtained in uniaxial tension tests of NiTi specimen plotted in Figure 3.4b show that extrapolated CC line martensite start temperature can be identified with good approximation with M_s^0 temperature; that is, start temperature of austenite into self-accommodating martensite p.t. obtained in DSC measurement. This conclusion is further confirmed by experimental data obtained for tubular NiTi specimen and shown in Figure 5.10a, in Chapter 5. This indicates that it is reasonable to accept that self-accommodating martensite start temperature and oriented martensite start temperature are the same $(M_s^{1-2} = M_s^0)$. The extrapolated CC line austenite start temperature at zero stress marked in Figure 3.4b with symbol A_s^{2-1} can be identified to be start temperature of oriented martensite to austenite phase transition—and being considerably different from conventional austenite start temperature A_s^0 of self-accommodating martensite into austenite p.t. obtained in the standard way with DSC $(A_s^{2-1} \neq A_s^0)$. Based on these observations, the following relations $(3.16)_{1,2}$ can be obtained from (3.15) upon taking advantage of the property that $g^{\mathrm{ult}}(\boldsymbol{\sigma} = 0) = 0$:

$$M_s^0 = \left(\Delta u_2^0 - \phi_{12}\right)/\left(\Delta s_2^0 - s_{12}\right), A_s^{2-1} = \left(\Delta u_2^0 + \phi_{12}\right)/\left(\Delta s_2^0 + s_{12}\right)$$
$$M_s^0 = \left(\Delta u_3^0 - \phi_{13}\right)/\left(\Delta s_3^0 - s_{13}\right), A_s^0 = \left(\Delta u_3^0 + \phi_{13}\right)/\left(\Delta s_3^0 + s_{13}\right), M_s^{1-2} = M_s^0 \tag{3.16}$$

Relations $(3.16)_{3,4}$ can be straightforwardly obtained upon taking advantage of the following reasoning. In full analogy to the development of austenite to oriented martensite phase transition theory $(\mathcal{A} \leftrightarrow \mathcal{M}_{\mathrm{or}})$, there can be developed the theory of austenite to self-accommodating martensite phase transition $(\mathcal{A} \leftrightarrow \mathcal{M}_{\mathrm{sa}})$. This last transition is temperature-induced at zero macroscopic stress $(\boldsymbol{\sigma} = 0)$. For that purpose, the expression for coherency energy $(3.3)_4$ must be replaced with expression

$\phi^{coh}(z_1,z_3)=z_3\phi_3^0+z_1z_3(\phi_{13}-Ts_{13})$; $\phi_3^0,\Delta s_3^{st},\phi_{13},s_{13}=\text{const}$. Then, the thermodynamic driving force of phase transition $\mathcal{A}\leftrightarrow\mathcal{M}_{sa}$ can be expressed with the formula $\pi^{1-3}\equiv-\partial g/\partial z_3=\pi_0^{1-3}(T)-(1-2z_3)(\phi_{13}-Ts_{13})$, where $\pi_0^{1-3}(T)=\Delta u_3^0-T\Delta s_3^0$, $\Delta u_3^0=\Delta u^0-\phi_3^0$, and $\Delta s_3^0=\Delta s^0-\Delta s_3^{st}$. Taking advantage of the conditions $\pi^{1-3}(T,z=0)=0$ and $\pi^{1-3}(T,z=1)=0$, relations $(3.16)_{3,4}$ are obtained.

As indicated in the text below, (3.15) experimental results show that it is reasonable to accept $s_{12}=0$. Then, it can be noticed from $(3.16)_{1,2}$ that the direct reason for splitting of *thermodynamic equilibrium temperature* between austenite and oriented martensite defined as $T_{or}^{eq}\equiv\Delta u_2^0/\Delta s_2^0$ into experimentally observed temperatures M_s^0 and A_s^{2-1} is interaction energy ϕ_{it} $(\phi_{it}=\phi_{12}-0\cdot T)$. When $s_{12}\ll\Delta s_2^C$ then $T_{or}^{eq}\cong(M_s^0+A_s^{2-1})/2$. When interaction energy would be zero, forward and reverse phase transformations $\mathcal{A}\leftrightarrow\mathcal{M}_{or}$ would start and proceed to their termination, at the same "equilibrium" transformation temperature T_{or}^{eq}.

No experimental information is available on term s_{13} entering interaction energy of self-accommodating martensite. However, in an analogy to oriented martensite the value of s_{12} can be expected to be close to zero ($s_{13}\cong 0\ll\Delta s_3^0$). Similar to the case of austenite and oriented martensite comparing $(3.16)_{3,4}$, it can noticed that the direct reason for splitting of equilibrium temperature between austenite and self-accommodating martensite, defined as $T_{sa}^{eq}\equiv\Delta u_3^0/\Delta s_3^0$, into experimentally observed temperatures M_s^0 and A_s^0, is again interaction energy $\phi_{it}=\phi_{13}-0\cdot T$.

When there are no incompatibilities of martensitic phase transformation eigenstrains, mesoscopic stresses originating from such incompatibilities would be zero $\breve{\boldsymbol{\sigma}}(\boldsymbol{x})=\boldsymbol{0}$, and thus coherency energies ϕ^{coh} ($\phi^{coh}(z_1,z_3)=0$, $\phi^{coh}(z_1,z_2)=0$) would be zero. In such a case, the following constants would take the zero values ($\Rightarrow\phi_{12}=s_{12}=\phi_{13}=s_{13}=\phi_2^0=\phi_3^0=\Delta s_2^{st}=\Delta s_3^{st}=0$) and consequentially all austenite, martensite "equilibrium" transformation temperatures will take the same value $T^{eq}=T_{or}^{eq}=T_{sa}^{eq}=M_s^0=A_s^0=A_s^{2-1}=\Delta u^0/\Delta s^0$.

The criteria (3.15) enable easy determination of critical temperatures of phase transformations induced by temperature in the presence of macroscopic stress $\boldsymbol{\sigma}=\text{const}$ from the formulas below.

$$M_s(\boldsymbol{\sigma})=M_s^0+\frac{g^{ult}(\boldsymbol{\sigma})}{\left(\Delta s_2^0-s_{12}\right)},A_s^{2-1}(\boldsymbol{\sigma})=A_s^{2-1}+\frac{g^{ult}(\boldsymbol{\sigma})}{\left(\Delta s_2^0+s_{12}\right)},$$

As potential $g^{ult}(\boldsymbol{\sigma})\geq 0$ is always nonnegative, then macroscopic external stresses can only bring about an increase of characteristic temperatures of phase transformation.

3.4.5 Internal material dissipation of work

The first principle of thermodynamics can be written in the rate form as follows:

$$\begin{aligned}&\dot{u}=\dot{w}-\dot{q},\dot{u}=\dot{\phi}+\dot{s}T+\dot{T}s\\&\dot{w}=(\boldsymbol{\sigma}/\rho)\cdot\dot{\boldsymbol{\varepsilon}},\end{aligned}\tag{3.17}$$

where $\dot{u}$ denotes rate of change of specific internal energy of macroelement, $\dot{w}$ is power of specific external mechanical work (done over macroelement), $\dot{q}$ is power of specific energy exchanged with environment in the form of heat ($\dot{q}$ has positive sign, when energy is removed from the macroelement), $\dot{\phi}$ denotes rate of variation of specific free energy of the macroelement, $\dot{s}$ is rate of variation of specific entropy, and $\dot{\boldsymbol{\varepsilon}}$ is rate of variation of total strain of macroelement.

The *Clausius-Duhem inequality* expressing the second law of thermodynamics states that entropy in any spontaneous thermodynamic process may not decrease. The inequality written for thermodynamic processes ongoing in a macroelement in thermally homogeneous conditions ($\nabla T \equiv 0$) takes the form

$$\dot{w}_{\text{Disp}} \equiv T\dot{s} + \dot{q} \geq 0, \dot{w}_{\text{Disp}} = \dot{w} - \dot{w}_{\text{r}}, \tag{3.18}$$

where $\dot{w}_{\text{Disp}}$ is power of specific (per unit of mass) mechanical *work dissipation*, and $\dot{s}$ is power of specific entropy. Alternatively, $\dot{w}_{\text{Disp}}$ can be expressed as a difference between the power of specific mechanical work $\dot{w}$, actually done over macroelement and power of specific mechanical work that would have to be done over macroelement in equivalent thermodynamically reversible process $\dot{w}_{\text{r}}$.

The *fundamental Gibbs equation* expressing variation of free energy of SMA macroelement in processes reversible thermodynamically written for free energy function ϕ postulated in model R_L takes the form

$$\dot{\phi}(\boldsymbol{\varepsilon}, T, z) = (\boldsymbol{\sigma}/\rho) \cdot \dot{\boldsymbol{\varepsilon}} - s\dot{T} - \pi^{1-2}\dot{z} \tag{3.19}$$

Upon elimination from equations $(3.17)_1$, $(3.17)_2$, and (3.19) of $\dot{q}$ and $\dot{s}$, and next substituting the obtained result into $(3.18)_1$, one obtains the following expression for power of *work dissipation:*

$$\dot{w}_{\text{Disp}} = \pi^{1-2}\dot{z} \geq 0 \tag{3.20}$$

In accordance with the second law of thermodynamics, it must be greater or equal to zero in any thermodynamic process. The above expression indicates that any active forward phase transformation $\dot{z} > 0$ (austenite$\rightarrow$martensite), can be initiated only from states in which $\pi^{1-2} \geq 0$, while any active reverse transformation $\dot{z} < 0$ (martensite$\rightarrow$austenite) can be initiated only from states in which $\pi^{1-2} \leq 0$.

Upon elimination of $\dot{w}_{\text{Disp}}$ from equations $(3.18)_1$ and (3.20), one obtains $\dot{q} = \pi^{1-2}\dot{z} - T\dot{s}$. Next, upon eliminating from such obtained relation the term $T\dot{s}$, and making use of material derivative $(3.13)_2$, one obtains the following relation for the power of heat, which must be exchanged with the environment of SMA macroelement during the progress of any thermodynamic process:

$$\begin{aligned} &\dot{q} = -c_{\text{p}}\dot{T} + \dot{q}_{\text{tr}}, \dot{q}_{\text{tr}} \equiv -T\boldsymbol{\alpha} \cdot \dot{\boldsymbol{\sigma}}/\rho + c_{\text{f}}^{*}\dot{z} + \pi^{1-2}\dot{z}, \\ &c_{\text{f}}^{*}(z) \equiv T\left[\Delta s_2^0 - (1-2z)s_{12}\right] \end{aligned} \tag{3.21}$$

The term $-c_{\text{p}}\dot{T}$ describes thermal effects connected with heat capacity of the material. Expression $\dot{q}_{\text{tr}}$ describes effects connected with internal (local) generation of heat in the material resulting from undergoing physical processes. The term $\pi^{1-2}\dot{z}$ describes dissipation of mechanical work, the term $T\boldsymbol{\alpha}\cdot\dot{\boldsymbol{\sigma}}/\rho$ describes piezocaloric effect, and the term $c_f^*\dot{z}$ describes heat power, which would be generated in a virtual infinitesimal thermodynamically reversible process of phase transformation as a result of the existence of latent heat of phase transition. Taking advantage of the equations $(3.13)_4$ and $(3.21)_3$, the thermodynamic driving force of phase transformation can be expressed as follows:

$$\pi^{1-2} = c_f - c_f^*, \; c_f(\boldsymbol{\sigma}, z) \equiv g^{ult}(\boldsymbol{\sigma}) + \Delta u_2^0 - (1-2z)\phi_{12},$$
$$l_f \equiv -\int_{z=1}^{z=0} c_f dz, \; l_f^* \equiv -\int_{z=1}^{z=0} c_f^*(z) dz, \; l_f(\boldsymbol{\sigma}=0) = \Delta u_2^0, \; l_f^* \equiv T\Delta s_2^0,$$
$$c_f^{sa}(\boldsymbol{\sigma}, z) = \Delta u_3^0 - (1-2z_3)\phi_{13}, \; l_f^{sa} = \int_{z_3=0}^{z_3=1} c_f^{sa} dz = \Delta u_3^0, \; l_f^{*\,sa} = T\Delta s_3^0. \quad (3.22)$$

The term $c_f\dot{z}$ describes actual heat power, which would be generated in thermodynamically irreversible, infinitesimal process of austenite to oriented martensite phase transformation. It results from the existence of latent heat of phase transformation and dissipation of mechanical work. The quantity l_f describes *actual latent heat of phase transformation*, in thermodynamically irreversible process. The quantity l_f^* describes so-called apparent or *reversible latent heat of phase transformation*; that is, heat that would be exchanged with the surroundings had the transformation been thermodynamically reversible. Direct integration of $(3.22)_3$ upon substitution of $(3.22)_2$ and accepting macroscopic stress equal to zero ($\boldsymbol{\sigma} = 0$) shows that the heat of complete p.t. from pure oriented martensite to pure austenite is simply equal to $-\Delta u_2^0$ (minus sign denotes that heat is absorbed from surroundings); see $(3.22)_5$. This thermal effect can also be experimentally measured with the DSC method. On account of the first law of thermodynamics, the heat absorbed during $\mathcal{M}_{or} \rightarrow \mathcal{A}$ phase transition at zero stress $\left(-\Delta u_2^0\right)$, must be exactly equal to the heat emitted during virtual opposite p.t. of pure austenite to pure oriented martensite $\mathcal{A} \rightarrow \mathcal{M}_{or}$ (Δu_2^0). This is so because during the closed thermal cycle at zero stress no work is executed ($W = 0$) and, thus, resultant heat exchange during the closed heating-cooling cycle must be zero $\int_{z=1}^{z=0} c_f dz + \int_{z=0}^{z=1} c_f dz = 0$. The $\mathcal{A} \rightarrow \mathcal{M}_{or}$ p.t. at zero macroscopic stress has been qualified as virtual because such transition is impossible to realize in practice. Upon lowering temperature at $\boldsymbol{\sigma} = 0$, only austenite to self-accommodating martensite p.t. can be effected $\mathcal{A} \rightarrow \mathcal{M}_{sa}$.

In full analogy to the above derivations taking advantage of the approach delineated below formula (3.16), it is straightforward to show that actual latent heat l_f^{sa} of complete phase transition from pure austenite to pure self-accommodating martensite $\mathcal{A} \rightarrow \mathcal{M}_{sa}$ must be equal to Δu_3^0 (see $(3.22)_8$), and reversible heat $l_f^{*\,sa}$ of such phase transformation is equal to $T\Delta s_3^0$ (see $(3.22)_9$).

Using formula $(3.21)_1$, the $\dot{q}_{tr}$ can be expressed as follows (see $(3.21)_2$):

$$\dot{q}_{tr} = -T\boldsymbol{\alpha}\cdot\dot{\boldsymbol{\sigma}}/\rho + c_f\dot{z} \quad (3.23)$$

In a majority of cases concerning shape memory alloys, power of heat $\dot{q}$ (positive when heat is removed to environment) exchanged between the SMA macroelement and the environment takes place through conduction. For isotropic SMA alloys, this leads to the following equation of heat exchange in the case of an initial-boundary value problem:

$$c_p \partial T(\boldsymbol{x},t)/\partial t + (1/\rho)\mathrm{div}(-\lambda_c \nabla T(\boldsymbol{x},t)) - \dot{q}_{tr}(\boldsymbol{x},t) = 0, \tag{3.24}$$

where λ_c denotes isotropic coefficient of heat conduction.

3.4.6 Phase transformation kinetics equations

In accordance with model R_L assumptions, phase transformations in SMA materials practically never do run through states of full thermodynamic equilibrium states. Hence, closing the specification of the constitutive SMA model in mathematical terms requires the proposition of so-called phase transformation kinetics relations describing quantitatively the volume fraction of martensitic phase evolution in response to external loading variations. The first element of the phase transformation kinetics rule in the R_L models family takes the form of two families of curves $\Psi^{(\alpha)}(\pi^{1-2}, z) = \mathrm{const}$, $\alpha = 1,2$ that are different for forward and reverse phase transitions. They define in $\pi^{1-2} - z$ space possible segments of active monotonic phase transitions ($\mathrm{d}z \neq 0$). At the same time they define quantitative relations between the thermodynamic driving force (t.d.f.) of phase transition π^{1-2} and conjugated with it thermodynamic flux; that is, a volume fraction of martensitic phase z that must be fulfilled during active p.t. These relations must be supplemented with conditions that have to be fulfilled in order for active p.t. to take place at all; for example, those resulting from the requirements of the second law of thermodynamics, that is, the requirement of nonnegative dissipation of work ($\pi^{1-2}\dot{z} \geq 0$). The other constraints can originate from additional experimental observations; for example, the existence of lag threshold values for the initiation of phase transformation. The following specific form of parametric functions $\Psi^{(\alpha)}$ have been selected to enable reproduction of a broad range of experimentally observed p.t. curves:

$$\begin{gathered}
\Psi^{(1)}\left(D, z; z_1^*\right) = \left[D^{m_1} + r_1 \ \ln(1-z) - p_1 \ z\right] = D_s^{(1)}\left(z_1^*\right) \\
\psi^{(2)}\left(D, z; z_2^*\right) = \left[(-D)^{m_2} + r_2 \ln(z) - p_2(1-z)\right] = \left(-D_s^{(2)}\left(z_2^*\right)\right) \\
D_s^{(1)}\left(z_1^*\right) = r_1 \ \ln\left(1-z_1^*\right) - p_1 z_1^*, \left(-D_s^{(2)}\left(z_2^*\right)\right) = r_2 \ \ln\left(z_2^*\right) - p_2 \ \left(1-z_2^*\right) \\
D \equiv \pi^{1-2}(\sigma, T, z)/\pi_N^{1-2}, \pi_N^{1-2} = \mathrm{const}, D_s^{(1)}\left(z_1^* = 0\right) = 0, D_s^{(2)}\left(z_2^* = 1\right) = 0
\end{gathered} \tag{3.25}$$

where $m_1, r_1, p_1, m_2, r_2, p_2$ are material constants of phase transformation kinetics rules. These constants should be identified individually for each shape memory alloy on the basis of experimental data. The z_α^*, $\alpha = 1,2$ denote values of volume fraction of martensitic phase at which starts the next segment of active, monotonous forward (reverse) phase transformation. The symbol D denotes thermodynamic driving force (t.d.f.) of phase transition normalized with some normalizing factor π_N^f Reasonable selection of specific normalizing factor π_N^f is an open scientific problem at present and requires further research. As a normalizing factor, the term ϕ_{12} can be accepted, characterizing the width of stress hysteresis loop of specific SMA material, or the term ϕ_{13}, characterizing the split between A_s^0 and M_s^0 temperatures (see (3.16)). Specific mathematical, functional form of expressions $D_s^{(1)}(z_1^*), D_s^{(2)}(z_2^*)$ in terms of starting value of volume fraction of product phase z_α^* defines the way of formation of internal hysteresis loops during partial phase transformations. The shape of the outermost hysteresis loop can be recreated when it is accepted $D_s^{(1)}(z_1^*=0)=D_s^{(2)}(z_2^*=1)=0$. It is easy to notice that when $m_1=1, p_1=0$ is substituted in $(3.25)_1$, then the classical exponential kinetics rule is obtained (Koistinen and Marburger, 1959).

For numerical computations, the most useful is rate form of phase transformation kinetics rules. The general relations of thermoelastic martensitic *R-L kinetics phase transformations rules* proposed by Raniecki and Lexcellent (1994), corresponding to integral relations (3.25), take the following form:

$$\dot{z}=\begin{cases}\lambda_{(1)}^{\sigma}\dot{\tilde{\pi}}_{\sigma}^{1-2} & \text{if } \left\{\pi^{1-2}(\sigma,T,z)\geq Y_{(1)}\geq 0 \text{ and } \dot{\tilde{\pi}}_{\sigma}^{1-2}(\dot{\sigma},\dot{T})>0\right\} \quad (\mathcal{A}\rightarrow\mathcal{M})\\ \lambda_{(2)}^{\sigma}\dot{\tilde{\pi}}_{\sigma}^{1-2} & \textit{if } \left\{\pi^{1-2}(\sigma,T,z)\leq -Y_{(2)}\leq 0 \text{ and } \dot{\tilde{\pi}}_{\sigma}^{1-2}(\dot{\sigma},\dot{T})<0\right\} \quad (\mathcal{M}\rightarrow\mathcal{A})\\ 0 & \text{otherwise}\end{cases} \tag{3.26}$$

They can be derived using constraints relations $\mathrm{d}\Psi^{(\alpha)}/\mathrm{d}t=\frac{\partial\Psi^{(\alpha)}}{\partial\pi^{1-2}}d\pi^{1-2}-\frac{\partial\Psi^{(\alpha)}}{\partial z}\dot{z}=0$, $\alpha=1,2$. The above relations can also be written in very compact form taking advantage of the properties of functions $\overline{H}(x)$ and $\langle x\rangle$

$$\dot{z}=\lambda_{(1)}^{\sigma}\overline{H}\left(\pi^{1-2}-Y_{(1)}\right)\left\langle\dot{\tilde{\pi}}_{\sigma}^{1-2}\right\rangle-\lambda_{(2)}^{\sigma}\overline{H}\left(-\pi^{1-2}-Y_{(2)}\right)\left\langle-\dot{\tilde{\pi}}_{\sigma}^{1-2}\right\rangle, \tag{3.27}$$

$$\overline{H}(x)=\begin{cases}1 & \text{if } x\geq 0\\ 0 & \text{if } x<0\end{cases}, \langle x\rangle=\begin{cases}x & \text{if } x\geq 0\\ 0 & \text{if } x<0\end{cases},$$

Here, $Y_{(\alpha)}(\boldsymbol{\sigma},T,z)\geq 0, \alpha=1,2$ denote lag threshold values functions. They operate in such a way that only when respective threshold values are reached by thermodynamic driving force, can active phase transformation start (see Figure 3.6f). The threshold functions fulfill conditions $Y_{(1)}\big|_{z=0}=Y_{(2)}\big|_{z=1}=0$. The power $\dot{\tilde{\pi}}_{\sigma}^{1-2}$ inducing forward $(\mathcal{A}\rightarrow\mathcal{M})$ or reverse $(\mathcal{M}\rightarrow\mathcal{A})$ active phase transformation is defined as follows (see (3.14)):

$$\dot{\tilde{\pi}}_{\sigma}^{1-2}\equiv\frac{\partial\pi^{1-2}}{\partial\boldsymbol{\sigma}}\cdot\dot{\boldsymbol{\sigma}}+\frac{\partial\pi^{1-2}}{\partial T}\dot{T}=(\boldsymbol{\kappa}/\rho)\cdot\dot{\boldsymbol{\sigma}}-\left[\Delta s_2^0-(1-2z)s_{12}\right]\dot{T} \tag{3.28}$$

The state functions $\lambda_{(\alpha)}^{\sigma}$ take the form

$$\lambda_{(1)}^{\sigma}=\frac{(m_1/2)(1-z)<D>^{m_1-1}}{\phi_{12}[p_1(1-z)+r_1]-m_1(1-z)\phi_{\mathrm{it}}(T)\langle D\rangle^{m_1-1}}, \tag{3.29}$$

$$\lambda^{\sigma}_{(2)} = \frac{(m_2/2)z(\langle -D\rangle)^{m_2-1}}{\phi_{12}[p_2z+r_2] - m_2z\phi_{\mathrm{it}}(T)\langle(-D)\rangle^{m_2-1}},$$

and they fulfill conditions $0 \leq \lambda^{\sigma}_{(\alpha)}(\boldsymbol{\sigma}, T, z) < \infty$ and $\lambda^{\sigma}_{(1)}z = 1 = \lambda^{\sigma}_{(2)}\Big|_{z=0} = 0$.

Phase transition kinetics rules can also be written in an alternative form; that is, dependent on rates of external parameters $(\dot{\boldsymbol{\varepsilon}}, \dot{T})$, and not parameters $(\dot{\boldsymbol{\sigma}}, \dot{T})$:

$$\dot{z} = \lambda^{\varepsilon}_{(1)}\overline{H}\left(\pi^{(1-2)} - Y_{(1)}\right)\left\langle \dot{\tilde{\pi}}^{1-2}_{\varepsilon}\right\rangle - \lambda^{\varepsilon}_{(2)}\overline{H}\left(-\pi^{1-2} - Y_{(2)}\right)\left\langle -\dot{\tilde{\pi}}^{1-2}_{\varepsilon}\right\rangle,$$

$$\lambda^{\varepsilon}_{(1)} = \frac{m_1(1-z)\langle D\rangle^{m_1-1}}{2\phi_{12}[p_1(1-z)+r_1] - (1-z)m_1\langle D\rangle^{m_1-1}[2\phi_{\mathrm{it}}(T) - \boldsymbol{\kappa}\cdot \boldsymbol{L}\boldsymbol{\kappa}/\rho]},$$

$$\lambda^{\varepsilon}_{(2)} = \frac{m_2z(\langle -D\rangle)^{m_2-1}}{2\phi_{12}[p_2z+r_2] - m_2z(\langle -D\rangle)^{m_2-1}[2\phi_{\mathrm{it}}(T) - \boldsymbol{\kappa}\cdot \boldsymbol{L}\boldsymbol{\kappa}/\rho]}, \tag{3.30}$$

$$\dot{\tilde{\pi}}^{1-2}_{\varepsilon} \equiv \frac{\partial \pi^{1-2}}{\partial \boldsymbol{\varepsilon}}\dot{\boldsymbol{\varepsilon}} + \frac{\partial \pi^{1-2}}{\partial T}\dot{T} = (\boldsymbol{\kappa}/\rho)\cdot \boldsymbol{L}\dot{\boldsymbol{\varepsilon}} - [\Delta s^* - (1-2z)s_{12}]\dot{T} + (\boldsymbol{\kappa}/\rho)\cdot \boldsymbol{L}\boldsymbol{\alpha}\dot{T},$$
$$D \equiv \pi^{1-2}(\boldsymbol{\varepsilon}, T, z)/\pi^{\mathrm{f}}_{\mathrm{N}}$$

More information on phase transition kinetics relations can be found in Chapter 6 of this book.

3.4.7 Complete set of incremental constitutive relations

Further $\mathrm{R_L}$ model specification below is limited to isotropic SMA materials. Thus, tensors of elastic modules and thermal expansion $\mathbf{L}$, $\boldsymbol{\alpha}$ have the following components:

$$\mathrm{L}_{ijkl} = \mu\left(\delta_{ik}\delta_{jl} + \delta_{il}\delta_{jk}\right) + \delta_{ij}\delta_{kl}(K - 2\mu/3); \quad (\alpha)_{ij} = \alpha\delta_{ij}, \tag{3.31}$$

where $\mu = E/2(1+\nu)$ is shear modulus of elasticity, $K = E/(3(1-2\nu))$ is bulk modulus of elasticity, and ν is Poisson coefficient.

The explicit form of *ultimate phase eigenstrains potential* $(3.9)_1$ defining tensor of ultimate phase eigenstrains $\boldsymbol{\kappa}$ is postulated in the form

$$\rho g^{\mathrm{ult}}(\overline{\boldsymbol{\sigma}}) = \gamma f(y)\sigma_{\mathrm{ef}}, \tag{3.32}$$

where γ is the amplitude of pseudoelastic flow in pure shear (treated as material constant), σ_{ef} denotes classical effective stress of Cauchy stress tensor, and $\overline{\boldsymbol{\sigma}}$ is Cauchy stress deviator.

$$\sigma_{\mathrm{ef}} \equiv \sqrt{3J_2} = \left[(3/2)\overline{\sigma}_{ij}\overline{\sigma}_{ij}\right]^{1/2}, \sigma' \equiv \sqrt{2J_2} = \left[\mathrm{tr}\left(\overline{\sigma}^2\right)\right]^{1/2} = \left[\overline{\sigma}_{ij}\overline{\sigma}_{ij}\right]^{1/2},$$
$$\overline{\sigma} = \mathrm{dev}(\sigma), \overline{\sigma}_{ij} = \sigma_{ij} - \frac{1}{3}\sigma_{ii}\delta_{ij}, \sigma_{\mathrm{ef}} = \sqrt{3/2}\,\sigma', \gamma = \sqrt{2/3}\,\eta, \gamma\sigma_{\mathrm{ef}} = \eta\sigma', \tag{3.33}$$

where J_2 denotes the second invariant of stress deviator. The function f(y) determines shape of limit surface of pseudoelastic flow in stress deviators space $\overline{\boldsymbol{\sigma}}$. In the sequel it is called *shape function*, and it is defined as follows:

$$f(y) \equiv \sigma^{AM}(0,T)/\sigma^{AM}(y,T) > 0, f(y=0) = 1, \tag{3.34}$$

where $\sigma^{AM}(0,T)$ denotes critical effective stress at which $\mathcal{A} \rightarrow \mathcal{M}$ phase transformation starts during pure shear, and $\sigma^{AM}(y,T)$ denotes critical effective stress at which $\mathcal{A} \rightarrow \mathcal{M}$ phase transformation starts when stress tensor has direction y. The function $f(y)$ is an input information for the R_L model and must be identified for particular SMA material from experimental data.

The formula $(3.13)_4$ for thermodynamic driving force of phase transformation π^{1-2} for isotropic SMA materials using expression for potential g^{ult} (3.32) and taking advantage of $(3.16)_1$ takes the following form:

$$\begin{aligned}
&\pi^{1-2}(\sigma_{ef}, T, z) = \left(\Delta s_2^0 - s_{12}\right) \cdot \left(M_s^0 - T\right) + 2(\phi_{12} - T s_{12}) \cdot z + \gamma f(y)\sigma_{ef}/\rho \\
&D = \pi^f(\sigma_{ef}, T, z)/\ \phi_{12} = D_T \cdot \left(M_s^0 - T\right) + D_\sigma \cdot f(y) \cdot \sigma_{ef} + D_z \cdot z, \\
&D_T = \left[\Delta s_0^2 - s_{12}(1-2z)\right]/\phi_{12}, D_\sigma = \gamma/(\phi_{12}\rho), D_z = 2\left[1 - M_s^0 s_{12}/\phi_{12}\right], \\
&\text{or} \\
&D_T = \left(\Delta s_0^2 - s_{12}\right)/\phi_{12}, D_\sigma = \gamma/(\phi_{12}\rho), D_z = 2[1 - T s_{12}/\phi_{12}]
\end{aligned} \tag{3.35}$$

when ϕ_{12} is accepted as a normalizing factor $(\pi_N^{1-2} = \phi_{12})$ in definition $(3.25)_5$ of normalized t.d.f. then it can be expressed with formulas $(3.35)_{2,3,4}$.

Experimental knowledge of function $f(y)$ enables the determination of modeling values of critical effective stresses of phase transformation σ^{AM} and σ^{MA} at which forward and reverse phase transformation starts when the stress tensor has "direction" y,

$$\begin{aligned}
&\pi^{1-2}(\boldsymbol{\sigma}, T, z=0) = 0 \Rightarrow \sigma_{ef} = \sigma^{AM}(y,T) = \rho\left(\Delta s_2^0 - s_{12}\right)\left(T - M_s^0\right)/[\gamma f(y)], \\
&\pi^{1-2}(\boldsymbol{\sigma}, T, z=1) = 0 \Rightarrow \sigma_{ef} = \sigma^{MA}(y,T) = \rho\left(\Delta s_2^0 + s_{12}\right)\left(T - A_s^{2-1}\right)/[\gamma f(y)]
\end{aligned} \tag{3.36}$$

These equations constitute correspondences of classical Clausius-Clapeyron relations describing sensitivity of critical pseudoelastic flow stresses on temperature. The domain of full (unconstrained) thermodynamic equilibrium states of two-phase SMA macroelement is defined with the equation $\pi^{1-2}(\sigma_{ef}, T, z) = 0$ and makes a line segment in stress-strain coordinates ($T = $const). Locations of *full thermodynamic equilibrium states* are shown with dashed lines and labeled $\pi^{1-2} = 0$ in the graphs shown in Figure **3.6**.

The argument y of shape function describing the direction of stress tensor in space of deviators is defined with the following formula (*Lode parameter*):

$$y = \cos(3\theta) = \frac{27 J_3}{2\sigma_{ef}^3} = \frac{3\sqrt{3}J_3}{2J_2^{3/2}}, -1 \le y \le 1, J^3 \equiv \tfrac{1}{3}\mathrm{tr}\left(\overline{\boldsymbol{\sigma}}^3\right) = \det\left(\overline{\boldsymbol{\sigma}}\right), \tag{3.37}$$

where J_3 denotes the third invariant of the stress deviator. The parameter y is simply one of the well-known Lode invariants, the other is σ'(see $(3.33)_2$), and θ represents the angle between the image of positive half-axis of principal stress σ_1 and the image of stress tensor on the octahedral plane. The y treated as a function of stress deviator is homogeneous function of degree zero. For example, value of argument $y=1 \quad (\theta=0^{\circ})$ corresponds to tension, $y=0 \quad (\theta=30^{\circ})$ denotes pure shear, while $y=-1 \quad (\theta=60^{\circ})$ corresponds to compression.

The relations between stress and strain take the following form:

$$\begin{aligned}
&\varepsilon_{\mathrm{v}}=\sigma_{\mathrm{m}}/K+3\alpha(T-T_0),\overline{\boldsymbol{\varepsilon}}=\overline{\boldsymbol{\sigma}}/2\mu+\boldsymbol{\varepsilon}^{\mathrm{pt}},\boldsymbol{\varepsilon}^{\mathrm{pt}}=z\boldsymbol{\kappa}, \\
&\sigma_{\mathrm{m}}=\frac{1}{3}\mathrm{tr}(\boldsymbol{\sigma}),\overline{\boldsymbol{\varepsilon}}=\mathrm{dev}(\boldsymbol{\varepsilon}),\varepsilon_{\mathrm{v}}=\mathrm{tr}(\boldsymbol{\varepsilon})
\end{aligned} \tag{3.38}$$

The *ultimate phase eigenstrains tensor* $\boldsymbol{\kappa}$ in accordance with definition $(3.9)_2$ takes the form

$$\begin{aligned}
&\boldsymbol{\kappa}=\partial(\gamma f(y)\sigma_{\mathrm{ef}})/\partial\boldsymbol{\sigma}=\overline{\boldsymbol{\kappa}}+\overline{\overline{\boldsymbol{\kappa}}}, \\
&\overline{\boldsymbol{\kappa}}=\left(\sqrt{3/2}\gamma\right)\mathrm{f}(y)\boldsymbol{n}^{\sigma},\overline{\overline{\boldsymbol{\kappa}}}=3\left(\sqrt{3/2}\gamma\right)\frac{\mathrm{d}f}{\mathrm{d}y}\left[\sqrt{6}\left((\boldsymbol{n}^{\sigma})^2-\mathbf{1}/3\right)-y\boldsymbol{n}^{\sigma}\right], \\
&\boldsymbol{n}^{\sigma}\equiv\overline{\boldsymbol{\sigma}}/\sigma',\mathrm{tr}(\boldsymbol{n}^{\sigma})=0,\mathrm{tr}\left((\boldsymbol{n}^{\sigma})^2\right)=1,\mathrm{tr}\left((\boldsymbol{n}^{\sigma})^4\right)=\tfrac{1}{2}, \\
&y=\sqrt{6}\mathrm{tr}\left((\boldsymbol{n}^{\sigma})^3\right)=3\sqrt{6}\det(\boldsymbol{n}^{\sigma}),(\boldsymbol{n}^{\sigma})^3=\tfrac{1}{2}\boldsymbol{n}^{\sigma}+\mathbf{1}\det(\boldsymbol{n}^{\sigma})=\tfrac{1}{2}\boldsymbol{n}^{\sigma}+\tfrac{1}{3}\mathrm{tr}((\boldsymbol{n}^{\sigma})^3\mathbf{1}
\end{aligned} \tag{3.39}$$

Relation $(3.39)_9$ follows directly from the Cayley-Hamilton theorem.

As can be easily verified, tensors $\overline{\boldsymbol{\kappa}}$ and $\overline{\overline{\boldsymbol{\kappa}}}$ are mutually orthogonal in the sense that $\mathrm{tr}(\overline{\overline{\boldsymbol{\kappa}}}\cdot\overline{\boldsymbol{\kappa}})\equiv 0$ as $\mathrm{tr}(\overline{\overline{\boldsymbol{\kappa}}}\cdot\boldsymbol{n}^{\sigma})\equiv 0$. An adopted form of ultimate eigenstrains potential g^{ult} brings the following about of phase transformation strain $\boldsymbol{\varepsilon}^{\mathrm{pt}}$ radius direction after radius direction $\boldsymbol{n}^{\sigma}$ of stress tensor, in accordance with experimental data shown in Figure 3.9. Also in accordance with experimental data phase transformation strain $\boldsymbol{\varepsilon}^{\mathrm{pt}}$ is traceless $\mathrm{tr}(\boldsymbol{\kappa})=0$, as volumetric changes resulting from thermoelastic phase transformation are in the first approximation negligible.

Equivalent strain connected with phase transformation $\varepsilon_{\mathrm{eq}}^{\mathrm{pt}}$ can computed with the formulas

$$\begin{aligned}
&\varepsilon_{\mathrm{eq}}^{\mathrm{pt}}=\left[(2/3)\boldsymbol{\varepsilon}^{\mathrm{pt}}\cdot\boldsymbol{\varepsilon}^{\mathrm{pt}}\right]^{1/2}=z\overline{\gamma}(y),\overline{\gamma}(y)=\varepsilon_{\mathrm{eq}}^{\mathrm{pt}}(y,z=1)=\gamma F(y), \\
&F(y)\equiv\left[(2/3)\mathrm{tr}(\boldsymbol{\kappa}^2)\right]^{1/2}=\left[f^2+9(1-y^2)\left(\mathrm{d}f^2/\mathrm{d}y\right)^2\right]^{1/2}, \\
&\varepsilon_{\mathrm{eq}}=\sqrt{2/3}\varepsilon',\varepsilon'\equiv\left[\bar{\varepsilon}_{ij}\bar{\varepsilon}_{ij}\right]^{1/2},\boldsymbol{n}^{\varepsilon}\equiv\overline{\boldsymbol{\varepsilon}}/\varepsilon'
\end{aligned} \tag{3.40}$$

$\overline{\gamma}(y)$ is an amplitude of effective macroscopic phase transition strain connected with complete $(z=0\rightarrow 1)$ phase transition associated with stress tensor direction y.

Please note that the formula for pseudoelastic flow amplitude in simple tension reduces to the following expression: $\gamma_{\mathrm{tn}} = \overline{\gamma}(y=1) = \gamma F(y=1) = \gamma f(y=1)$. Thanks to this property, R_L model usage in uniaxial tension cases does not require knowledge of a shape function. It only requires knowledge of the value of *pseudoelastic flow amplitude* in simple tension γ_{tn}. Then, in all formulas of the R_L model requiring product $\gamma f(y=1)$ (e.g., in (3.32)), directly γ_{tn} must be substituted.

Intensity of Cauchy stress σ' can be expressed with the formula (see $(3.33)_2$)

$$\sigma'/2\mu = \left[(\boldsymbol{\varepsilon}')^2 - (3\eta z \mathrm{d}f(y)/\mathrm{d}y)^2(1-y^2)\right]^{1/2} - \eta f(y) z,$$
$$(\varepsilon')^2 = [(\sigma'/2\mu) + \eta f(y) z]^2 + (3\eta z \mathrm{d}f(y)/\mathrm{d}y)^2(1-y^2) \tag{3.41}$$

The direction tensor $\boldsymbol{n}^\sigma$ and direction tensor $\boldsymbol{n}^\varepsilon$ are coaxial; they commute $\boldsymbol{n}^\sigma \boldsymbol{n}^\varepsilon = \boldsymbol{n}^\varepsilon \boldsymbol{n}^\sigma$. When shape function is identically equal to one $f(y) \equiv 1$, relation (3.41) simplifies to $\sigma' = 2\mu[\varepsilon' - \eta z]$.

The power $\dot{\tilde{\pi}}_\sigma^{1-2}, \dot{\tilde{\pi}}_\varepsilon^{1-2}$ (see (3.28), $(3.30)_4$), appearing in the relation of phase transformation kinetics inducing process of active phase transformation, can be now expressed for isotropic SMA alloys as

$$\dot{\tilde{\pi}}_\sigma^{1-2} = (1/\rho)\left(\overline{\boldsymbol{\kappa}} + \overline{\overline{\boldsymbol{\kappa}}}\right)\cdot\dot{\boldsymbol{\sigma}} - \left[\Delta s_2^0 - (1-2z)s_{12}\right]\dot{T},$$
$$\dot{\tilde{\pi}}_\varepsilon^{1-2} = (1/\rho)\left(\overline{\boldsymbol{\kappa}} + \overline{\overline{\boldsymbol{\kappa}}}\right)\cdot\mathbf{L}\dot{\boldsymbol{\varepsilon}} - \left[\Delta s_2^0 - (1-2z)s_{12}\right]\dot{T} \tag{3.42}$$

When $f(y) \equiv 1$, then the formula for $\dot{\tilde{\pi}}_\varepsilon^{1-2}$ simplifies to $\dot{\tilde{\pi}}_\varepsilon^{1-2} = (2\mu\eta/\rho)\dot{\varepsilon}' - \left[\Delta s_2^0 - (1-2z)s_{12}\right]\dot{T}$.

The formula describing the magnitude of heat effects connected with phase transformation and piezocaloric effect takes the form (see (3.23))

$$\dot{q}_{\mathrm{tr}} \equiv \left[\gamma f(y)\sigma_{\mathrm{ef}} + \Delta u_2^0 - (1-2z)\phi_{12}\right]\dot{z} - (3T\alpha/\rho)\dot{\sigma}_m \tag{3.43}$$

The mathematically closed set of rate constitutive equations of R_L model for isotropic SMA alloys, valid in the pseudoelastic range of their behavior, is composed of differentiated, with respect to time, equations (3.38), (3.39), $(3.13)_3$, and (3.26) together with auxiliary equations (3.35), (3.42), and (3.43).

Full specification of R_L model for isotropic SMA requires for each such material identification of the following set of data on the base of experimental tests results (see (3.11)):

$$\left.\begin{array}{l}(\alpha \to \boldsymbol{\alpha}), (E, \nu \to \boldsymbol{L}), \phi_{12}, s_{12}, \Delta u_2^0, \Delta s_2^0, \rho, c_p, \lambda_c \\ \gamma, f(y) \to g^{\mathrm{ult}}(\boldsymbol{\sigma}, T)\end{array}\right\} \to \rho g(\boldsymbol{\sigma}, T, z) \tag{3.44}$$

$m_1, m_2, r_1, r_2, p_1, p_2$ — cf. (3.29).

Macroscopic free energy function of two-phase SMA material macroelement-mesomechanical studies

4

4.1 Introduction

In order to enable a more precise and more reliable description of SMA materials behavior it is desirable and useful to make attempts to validate the specific form of macroscopic free energy function posed heuristically with mesomechanical studies. This is done in this chapter upon taking advantage and relevant processing of metallurgical information on the mesoscopic behavior of SMA materials, recalled from Sections 2.1 and 2.2 of Chapter 2. It is accepted here that, in the most general case, the SMA macroelement in mesoscale is a body composed of three linear, thermoelastic media with eigenstrains. It is accepted that the body fulfills conditions that allow it to be treated as a so-called representative volume element (RVE). A broad discussion concerning the concept of RVE, its properties and selection criteria, can be found in a monograph by Nemat-Naser and Hori (1993) (see also **Annex 4.2** later in this chapter).

In order to estimate the elastic energy stored in the macroelement (RVE) of the shape memory alloy, the boundary value problem of linear theory of elasticity is formulated for a body with fixed mesostructure, heterogeneous elastic properties, and a heterogeneous field of eigenstrains. Prescribed on the surface of the body force, boundary conditions fulfill Hill's postulate by assumption, which allows meaningfully introducing the concept of so-called effective properties of the material. Accepting that SMA material RVE can be treated in mesoscale as a multiphase system, remaining in mechanical equilibrium, composed of linear thermoelastic solids obeying local constitutive relations of elasticity with eigenstrains $\breve{\boldsymbol{\varepsilon}}_{(\alpha)}(\boldsymbol{x}) = \breve{\boldsymbol{M}}(\boldsymbol{x})\breve{\boldsymbol{\sigma}}_{(\alpha)}(\boldsymbol{x}) + \breve{\boldsymbol{\Gamma}}_{(\alpha)}(\boldsymbol{x}) \;\; \boldsymbol{x} \in V_{\alpha}$, and loaded with boundary surface tractions $\boldsymbol{t}^{(n)}(\boldsymbol{x}) = \boldsymbol{\sigma}(\boldsymbol{x})\boldsymbol{n}(\boldsymbol{x}) \;\; \boldsymbol{x} \in \partial V$ leading to fulfillment of Hill's postulate, we are interested in how individual elements of the problem of elastic mechanical equilibrium influence the elastic energy of the system regardless of the specific mesostructure that the system can take. For that purpose, the original problem of linear elastic mechanical equilibrium is divided into three separate component problems in which the "loadings" are (a) surface boundary tractions only, marked with subscript "ex"; (b) uniform average phase eigenstrains strains $\boldsymbol{\Gamma}_{(\alpha)}$ only, marked with subscript "a-in"; and (c) phase fluctuating eigenstrains fields $\breve{\boldsymbol{\Gamma}}^{\mathrm{f}}_{(\alpha)}(\boldsymbol{x})$ only, marked with subscript "f-in." Eigenstrains are separated into a phase average part and a phase

Pseudoelasticity of Shape Memory Alloys. http://dx.doi.org/10.1016/B978-0-12-801697-8.00004-1

fluctuating part: $\breve{\boldsymbol{\Gamma}}(\boldsymbol{x}) = \boldsymbol{\Gamma}_{(\alpha)} + \breve{\boldsymbol{\Gamma}}^{\mathrm{f}}_{(\alpha)}(\boldsymbol{x})$, $\left(\boldsymbol{\Gamma}_{(\alpha)} = \left\langle \breve{\boldsymbol{\Gamma}}(\boldsymbol{x}) \right\rangle_{V_\alpha} \equiv \left(\frac{1}{V_\alpha}\right)\int_{V_\alpha} \breve{\boldsymbol{\Gamma}}(\boldsymbol{x})\mathrm{d}\boldsymbol{x}\right)$. The symbolic solutions are then combined thanks to the linearity of the component problems. Some results are obtained for the most general case of an inhomogeneous elastic body, but later in the sequence attention is focused on the two-phase body.

The expression for elastic energy, obtained as a result of the above-mentioned operations, reveals the appearance of certain mathematical terms, but still the elastic energy function is dependent on a number of internal parameters h_K characterizing martensitic mesostructures, not specified here in an explicit way. In experimental studies devoted to martensitic transformations it is observed that a martensitic mesostructure evolves, adapting to changing thermomechanical loading conditions. This observation allowed for the formulation of the optimum rearrangement rule, that is, the tendency to reach mesostructural thermodynamic equilibrium with respect to actual thermomechanical loading. This hypothesis is used to obtain a more useful form of the formula for elastic energy of the SMA macroelement, through diminishing the number of internal parameters appearing in the elastic energy function. Symbolic assessment of the average elastic energy stored in the SMA macroelement delivers mesomechanical justification for heuristic suppositions regarding the macroscopic form of SMA materials' free energy function.

Figure 4.1 shows schematically the *mesoscopic structure* of a macroscopic RVE of the SMA material accepted for the present investigation. It is assumed that in the most general case the SMA RVE can consist of three phases: austenite $\mathcal{A}$ (phase 1), oriented martensite $\mathcal{M}_{\mathrm{or}}$ (phase 2), and self-accommodating martensite $\mathcal{M}_{\mathrm{sa}}$ (phase 3). The difference in mesostructural properties of those phases is schematically illustrated in Figure 4.1b. Austenite has a zero mesoscopic eigenstrains field $\breve{\boldsymbol{\Gamma}}_{(1)}(\boldsymbol{x}) \equiv \boldsymbol{0} \Leftrightarrow \boldsymbol{\Gamma}_{(1)} = \breve{\boldsymbol{\Gamma}}^{\mathrm{f}}_{(1)}(\boldsymbol{x}) = \boldsymbol{0}$, where

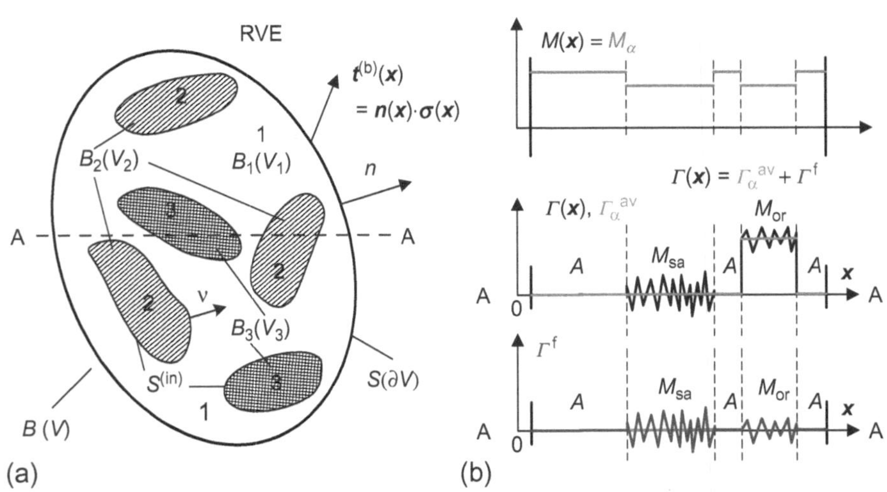

Figure 4.1 (a) Schematic illustration of mesoscopic structure of macroscopic representative volume element of SMA material; (b) Characteristics of elastic properties and eigenstrain fields in austenitic phase, self-accommodating martensite and oriented martensite.

$\boldsymbol{\Gamma}_{(1)}$ is the phase average eigenstrain of austenite, and $\breve{\boldsymbol{\Gamma}}^{\mathrm{f}}_{(1)}(\boldsymbol{x})$ is the phase eigenstrain fluctuation of austenite. The oriented martensite has nonzero phase average eigenstrain $\boldsymbol{\Gamma}_{(2)} \neq 0$ and nonzero fluctuation phase eigenstrain field $\breve{\boldsymbol{\Gamma}}^{\mathrm{f}}_{(2)}(\boldsymbol{x}) \neq 0$, and self-accommodating martensite has zero phase average eigenstrain $\boldsymbol{\Gamma}_{(3)} = 0$ and nonzero phase fluctuation eigenstrain field $\breve{\boldsymbol{\Gamma}}^{\mathrm{f}}_{(3)}(\boldsymbol{x}) \neq 0$. The following assumption about the piecewise uniform field of elastic properties is reasonable and frequently encountered in the literature devoted to SMA materials: $\breve{\mathbf{M}}(\boldsymbol{x}) = \mathbf{M}_1$, $\boldsymbol{x} \in V_1$, $\breve{\mathbf{M}}(\boldsymbol{x}) = \mathbf{M}_2 = \mathbf{M}_3$, $\boldsymbol{x} \in V_2 + V_3$ $(\alpha = 1,2,3)$. Here, in the sequel below investigation is actually limited to two-phase mesostructures of austenite and oriented martensite only. This is so because thermomechanical loadings investigated in this book are limited to the ones invoking pseudoelastic behavior of SMA materials.

The line of approach and some results presented in this chapter targeted at mesoscopic justification of the adopted macroscopic free energy function for SMA materials when RVE is treated as a two-phase mixture were published in a paper by Ziółkowski and Raniecki (1999). Independent research results on the subject located in the same stream were published by Bernardini (2001). Bernardini elaborated on a mesomechanics-based method of the combined, consistent estimation of three terms: elastic energy g^{el}, macroscopic phase transformation strain $\boldsymbol{\varepsilon}^{\mathrm{pt}}$, and coherency energy ϕ^{coh} for two-phase SMA material RVE. Bernardini took advantage of the fact that, in the case of a two-phase medium, unique connections exist between fourth-order dimensionless mechanical *stress concentration factor tensors*, commonly denoted by the symbol $\boldsymbol{\mathcal{B}}_{(\alpha)}$ (see, e.g., Dvorak and Benveniste (1992) and, e.g., effective elastic compliance $\boldsymbol{M}^{\mathrm{ef}}$). Thus, at fixed mesostructure and phase eigenstrains, field stress concentration tensors $\boldsymbol{\mathcal{B}}_{(\alpha)}$, $\boldsymbol{\varepsilon}^{\mathrm{pt}}$, and ϕ^{coh} can be uniquely expressed as functions of $\boldsymbol{M}^{\mathrm{ef}}$. Thus, the quality of estimate of $\boldsymbol{M}^{\mathrm{ef}}$ entails the quality of estimates of $\boldsymbol{\varepsilon}^{\mathrm{pt}}$ and ϕ^{coh}. Nontrivial extension of the Bernardini studies to three phases, essential for modeling of one-way memory effect, is delivered in Raniecki and Ziółkowski (2015).

4.2 Mesomechanics of thermoelastic martensitic structures

4.2.1 *Mechanical equilibrium of macroelement with nonhomogeneous elastic properties and eigenstrains*

In order to evaluate the value of elastic energy stored in the macroelement RVE of the shape memory alloy let us start with the very general assumption that it can be treated as a linearly elastic body $\mathcal{B}$, with fixed mesostructure, in which inhomogeneous field of elastic compliance modules $\breve{\mathbf{M}}(\boldsymbol{x})$ $\left(\breve{\mathbf{M}}(\boldsymbol{x}) = \breve{\mathbf{L}}^{-1}(\boldsymbol{x})\right)$ and inhomogeneous field of eigenstrains $\breve{\boldsymbol{\Gamma}}(\boldsymbol{x})$ are known. These fields are such that, by assumption, they assure the existence of continuous in the whole volume of the body $\mathcal{B}$ displacement field $\breve{\boldsymbol{u}}(\boldsymbol{x})$, $\boldsymbol{x} \in V$, constituting solution of the boundary value problem of static mechanical equilibrium, at the assumption of zero volume forces. The requirement for continuity of the displacement field comes from the experimentally found property of coherency of martensitic transformation. Fields of elastic properties and eigenstrains are, by assumption, piecewise continuous. They can be subject to jump discontinuities on certain surfaces and interphase boundaries $\mathcal{S}^{(\text{in})}$ (see Figure 4.1a). Due to this conjecture the field of local total strains $\breve{\boldsymbol{\varepsilon}}(\boldsymbol{x})$ generated by displacement field $\breve{\boldsymbol{u}}(\boldsymbol{x})$ and local stress field $\breve{\boldsymbol{\sigma}}(\boldsymbol{x})$ constituting solution of quasi-static mechanical equilibrium problem of SMA macroelement, can undergo jump changes on surfaces of strong discontinuities $\mathcal{S}^{(\text{in})}$. In such a case they must fulfill static and geometrical conditions of compatibility, respectively.

$$[\![\breve{\boldsymbol{\sigma}}]\!]\cdot\boldsymbol{\nu}(\boldsymbol{x}) = \left[\breve{\boldsymbol{\sigma}}_{(\alpha)}(\boldsymbol{x}) - \breve{\boldsymbol{\sigma}}_{(\beta)}(\boldsymbol{x})\right]\Big|_{\boldsymbol{x}\in\mathcal{S}^{(\text{in})}}\cdot\boldsymbol{\nu}(\boldsymbol{x}) = \boldsymbol{0} \quad \alpha,\beta = 1,2,\ldots$$

$$[\![\breve{\boldsymbol{\varepsilon}}]\!] = \left[\breve{\boldsymbol{\varepsilon}}_{(\alpha)}(\boldsymbol{x}) - \breve{\boldsymbol{\varepsilon}}_{(\beta)}(\boldsymbol{x})\right]\Big|_{\boldsymbol{x}\in\mathcal{S}^{(\text{in})}} = 0.5[\boldsymbol{\lambda}(\boldsymbol{x})\otimes\boldsymbol{\nu}(\boldsymbol{x}) + \boldsymbol{\nu}(\boldsymbol{x})\otimes\boldsymbol{\lambda}(\boldsymbol{x})], \tag{4.1}$$

where $\boldsymbol{\nu}(\boldsymbol{x})$ is unit vector normal to interphase boundary $\mathcal{S}^{(\text{in})}$ directed as shown in Figure 4.1, and vector $\boldsymbol{\lambda}$ determines the amount of strains jump. Stresses $\breve{\boldsymbol{\sigma}}_{(\alpha)}(\boldsymbol{x})$, $\breve{\boldsymbol{\sigma}}_{(\beta)}(\boldsymbol{x})$ are limit values of stress field $\breve{\boldsymbol{\sigma}}(\boldsymbol{x})$ when approaching the point $\boldsymbol{x}$ lying on the surface $\mathcal{S}^{(\text{in})}$ while remaining in the domain 1 and remaining in the domain 2, respectively. Similarly, it is in the case of strain fields $\breve{\boldsymbol{\varepsilon}}_{(\alpha)}(\boldsymbol{x})$, $\breve{\boldsymbol{\varepsilon}}_{(\beta)}(\boldsymbol{x})$. The condition $(4.1)_1$ assures mechanical equilibrium on interphase surfaces $\mathcal{S}^{(\text{in})}$, while condition $(4.1)_2$ assures that strain field originates from continuous (coherent) displacement field.

Here, conjecture is made that an eigenstrains field $\breve{\boldsymbol{\Gamma}}(\boldsymbol{x})$ is generated only as a result of progress of the martensitic phase transformation. In a more general case, this field can have various physical origins. For example, it can result from thermal expansion or permanent plastic strains generated after the process of "training" of the SMA alloy, leading to two-way shape memory effect. Hence, the studies conducted here can have broader areas of application than only to thermoelastic martensitic transformation in the range of pseudoelastic behavior of SMA alloys.

It is assumed that mesostructure of the body $\mathcal{B}$ remains fixed at all times, due to the fact that fields $\breve{\mathbf{M}}(\boldsymbol{x})$ and $\breve{\boldsymbol{\Gamma}}(\boldsymbol{x})$ are treated as constant. It is also assumed that SMA macroelement remains in full thermal equilibrium all the time and that the temperature field in it is uniform $T = \text{const}$. As a result, in the sequel below temperature is treated as a fixed parameter and in the majority of formulas it will not appear explicitly. During real martensitic transformation temperature and microstructure evolves in time. Hence, the boundary value problem of linear theory of elasticity studied here can be treated as certain instantaneous "frozen" state appearing during real thermoelastic martensitic phase transformation.

Let us investigate the following problem of mechanical equilibrium of a body $\mathcal{B}$ for fields $\breve{\boldsymbol{u}}(\boldsymbol{x})$, $\breve{\boldsymbol{\varepsilon}}(\boldsymbol{x})$, $\breve{\boldsymbol{\sigma}}(\boldsymbol{x})$; a problem of linear elastostatics without the presence of volume forces:

$$\nabla\cdot\breve{\boldsymbol{\sigma}} = 0, \quad \breve{\boldsymbol{\sigma}} = \breve{\boldsymbol{\sigma}}^{T}, \quad \boldsymbol{x}\in V, \tag{4.2}$$

with prescribed on the whole boundary of the body $\mathcal{B}$ force boundary conditions

$$\breve{\boldsymbol{t}}^{(\mathrm{b})}(\boldsymbol{x})=\breve{\boldsymbol{n}}(\boldsymbol{x})\breve{\boldsymbol{\sigma}}(\boldsymbol{x}),\ \boldsymbol{x}\in\partial V,\ \boldsymbol{\mathcal{F}}=\oint_{\partial \mathrm{V}}\breve{\boldsymbol{t}}^{(b)}\mathrm{d}s=\mathbf{0},\ \boldsymbol{\mathcal{M}}=\oint_{\partial \mathrm{V}}\boldsymbol{x}\times\breve{\boldsymbol{t}}^{(\mathrm{b})}\mathrm{d}s=\mathbf{0},\tag{4.3}$$

the field $\breve{\boldsymbol{t}}^{(\mathrm{b})}(\boldsymbol{x})$, by assumption, is a self-equilibrating field, so that the body $\mathcal{B}$ can remain in static mechanical equilibrium. Discussion is limited here to all-round surface tractions type boundary conditions for clarity of conduct. It does not change the general conclusions obtained here.

Local field of total strains $\breve{\boldsymbol{\varepsilon}}(\boldsymbol{x})$ is connected with local field of stress $\breve{\boldsymbol{\sigma}}(\boldsymbol{x})$ by linear constitutive relations

$$\begin{aligned}&\breve{\boldsymbol{\varepsilon}}(\boldsymbol{x})=\breve{\mathbf{M}}(\boldsymbol{x})\breve{\boldsymbol{\sigma}}(\boldsymbol{x})+\breve{\boldsymbol{\Gamma}}(\boldsymbol{x}),\ \breve{\boldsymbol{\varepsilon}}(\boldsymbol{x})=0.5\left[\left(\nabla\otimes\breve{\boldsymbol{u}}\right)+\nabla\otimes\breve{\boldsymbol{u}}\right)^{\mathrm{T}}],\ \boldsymbol{x}\in V,\\&\breve{\boldsymbol{\varepsilon}}^{(\mathrm{e})}(\boldsymbol{x})=\breve{\mathbf{M}}(\boldsymbol{x})\breve{\boldsymbol{\sigma}}(\boldsymbol{x}),\ \breve{\boldsymbol{\varepsilon}}^{(\mathrm{e})}(\boldsymbol{x})=\breve{\boldsymbol{\varepsilon}}(\boldsymbol{x})-\breve{\boldsymbol{\Gamma}}(\boldsymbol{x}),\end{aligned}\tag{4.4}$$

where symbol $\breve{\boldsymbol{\varepsilon}}^{(\mathrm{e})}(\boldsymbol{x})$ denotes field of small elastic mesoscopic strains.

In view of the admissibility of jump changes of values of eigenstrains field $\breve{\boldsymbol{\Gamma}}(\boldsymbol{x})$ on surfaces of strong discontinuities $\mathcal{S}^{(\mathrm{in})}$, the fields $\breve{\boldsymbol{\sigma}}(\boldsymbol{x})$ and $\breve{\boldsymbol{\varepsilon}}(\boldsymbol{x})$ also can experience jump changes. However, by assumption they fulfill conditions of static and geometric compatibility defined by formula (4.1). Nonhomogeneous fields of elastic properties and eigenstrains are, by assumption, such that there exists a continuous displacement field $\breve{\boldsymbol{u}}(\boldsymbol{x})$ constituting a solution of the mechanical balance equation, connected with the strains field by relation $(4.4)_2$, valid within small strains theory.

The variables that appear explicitly on macroscopic level of description of material behavior are macroscopic stress $\boldsymbol{\sigma}$ and macroscopic strain $\boldsymbol{\varepsilon}$. They originate from mesoscopic fields by averaging respective local fields $\breve{\boldsymbol{\sigma}}(\boldsymbol{x})$, $\breve{\boldsymbol{\varepsilon}}(\boldsymbol{x})$, constituting the solution of mechanical equilibrium problem (4.2)–(4.4) over the whole volume of macroelement (RVE). All the other mesoscopic information expresses itself on a macroscopic level only indirectly embedded, hidden in so-called effective material properties. Taking into account that the mesoscopic stress field is statically admissible, that is, fulfills conditions of mechanical equilibrium, it can be shown that the value of macroscopic stress is completely determined by the boundary tractions $\breve{\boldsymbol{t}}^{(\mathrm{b})}(\boldsymbol{x})$ prescribed on the surface of the body $\mathcal{B}$ (RVE) in accordance with the formula

$$\boldsymbol{\sigma}=\left\langle\breve{\boldsymbol{\sigma}}(\boldsymbol{x})\right\rangle_V=\frac{1}{V}\oint_{\partial V}\boldsymbol{x}\otimes\breve{\boldsymbol{t}}^{(\mathrm{b})}(\boldsymbol{x})+\breve{\boldsymbol{t}}^{(\mathrm{b})}(\boldsymbol{x})\otimes\boldsymbol{x}\mathrm{d}s\tag{4.5}$$

Similarly, a value of macroscopic strain is completely determined by the boundary values of continuous in the whole body $\mathcal{B}$ (RVE) displacements field $\breve{\boldsymbol{u}}(\boldsymbol{x})$, constituting a solution of the mechanical equilibrium problem (4.2)–(4.4) (i.e., kinematically admissible displacement field), in accordance with formula

$$\boldsymbol{\varepsilon}=\left\langle\breve{\boldsymbol{\varepsilon}}(x)\right\rangle_V=\frac{1}{V}\oint_{\partial V}0.5\left[\breve{\boldsymbol{n}}\otimes\breve{\boldsymbol{u}}^{(\mathrm{b})}+\breve{\boldsymbol{u}}^{(\mathrm{b})}\otimes\breve{\boldsymbol{n}}\right]\mathrm{d}s\tag{4.6}$$

Different approaches exist regarding definition of macroscopic variables; some authors take volume averaging and others take surface averaging as the defining operation. Equivalence of the two approaches is shown in **Annex 4.2**.

Specific total work done on the body $\mathcal{B}$ as a result of prescribed, constant in time, boundary force loadings $\boldsymbol{t}^{(b)}(\boldsymbol{x})$ on respective boundary displacements $\breve{\boldsymbol{u}}^{(b)}(\boldsymbol{x})$ can be expressed with the formula (see (A4.23) and (A4.24) in Annex 4.2)

$$\rho W^{(\mathrm{Tot})} = \langle \breve{\boldsymbol{\sigma}} \cdot \breve{\boldsymbol{\varepsilon}} \rangle_V = \langle \breve{\boldsymbol{\sigma}} \rangle_V \langle \breve{\boldsymbol{\varepsilon}} \rangle_V + \rho W^{\mathrm{f}},$$

$$\rho W^{\mathrm{f}} = \frac{1}{V} \oint_{\partial V} \breve{\boldsymbol{t}}^{\mathrm{f}} \cdot \breve{\boldsymbol{u}}^{\mathrm{f}} \, \mathrm{d}s, \tag{4.7}$$

where ρW^{f} denotes specific work of the surface tractions fluctuations field on corresponding to it the field of surface displacements fluctuations (see also (A4.15) and (A4.20) in Annex 4.2). Formula (4.7) is valid regardless of the constitutive law of the material from which the body $\mathcal{B}$ is made. As it comes from this formula, the specific work $\rho W^{(\mathrm{Tot})}$ depends not only on the average values of stress and strain fields (macroscopic variables), but also on specific work of fluctuations ρW^{f} independent from these variables, that is, independent mesoscopic information. This situation precludes meaningful introduction of macroscopic description of material behavior. In order to overcome this difficulty it is accepted in the sequence below that surface boundary tractions $\breve{\boldsymbol{t}}^{(\mathrm{b})}(\boldsymbol{x})$ (see (4.3)) are such that they are leading to fulfillment of *Hill's postulate* as follows:

$$\rho W^{\mathrm{f}} = \frac{1}{V} \oint_{\partial V} \breve{\boldsymbol{t}}^{\mathrm{f}} \cdot \breve{\boldsymbol{u}}^{\mathrm{f}} \, \mathrm{d}s = 0 \Rightarrow \rho W^{(\mathrm{Tot})} = \langle \breve{\boldsymbol{\sigma}} \rangle_V \cdot \langle \breve{\boldsymbol{\varepsilon}} \rangle_V \tag{4.8}$$

The Hill's postulate, also called the postulate of mechanical work compatibility, in fact imposes constraints on boundary conditions. When surface fluctuation fields fulfill Hill's postulate then specific total work done on the body $\mathcal{B}$, in the case of linearly elastic body equal to double elastic energy stored in the body $\mathcal{B}\left(\rho W^{(\mathrm{Tot})} = 2\rho W\right)$, depends in a parametric way on average stress and strain only $(\boldsymbol{\sigma}, \boldsymbol{\varepsilon})$ in accordance with $(4.8)_2$. This enables meaningful introduction of the concept of effective elastic properties of a material and transition into macroscopic description. Derivation of formula (4.7) and broader discussion of consequences of the Hill's postulate (4.8) is delivered in Annex 4.2.

4.2.2 Elastic energy of heterogeneous macroelement in the presence of eigenstrains

In this section it is assumed that the fields $\breve{\boldsymbol{u}}(\boldsymbol{x})$, $\breve{\boldsymbol{\varepsilon}}(\boldsymbol{x})$, $\breve{\boldsymbol{\sigma}}(\boldsymbol{x})$ constituting solution of the problem of linear elasticity (4.2)–(4.4) are known. In view of the fact that the studied problem is a problem of the linear theory of elasticity, the solution of this problem may be expressed as a sum of two solutions of other problems of elasticity constituting together the original problem. Here, special decomposition of the original problem into two equivalent problems is done. Solution of the first problem describes effects connected with "external" boundary conditions only while the solution of the second problem describes effects appearing as a result of the existence of a nonzero, incompatible eigenstrains field with zero surface tractions boundary conditions. The fields $\breve{\boldsymbol{\sigma}}(\boldsymbol{x})$ and $\breve{\boldsymbol{\varepsilon}}(\boldsymbol{x})$, constituting the solution of the original problem, can be expressed as follows:

$$\breve{\boldsymbol{\varepsilon}}(\boldsymbol{x}) = \breve{\boldsymbol{\varepsilon}}^{(\mathrm{ex})}(\boldsymbol{x}) + \breve{\boldsymbol{\varepsilon}}^{(\mathrm{in})}(\boldsymbol{x}),\ \breve{\boldsymbol{\sigma}}(\boldsymbol{x}) = \breve{\boldsymbol{\sigma}}^{(\mathrm{ex})}(\boldsymbol{x}) + \breve{\boldsymbol{\sigma}}^{(\mathrm{in})}(\boldsymbol{x}) \tag{4.9}$$

The pair $I = \left\{\breve{\boldsymbol{\sigma}}^{(\mathrm{ex})}, \breve{\boldsymbol{\varepsilon}}^{(\mathrm{ex})}\right\}$, by assumption, fulfills mechanical equilibrium equations, linear Hooke's law without eigenstrains, and external force boundary conditions of the primary problem.

$$\begin{aligned}
&\breve{\boldsymbol{\varepsilon}}^{(\mathrm{ex})}(\boldsymbol{x}) = \breve{\mathbf{M}}(\boldsymbol{x})\breve{\boldsymbol{\sigma}}^{(\mathrm{ex})}(\boldsymbol{x}) \quad \boldsymbol{x} \in V,\\
&\breve{\boldsymbol{t}}^{(\mathrm{b})}(\boldsymbol{x}) = \breve{\boldsymbol{n}}(\boldsymbol{x})\boldsymbol{\sigma} + \breve{\boldsymbol{t}}^{\mathrm{f}}(\boldsymbol{x}) = \breve{\boldsymbol{n}}(\boldsymbol{x})\breve{\boldsymbol{\sigma}}^{(\mathrm{ex})}(\boldsymbol{x}), \quad \boldsymbol{x} \in \partial V,\\
&\boldsymbol{\sigma}^{(\mathrm{ex})} = \left\langle \breve{\boldsymbol{\sigma}}^{(\mathrm{ex})}(\boldsymbol{x}) \right\rangle_V = \frac{1}{V}\oint_{\partial V} \boldsymbol{x} \otimes \breve{\boldsymbol{t}}^{(\mathrm{b})}(\boldsymbol{x}) + \breve{\boldsymbol{t}}^{(\mathrm{b})}(\boldsymbol{x}) \otimes \boldsymbol{x}\,\mathrm{d}s = \boldsymbol{\sigma}.
\end{aligned} \tag{4.10}$$

The pair $\mathrm{II} = \left\{\breve{\boldsymbol{\sigma}}^{(\mathrm{in})}, \breve{\boldsymbol{\varepsilon}}^{(\mathrm{in})}\right\}$, by assumption, fulfills mechanical equilibrium equations, linear Hooke's law with eigenstrains, and zero surface tractions boundary conditions.

$$\begin{aligned}
&\breve{\boldsymbol{\varepsilon}}^{(\mathrm{in})}(\boldsymbol{x}) = \breve{\mathbf{M}}(\boldsymbol{x})\breve{\boldsymbol{\sigma}}^{(\mathrm{in})}(\boldsymbol{x}) + \breve{\boldsymbol{\Gamma}}(\boldsymbol{x}), \quad \boldsymbol{x} \in V\\
&\mathbf{0} = \breve{\boldsymbol{t}}^{(\mathrm{b})(\mathrm{in})}(\boldsymbol{x}) = \breve{\boldsymbol{n}}(\boldsymbol{x})\breve{\boldsymbol{\sigma}}^{(\mathrm{in})}(\boldsymbol{x}), \ \left(\breve{\boldsymbol{t}}^{(\mathrm{u})(\mathrm{in})}(\boldsymbol{x}) = \breve{\boldsymbol{n}}(\boldsymbol{x})\boldsymbol{\sigma}^{(\mathrm{in})} = \breve{\boldsymbol{t}}^{\mathrm{f}(\mathrm{in})}(\boldsymbol{x}) = \mathbf{0}\right), \quad \boldsymbol{x} \in \partial V,\\
&\boldsymbol{\sigma}^{(\mathrm{in})} \equiv \left\langle \breve{\boldsymbol{\sigma}}^{(\mathrm{in})}(\boldsymbol{x}) \right\rangle_V = \frac{1}{V}\oint_{\partial V} \boldsymbol{x} \otimes \breve{\boldsymbol{t}}^{(\mathrm{b})(\mathrm{in})}(\boldsymbol{x}) + \breve{\boldsymbol{t}}^{(\mathrm{b})(\mathrm{in})}(\boldsymbol{x}) \otimes \boldsymbol{x}\,\mathrm{d}s = \mathbf{0}.
\end{aligned} \tag{4.11}$$

Hence, self-equilibrating field of internal stresses $\breve{\boldsymbol{\sigma}}^{(\mathrm{in})}$, with zero average, is generated solely by incompatible eigenstrains field $\breve{\boldsymbol{\Gamma}}(\boldsymbol{x})$. It is identically equal to zero when the eigenstrains field does not appear $\left(\breve{\boldsymbol{\Gamma}}(\boldsymbol{x}) = \mathbf{0} \Rightarrow \breve{\boldsymbol{\sigma}}^{(\mathrm{in})}(\boldsymbol{x}) = \mathbf{0}\right)$ or when the field $\breve{\boldsymbol{\Gamma}}(\boldsymbol{x})$ is compatible. Because in this last situation continuous displacement field exists generating such strain field that $\breve{\boldsymbol{\varepsilon}}^{(\mathrm{in})}(\boldsymbol{x}) = \breve{\boldsymbol{\Gamma}}(\boldsymbol{x})$.

By assumption, the external loading field $\breve{\boldsymbol{t}}^{(\mathrm{b})}(\boldsymbol{x})$ is such that for a pair of external fields $I = \left(\breve{\boldsymbol{\sigma}}^{(\mathrm{ex})}, \breve{\boldsymbol{\varepsilon}}^{(\mathrm{ex})}\right)$ the postulate of mechanical energy compatibility, Hill's postulate, is fulfilled.

$$\frac{1}{V}\oint_{\partial V} \breve{\boldsymbol{t}}^{\mathrm{f}(\mathrm{ex})} \otimes \breve{\boldsymbol{u}}^{\mathrm{f}(\mathrm{ex})}\,\mathrm{d}s = \frac{1}{V}\oint_{\partial V} \breve{\boldsymbol{t}}^{\mathrm{f}} \otimes \breve{\boldsymbol{u}}^{\mathrm{f}(\mathrm{ex})}\,\mathrm{d}s \equiv \mathbf{0} \tag{4.12}$$

As a result of this assumption, the elastic energy stored in body $\mathcal{B}$ depends only on the average values of external fields of stress $\boldsymbol{\sigma}$ and strain $\boldsymbol{\varepsilon}$. The external stress tensor field can be expressed with *stress concentration tensor* as the linear function of average (macroscopic) stress as follows: $\breve{\boldsymbol{\sigma}}^{(\mathrm{ex})}(\boldsymbol{x})=\mathcal{B}^{(\mathrm{ex})}(\boldsymbol{x})\boldsymbol{\sigma}$. This enables introduction of the concept of effective modules of elasticity for the external fields (see $(4.10)_1$).

$$\boldsymbol{\varepsilon}^{\mathrm{E}} \equiv \left\langle \breve{\boldsymbol{\varepsilon}}^{(\mathrm{ex})}(\boldsymbol{x})\right\rangle_V = \left\langle \breve{\mathbf{M}}(\boldsymbol{x})\breve{\boldsymbol{\sigma}}^{(\mathrm{ex})}(\boldsymbol{x})\right\rangle_V = \mathbf{M}^{\mathrm{ef}}\boldsymbol{\sigma}^{(\mathrm{ex})} = \mathbf{M}^{\mathrm{ef}}\boldsymbol{\sigma},$$

$$\mathbf{M}^{\mathrm{ef}} = \left\langle \breve{\mathbf{M}}(\boldsymbol{x})\mathcal{B}^{(\mathrm{ex})}(\boldsymbol{x})\right\rangle_V \tag{4.13}$$

where $\mathbf{M}^{\mathrm{ef}}$ denotes the tensor of *effective modules of elastic compliance* $\mathbf{M}^{\mathrm{ef}}=\left(\mathbf{L}^{\mathrm{ef}}\right)^{-1}$.

By averaging formulas $(4.9)_1$ and $(4.11)_1$ over the whole volume of the body $\mathcal{B}$, one obtains relations for macroscopic elastic strain $\boldsymbol{\varepsilon}^{\mathrm{E}}$ and *macroscopic phase strain* $\boldsymbol{\varepsilon}^{\mathrm{pt}}$

$$\boldsymbol{\varepsilon}^{\mathrm{E}} = \boldsymbol{\varepsilon} - \boldsymbol{\varepsilon}^{\mathrm{pt}},\ \boldsymbol{\varepsilon}^{\mathrm{pt}} \equiv \left\langle \breve{\boldsymbol{\varepsilon}}^{(\mathrm{in})}(\boldsymbol{x})\right\rangle_V = \frac{1}{2V}\oint_{\partial V} \breve{\boldsymbol{n}}\otimes\breve{\boldsymbol{u}}^{(\mathrm{in})} + \breve{\boldsymbol{u}}^{(\mathrm{in})}\otimes\breve{\boldsymbol{n}}\,\mathrm{d}s,$$

$$\boldsymbol{\varepsilon}^{\mathrm{pt}} = \left\langle \breve{\mathbf{M}}(\boldsymbol{x})\breve{\boldsymbol{\sigma}}^{(\mathrm{in})}(\boldsymbol{x})\right\rangle_V + \boldsymbol{\Gamma},\ \boldsymbol{\Gamma} \equiv \left\langle \breve{\boldsymbol{\Gamma}}(\boldsymbol{x})\right\rangle_V \tag{4.14}$$

As it comes from (4.14), macroscopic phase strain $\boldsymbol{\varepsilon}^{\mathrm{pt}}$ connected with the field of eigenstrains $\breve{\boldsymbol{\Gamma}}(\boldsymbol{x})$ is equal to average value of the eigenstrain field $\boldsymbol{\Gamma}(\boldsymbol{\varepsilon}^{\mathrm{pt}}=\boldsymbol{\Gamma})$ when the field $\breve{\boldsymbol{\Gamma}}(\boldsymbol{x})$ is geometrically compatible, because in such a case $\breve{\boldsymbol{\sigma}}^{(\mathrm{in})}(\boldsymbol{x})=\mathbf{0}$, or when the field of modules of compliance is uniform $\left(\breve{\mathbf{M}}(\boldsymbol{x})=\mathbf{M}=\mathrm{const}\right)$, because then $\left\langle \breve{\mathbf{M}}(\boldsymbol{x})\breve{\boldsymbol{\sigma}}^{(\mathrm{in})}\right\rangle_V = \mathbf{M}\left\langle \breve{\boldsymbol{\sigma}}^{(\mathrm{in})}\right\rangle_V = \mathbf{0}$, in view of zero, uniform boundary tractions of "in" problem $\breve{\boldsymbol{t}}^{(\mathrm{b})(\mathrm{in})}(\boldsymbol{x})=\mathbf{0}$ (see $(4.11)_2$).

By averaging $(4.4)_{3\text{–}4}$ and using (4.14) one obtains the following relations:

$$\boldsymbol{\varepsilon}^{(\mathrm{e})} = \left\langle \breve{\mathbf{M}}\breve{\boldsymbol{\sigma}}\right\rangle_V = \left\langle \breve{\mathbf{M}}\breve{\boldsymbol{\sigma}}^{(\mathrm{ex})} + \breve{\mathbf{M}}\breve{\boldsymbol{\sigma}}^{(\mathrm{in})}\right\rangle_V = \boldsymbol{\varepsilon}^{\mathrm{E}} + \left\langle \breve{\mathbf{M}}\breve{\boldsymbol{\sigma}}^{(\mathrm{in})}\right\rangle_V,$$

$$\boldsymbol{\varepsilon}^{(\mathrm{e})} \equiv \left\langle \breve{\boldsymbol{\varepsilon}}^{(\mathrm{e})}\right\rangle_V = \left\langle \breve{\boldsymbol{\varepsilon}}(\boldsymbol{x}) - \breve{\boldsymbol{\Gamma}}(\boldsymbol{x})\right\rangle_V = \boldsymbol{\varepsilon} - \boldsymbol{\Gamma} = \boldsymbol{\varepsilon}^{\mathrm{E}} + \boldsymbol{\varepsilon}^{\mathrm{pt}} - \boldsymbol{\Gamma}. \tag{4.15}$$

Thus, *macroscopic total strain* $\boldsymbol{\varepsilon}=\boldsymbol{\varepsilon}^{\mathrm{E}}+\boldsymbol{\varepsilon}^{\mathrm{pt}}$ and macroscopic constitutive relations take the form $\boldsymbol{\varepsilon}=\mathbf{M}^{\mathrm{ef}}\boldsymbol{\sigma}+\boldsymbol{\varepsilon}^{\mathrm{pt}}$.

In the case of a linearly elastic body, specific total work $(4.8)_2$ equals double elastic energy stored in the body $\mathcal{B}\left(\rho W^{(\mathrm{Tot})}=2\rho W\right)$. Taking advantage of the decomposition (4.9) into external and internal fields, the elastic energy stored in the linear elastic solid body $\mathcal{B}$ (RVE) can be expressed as follows:

$$\begin{aligned}
2\rho W &= \left\langle \breve{\boldsymbol{\sigma}}\cdot\breve{\boldsymbol{\varepsilon}}^{(\mathrm{e})}\right\rangle_V = \left\langle \breve{\boldsymbol{\varepsilon}}^{(\mathrm{e})}\breve{\mathbf{L}}\breve{\boldsymbol{\varepsilon}}^{(\mathrm{e})}\right\rangle_V = \left\langle \breve{\boldsymbol{\sigma}}\cdot\breve{\mathbf{M}}\breve{\boldsymbol{\sigma}}\right\rangle_V \\
&= \left\langle \left(\breve{\boldsymbol{\sigma}}^{(\mathrm{ex})} + \breve{\boldsymbol{\sigma}}^{(\mathrm{in})}\right)\cdot\left(\breve{\boldsymbol{\varepsilon}}^{(\mathrm{ex})} + \breve{\mathbf{M}}\breve{\boldsymbol{\sigma}}^{(\mathrm{in})}\right)\right\rangle_V = \left\langle \breve{\boldsymbol{\sigma}}^{(\mathrm{ex})}\cdot\breve{\boldsymbol{\varepsilon}}^{(\mathrm{ex})}\right\rangle_V + \left\langle \breve{\boldsymbol{\sigma}}^{(\mathrm{in})}\cdot\breve{\boldsymbol{\varepsilon}}^{(\mathrm{ex})}\right\rangle_V \\
&+ \left\langle \breve{\boldsymbol{\sigma}}^{(\mathrm{ex})}\cdot\breve{\mathbf{M}}\breve{\boldsymbol{\sigma}}^{(\mathrm{in})}\right\rangle_V + \left\langle \breve{\boldsymbol{\sigma}}^{(\mathrm{in})}\cdot\breve{\mathbf{M}}\breve{\boldsymbol{\sigma}}^{(\mathrm{in})}\right\rangle_V = \left\langle \breve{\boldsymbol{\sigma}}^{(\mathrm{ex})}\cdot\breve{\boldsymbol{\varepsilon}}^{(\mathrm{ex})}\right\rangle_V + \left\langle \breve{\boldsymbol{\sigma}}^{(\mathrm{in})}\cdot\breve{\mathbf{M}}\breve{\boldsymbol{\sigma}}^{(\mathrm{in})}\right\rangle_V.
\end{aligned} \tag{4.16}$$

Using formula $(4.7)_1$ to the fields $\breve{\boldsymbol{\sigma}}^{(\text{in})}, \breve{\boldsymbol{\varepsilon}}^{(\text{ex})}$ and taking advantage of the condition $(4.11)_3$, it is easy to show the term $\left\langle \breve{\boldsymbol{\sigma}}^{(\text{in})} \cdot \breve{\boldsymbol{\varepsilon}}^{(\text{ex})} \right\rangle_V$ has zero value

$$\left\langle \breve{\boldsymbol{\sigma}}^{(\text{in})} \cdot \breve{\boldsymbol{\varepsilon}}^{(\text{ex})} \right\rangle_V = \underbrace{\left\langle \breve{\boldsymbol{\sigma}}^{(\text{in})} \right\rangle_V}_{=\mathbf{0}} \cdot \left\langle \breve{\boldsymbol{\varepsilon}}^{(\text{ex})} \right\rangle_V + \oint_{\partial V} \underbrace{\breve{\boldsymbol{t}}^{\text{f(in)}}}_{=\mathbf{0}} \cdot \breve{\boldsymbol{u}}^{\text{f(ex)}} \mathrm{d}s = 0, \tag{4.17}$$

the term $\left\langle \breve{\boldsymbol{\sigma}}^{(\text{ex})} \cdot \breve{\mathbf{M}} \breve{\boldsymbol{\sigma}}^{(\text{in})} \right\rangle_V$ is zero due to the assumed symmetry of tensor $\breve{\mathbf{M}}$, $(4.10)_1$ and (4.17).

$$\left\langle \breve{\boldsymbol{\sigma}}^{(\text{ex})} \cdot \breve{\mathbf{M}} \breve{\boldsymbol{\sigma}}^{(\text{in})} \right\rangle_V = \left\langle \breve{\boldsymbol{\sigma}}^{(\text{in})} \cdot \breve{\mathbf{M}} \breve{\boldsymbol{\sigma}}^{(\text{ex})} \right\rangle_V = \left\langle \breve{\boldsymbol{\sigma}}^{(\text{in})} \cdot \breve{\boldsymbol{\varepsilon}}^{(\text{ex})} \right\rangle_V = 0 \tag{4.18}$$

Hence, the total elastic energy stored in macroelement $\mathcal{B}$ can be expressed as follows (see (4.16)):

$$W = W^{(\text{ex})}\left(\boldsymbol{\varepsilon}^{\text{E}}\right) + W^{(\text{in})}, \tag{4.19}$$

where the macroscopic elastic energy $W^{(\text{ex})}(\boldsymbol{\varepsilon}^{\text{E}}) = W^{(\text{ex})}(\boldsymbol{\sigma})$ can be expressed in several equivalent forms.

$$\begin{aligned} 2\rho W^{(\text{ex})}(\boldsymbol{\varepsilon}^{\text{E}}) &\equiv \left\langle \breve{\boldsymbol{\sigma}}^{(\text{ex})} \cdot \breve{\boldsymbol{\varepsilon}}^{(\text{ex})} \right\rangle_V = \boldsymbol{\sigma} \cdot \boldsymbol{\varepsilon}^{(\text{ex})} \\ &= (\boldsymbol{\varepsilon} - \boldsymbol{\varepsilon}^{\text{pt}}) \cdot \mathbf{L}^{\text{ef}} (\boldsymbol{\varepsilon} - \boldsymbol{\varepsilon}^{\text{pt}}) = \boldsymbol{\sigma} \cdot \mathbf{M}^{\text{ef}} \boldsymbol{\sigma} = \boldsymbol{\varepsilon}^{\text{E}} \cdot \mathbf{L}^{\text{ef}} \boldsymbol{\varepsilon}^{\text{E}} \geq 0. \end{aligned} \tag{4.20}$$

This energy originates only from external loading and can play a role of potential for macroscopic stress $\boldsymbol{\sigma}$ (see (4.13)).

$$\boldsymbol{\sigma} = \rho \partial W^{(\text{ex})}\left(\boldsymbol{\varepsilon}^{\text{E}}\right) / \partial \boldsymbol{\varepsilon}^{\text{E}} = \rho \partial W^{(\text{ex})}(\boldsymbol{\varepsilon} - \boldsymbol{\varepsilon}^{\text{pt}}) / \partial \boldsymbol{\varepsilon} \tag{4.21}$$

The energy $W^{(\text{in})}\left(\breve{\boldsymbol{\Gamma}}(\boldsymbol{x})\right)$, as defined below,

$$2\rho W^{(\text{in})} \equiv \left\langle \breve{\boldsymbol{\sigma}}^{(\text{in})} \cdot \breve{\mathbf{M}} \breve{\boldsymbol{\sigma}}^{(\text{in})} \right\rangle_V = -\left\langle \breve{\boldsymbol{\sigma}}^{(\text{in})} \breve{\boldsymbol{\Gamma}} \right\rangle_V \geq 0, \tag{4.22}$$

depends only on an incompatible eigenstrains field and disappears when the field $\breve{\boldsymbol{\Gamma}}(\boldsymbol{x})$ is geometrically compatible. For that reason it is frequently called the energy of incompatibility of eigenstrains field, configurational energy, misfit energy, or "coherency energy." The last equality in (4.22) originates from the following relations (see (4.11)):

$$\begin{aligned} \left\langle \breve{\boldsymbol{\sigma}}^{(\text{in})} \cdot \breve{\boldsymbol{\varepsilon}}^{(\text{in})} \right\rangle_V &= \underbrace{\left\langle \breve{\boldsymbol{\sigma}}^{(\text{in})} \right\rangle_V}_{=\mathbf{0}} \cdot \left\langle \breve{\boldsymbol{\varepsilon}}^{(\text{in})} \right\rangle_V + \oint_{\partial V} \underbrace{\breve{\boldsymbol{t}}^{\text{f(in)}}}_{=\mathbf{0}} \cdot \breve{\boldsymbol{u}}^{\text{f(in)}} \mathrm{d}s = 0 \\ \left\langle \breve{\boldsymbol{\sigma}}^{(\text{in})} \cdot \breve{\boldsymbol{\varepsilon}}^{(\text{in})} \right\rangle_V &= \left\langle \breve{\boldsymbol{\sigma}}^{(\text{in})} \cdot \breve{\mathbf{M}} \breve{\boldsymbol{\sigma}}^{(\text{in})} \right\rangle_V + \left\langle \breve{\boldsymbol{\sigma}}^{(\text{in})} \cdot \breve{\boldsymbol{\Gamma}} \right\rangle_V = 0. \end{aligned} \tag{4.23}$$

In fact, the energy $W^{(\text{in})}$ does not depend on the average value of the eigenstrains field $\boldsymbol{\Gamma}$ but on its fluctuation only $\breve{\boldsymbol{\Gamma}}^{\text{f}}(\underline{x}) = \breve{\boldsymbol{\Gamma}}(\underline{x}) - \boldsymbol{\Gamma}$ as

$$-2\rho W^{(\text{in})} = \left\langle \breve{\boldsymbol{\sigma}}^{(\text{in})} \cdot \breve{\boldsymbol{\Gamma}} \right\rangle_V = \underbrace{\left\langle \breve{\boldsymbol{\sigma}}^{(\text{in})} \right\rangle_V}_{=\mathbf{0}} \cdot \boldsymbol{\Gamma} + \left\langle \breve{\boldsymbol{\sigma}}^{(\text{in})} \cdot \breve{\boldsymbol{\Gamma}}^{\text{f}} \right\rangle_V = \left\langle \breve{\boldsymbol{\sigma}}^{(\text{in})} \cdot \breve{\boldsymbol{\Gamma}}^{\text{f}} \right\rangle_V. \quad (4.24)$$

4.2.3 Elastic energy of a two-phase macroelement

In this section, it is assumed that the body $\mathcal{B}$ constitutes a two-phase continuum; that is, that the whole volume of the body $\mathcal{B}$ can be divided into two subdomains with volumes V_1, V_2 ($V_1 + V_2 = V$); (see Figure 4.1 with volume $V_3 = 0$). The parent phase—austenite—is denoted in the sequence with index $\alpha = 1$, while the product phase—oriented martensite—is denoted with index $\alpha = 2$. The magnitude of these domains is characterized by volume fractions $z_\alpha = \mathrm{V}_\alpha/\mathrm{V}$ ($z_1 + z_2 = 1$). Besides the assumption that particular phases can be distinguished in macroelement (RVE) and their volume fractions, the other features of macroelement microstructure remain unspecified; for example, shape of inclusions of one phase in the other, their mutual orientation, or other geometric parameters characterizing microstructure. The microstructure, by assumption, can be characterized as a set of internal parameters of deformational character h_K, $K = 1, 2, \ldots N$; here, not given explicitly. It is assumed that fields of elastic properties and eigenstrains of each phase can be inhomogeneous. Thus, elastic properties and eigenstrains fields in the whole body $\mathcal{B}$ can be expressed as follows:

$$\breve{\mathbf{M}}(\boldsymbol{x}) = \begin{matrix} \breve{\mathbf{M}}_{(1)}(\boldsymbol{x}) & \boldsymbol{x} \in V_1 \\ \breve{\mathbf{M}}_{(2)}(\boldsymbol{x}) & \boldsymbol{x} \in V_2 \end{matrix}, \quad \breve{\boldsymbol{\Gamma}}(\boldsymbol{x}) = \begin{matrix} \breve{\boldsymbol{\Gamma}}_{(1)}(\boldsymbol{x}) & \boldsymbol{x} \in V_1 \\ \breve{\boldsymbol{\Gamma}}_{(2)}(\boldsymbol{x}) & \boldsymbol{x} \in V_2 \end{matrix} \quad (4.25)$$

It is assumed that field of eigenstrains $\breve{\boldsymbol{\Gamma}}(\boldsymbol{x})$ experience jump changes on interphase boundaries $\mathcal{S}^{(\text{in})}$ (see Figure 4.1), while fields of elastic properties may or may not experience jump changes on these surfaces. Frequently encountered in the literature devoted to SMA materials is the assumption on piecewise uniform field of elastic properties $\breve{\mathbf{M}}(\boldsymbol{x}) = \mathbf{M}_\alpha$, $\boldsymbol{x} \in V_\alpha$.

By using phase averages and phase fluctuations and their relations with average and fluctuations over the whole volume of the body $\mathcal{B}$(see formulas (A4.11)–(A4.13) in Annex 4.1), to the field of eigenstrains $\breve{\boldsymbol{\Gamma}}(\boldsymbol{x})$, the following formulas are obtained:

$$\boldsymbol{\Gamma} = z_1 \boldsymbol{\Gamma}_{(1)} + z_2 \boldsymbol{\Gamma}_{(2)}, \quad \Delta\boldsymbol{\Gamma} = \boldsymbol{\Gamma}_{(2)} - \boldsymbol{\Gamma}_{(1)}$$

$$\breve{\boldsymbol{\Gamma}}^{\text{f}}(\boldsymbol{x}) \equiv \breve{\boldsymbol{\Gamma}}(\boldsymbol{x}) - \boldsymbol{\Gamma} = \begin{cases} -z_2 \Delta\boldsymbol{\Gamma} + \breve{\boldsymbol{\Gamma}}^{\text{f}}_{(1)}(\boldsymbol{x}) & \boldsymbol{x} \in V_1 \\ z_1 \Delta\boldsymbol{\Gamma} + \breve{\boldsymbol{\Gamma}}^{\text{f}}_{(2)}(\boldsymbol{x}) & \boldsymbol{x} \in V_2 \end{cases},$$

$$\breve{\boldsymbol{\Gamma}}^{\text{f}}(\boldsymbol{x}) = \breve{\boldsymbol{\Gamma}}(\boldsymbol{x}) - \left\langle \breve{\boldsymbol{\Gamma}} \right\rangle_V, \quad \left\langle \breve{\boldsymbol{\Gamma}}^{\text{f}}(\boldsymbol{x}) \right\rangle_V = \mathbf{0},$$

$$\breve{\boldsymbol{\Gamma}}^{\text{f}}_{(\alpha)}(\boldsymbol{x}) = \breve{\boldsymbol{\Gamma}}(\boldsymbol{x}) - \left\langle \breve{\boldsymbol{\Gamma}} \right\rangle_{V_\alpha}, \quad \left\langle \breve{\boldsymbol{\Gamma}}^{\text{f}}_{(\alpha)}(\boldsymbol{x}) \right\rangle_{V_\alpha} = \mathbf{0}. \quad (4.26)$$

In order to be able to evaluate the value of coherency energy $W^{(\text{in})}$, there must be a known self-equilibrating, residual stress field $\breve{\boldsymbol{\sigma}}^{(\text{in})}(\boldsymbol{x})\left(\left\langle\breve{\boldsymbol{\sigma}}^{(\text{in})}(\boldsymbol{x})\right\rangle_V=0\right)$ arising due to geometrical incompatibility of eigenstrains field $\breve{\boldsymbol{\Gamma}}(\boldsymbol{x})$.

The original constitutive relations involving internal stresses $\breve{\boldsymbol{\sigma}}^{(\text{in})}$ can be rewritten using eigenstrains phase averages and eigenstrains phase fluctuations (see (4.11)).

$$\begin{aligned}
\breve{\boldsymbol{\varepsilon}}^{(\text{in})}_{(1)}(\boldsymbol{x}) &= \breve{\mathbf{M}}(\boldsymbol{x})\breve{\boldsymbol{\sigma}}^{(\text{in})}_{(1)}(\boldsymbol{x}) + \boldsymbol{\Gamma}_1 + \breve{\boldsymbol{\Gamma}}^{\text{f}}_{(1)}(\boldsymbol{x}) \quad \boldsymbol{x}\in V_1,\\
\breve{\boldsymbol{\varepsilon}}^{(\text{in})}_{(2)}(\boldsymbol{x}) &= \breve{\mathbf{M}}(\boldsymbol{x})\breve{\boldsymbol{\sigma}}^{(\text{in})}_{(2)}(\boldsymbol{x}) + \boldsymbol{\Gamma}_2 + \breve{\boldsymbol{\Gamma}}^{\text{f}}_{(2)}(\boldsymbol{x}) \quad \boldsymbol{x}\in V_2
\end{aligned} \tag{4.27}$$

Then, the problem (4.11), for the field of internal stresses $\breve{\boldsymbol{\sigma}}^{(\text{in})}$, can be further decomposed into two equivalent problems "(a-in)" and "(f-in)" for a stress fields $\breve{\boldsymbol{\sigma}}^{(\text{a-in})}$ and $\breve{\boldsymbol{\sigma}}^{(\text{f-in})}$.

$$\breve{\boldsymbol{\varepsilon}}^{(\text{in})} = \breve{\boldsymbol{\varepsilon}}^{(\text{a-in})} + \breve{\boldsymbol{\varepsilon}}^{(\text{f-in})}, \quad \breve{\boldsymbol{\sigma}}^{(\text{in})} = \breve{\boldsymbol{\sigma}}^{(\text{a-in})} + \breve{\boldsymbol{\sigma}}^{(\text{f-in})} \tag{4.28}$$

In the problem "(a-in)," only eigenstrains phase averages are present:

$$\begin{aligned}
\breve{\boldsymbol{\varepsilon}}^{(\text{a-in})}_{(\alpha)}(\boldsymbol{x}) &= \breve{\mathbf{M}}(\boldsymbol{x})\breve{\boldsymbol{\sigma}}^{(\text{a-in})}_{(\alpha)}(\boldsymbol{x}) + \boldsymbol{\Gamma}_{(\alpha)}, \quad \boldsymbol{x}\in V_\alpha, \quad \alpha=1,2\\
\breve{\boldsymbol{t}}^{(\text{b})(\text{a-in})}(\boldsymbol{x}) &= \breve{\boldsymbol{n}}(\boldsymbol{x})\breve{\boldsymbol{\sigma}}^{(\text{a-in})}(\boldsymbol{x}) = \mathbf{0}, \quad \boldsymbol{x}\in\partial V
\end{aligned} \tag{4.29}$$

In the problem "(f-in)," only phase fluctuations $\breve{\boldsymbol{\Gamma}}^{\text{f}}_{(\alpha)}(\boldsymbol{x})$ of the originally prescribed field of eigenstrains are present, in each phase.

$$\begin{aligned}
\breve{\boldsymbol{\varepsilon}}^{(\text{f-in})}_{(\alpha)}(\boldsymbol{x}) &= \breve{\mathbf{M}}(\boldsymbol{x})\breve{\boldsymbol{\sigma}}^{(\text{f-in})}_{(\alpha)}(\boldsymbol{x}) + \breve{\boldsymbol{\Gamma}}^{\text{f}}_{(\alpha)}(\boldsymbol{x}), \quad \boldsymbol{x}\in V_\alpha\\
\breve{\boldsymbol{t}}^{(\text{b})(\text{f-in})}(\boldsymbol{x}) &= \breve{\boldsymbol{n}}(\boldsymbol{x})\breve{\boldsymbol{\sigma}}^{(\text{f-in})}(\boldsymbol{x}) = \mathbf{0}, \quad \boldsymbol{x}\in\partial V
\end{aligned} \tag{4.30}$$

The fields of phase fluctuations $\breve{\boldsymbol{\Gamma}}^{\text{f}}_{(\alpha)}(\boldsymbol{x})$ are independent from phase averages $\boldsymbol{\Gamma}_{(\alpha)}$.

The fields of stresses constituting solutions of auxiliary problems "(a-in)" and "(f-in)" are self-equilibrating, residual fields $\left\langle\breve{\boldsymbol{\sigma}}^{(\text{a-in})}(\boldsymbol{x})\right\rangle_V=0$, $\left\langle\breve{\boldsymbol{\sigma}}^{(\text{f-in})}(\boldsymbol{x})\right\rangle_V=0$. Hence, their phase averages in accordance with formula (A4.14) can be expressed as follows:

$$\begin{aligned}
\boldsymbol{\sigma}^{(\text{a-in})}_{(1)} &= z_2\boldsymbol{\Sigma}^{(\text{a-in})}, \quad \boldsymbol{\sigma}^{(\text{a-in})}_{(2)} = -z_1\boldsymbol{\Sigma}^{(\text{a-in})}, \quad \boldsymbol{\Sigma}^{(\text{a-in})} \equiv -\left(\boldsymbol{\sigma}^{(\text{a-in})}_{(2)} - \boldsymbol{\sigma}^{(\text{a-in})}_{(1)}\right),\\
\boldsymbol{\sigma}^{(\text{f-in})}_{(1)} &= z_2\boldsymbol{\Sigma}^{(\text{f-in})}, \quad \boldsymbol{\sigma}^{(\text{f-in})}_{(2)} = -z_1\boldsymbol{\Sigma}^{(\text{f-in})}, \quad \boldsymbol{\Sigma}^{(\text{f-in})} \equiv -\left(\boldsymbol{\sigma}^{(\text{f-in})}_{(2)} - \boldsymbol{\sigma}^{(\text{f-in})}_{(1)}\right).
\end{aligned} \tag{4.31}$$

Notice that if the stress field $\breve{\boldsymbol{\sigma}}^{(\text{a-in})}_{(\alpha)}(\boldsymbol{x})$ fulfills mechanical equilibrium equations with zero boundary tractions conditions and constitutive relations (4.29), then the same field will also fulfill mechanical equilibrium equations with zero surface tractions boundary conditions and constitutive relations for strain field as follows: $\breve{\boldsymbol{\varepsilon}}^{0(\text{a-in})}_{(\alpha)}(\boldsymbol{x}) = \breve{\boldsymbol{\varepsilon}}^{(\text{a-in})}_{(\alpha)}(\boldsymbol{x}) - \boldsymbol{\Gamma}_1 \left(\breve{\boldsymbol{\varepsilon}}^{0(\text{a-in})}_{(1)}(\boldsymbol{x}) = \breve{\mathbf{M}}(\boldsymbol{x})\breve{\boldsymbol{\sigma}}^{(\text{a-in})}_{(1)}(\boldsymbol{x}), \breve{\boldsymbol{\varepsilon}}^{0(\text{a-in})}_{(2)}(\boldsymbol{x}) = \breve{\mathbf{M}}(\boldsymbol{x})\breve{\boldsymbol{\sigma}}^{(\text{a-in})}_{(2)}(\boldsymbol{x}) + \Delta\boldsymbol{\Gamma} \right)$.

Hence, in order to find the stress field $\breve{\boldsymbol{\sigma}}^{(\text{a-in})}_{(\alpha)}(\boldsymbol{x})$ instead of solving problem (4.29) for the pair of fields $\breve{\boldsymbol{\sigma}}^{(\text{a-in})}_{(\alpha)}(\boldsymbol{x}) \leftrightarrow \breve{\boldsymbol{\varepsilon}}^{(\text{a-in})}_{(\alpha)}(\boldsymbol{x})$, it is sufficient to solve the boundary value problem for the pair of fields $\breve{\boldsymbol{\sigma}}^{(\text{a-in})}_{(\alpha)}(\boldsymbol{x}) \leftrightarrow \breve{\boldsymbol{\varepsilon}}^{0(\text{a-in})}_{(\alpha)}(\boldsymbol{x})$. However, the solution $\breve{\boldsymbol{\sigma}}^{(\text{a-in})}_{(\alpha)}(\boldsymbol{x})$ for this last pair does not depend on $\boldsymbol{\Gamma}_1$. Because the same field of stress constitutes the solution of problem (4.29), hence it can be immediately concluded that the solution of the problem "(a-in)" does not depend on $\boldsymbol{\Gamma}_1$, and it must be the linear function of $\Delta\boldsymbol{\Gamma}$. This solution can be symbolically expressed as follows:

$$\breve{\boldsymbol{\sigma}}^{(\text{a-in})}_{(1)}(\boldsymbol{x}) = z_2 \breve{\mathcal{H}}^{(\text{a-in})}(\boldsymbol{x})\Delta\boldsymbol{\Gamma},\ \breve{\boldsymbol{\sigma}}^{(\text{a-in})}_{(2)}(\boldsymbol{x}) = -z_1 \breve{\mathcal{H}}^{(\text{a-in})}(\boldsymbol{x})\Delta\boldsymbol{\Gamma},$$

$$\boldsymbol{\sigma}^{(\text{a-in})}_{(1)} = z_2 \mathcal{H}^{(\text{a-in})}\Delta\boldsymbol{\Gamma},\ \boldsymbol{\sigma}^{(\text{a-in})}_{(2)} = -z_1 \mathcal{H}^{(\text{a-in})}\Delta\boldsymbol{\Gamma}, \tag{4.32}$$

where $\boldsymbol{\Sigma}^{(\text{a-in})} = \mathcal{H}^{(\text{a-in})}\Delta\boldsymbol{\Gamma}$, and $\breve{\mathcal{H}}^{(\text{a-in})}(\boldsymbol{x})$ can be identified with Hill's stress concentration tensor (see Hill (1963)).

4.2.4 Macroscopic coherency energy $W^{(in)}$

The energy $W^{(\text{in})}$ resulting from incompatibility of the field of eigenstrains can be now expressed as follows (see (4.22) and (4.24)):

$$W^{(\text{in})} = -\frac{1}{2\rho}\left\langle \breve{\boldsymbol{\sigma}}^{(\text{in})} \breve{\boldsymbol{\Gamma}}^{\text{f}} \right\rangle_V = W^{(\text{a-in})} + W^{(\text{af-in})} + W^{(\text{f-in})} \geq 0 \tag{4.33}$$

Taking advantage of the decomposition (4.28) and relation (4.26) one obtains,

$$W^{(\text{in})} = -\tfrac{1}{2\rho} z_1 \left\langle \left(\breve{\boldsymbol{\sigma}}^{(\text{a-in})}_{(1)} + \breve{\boldsymbol{\sigma}}^{(\text{f-in})}_{(1)} \right) \cdot \left(\breve{\boldsymbol{\Gamma}}^{\text{f}}_{(1)} - z_2 \Delta\boldsymbol{\Gamma} \right) \right\rangle_{V_1} - \tfrac{1}{2\rho} z_2 \left\langle \left(\breve{\boldsymbol{\sigma}}^{(\text{a-in})}_{(2)} + \breve{\boldsymbol{\sigma}}^{(\text{f-in})}_{(2)} \right) \cdot \left(\breve{\boldsymbol{\Gamma}}^{\text{f}}_{(2)} + z_1 \Delta\boldsymbol{\Gamma} \right) \right\rangle_{V_2}.$$

Defined below energy $W^{(\text{a-in})}$, always nonnegative, describes elastic energy stored in the system "(a-in)". This energy is generated due to piecewise continuous, uniform eigenstrains field, such that in the domain of the first phase it takes the constant value $\boldsymbol{\Gamma}_1$ of first-phase average eigenstrain and in the domain of second phase it takes the constant value $\boldsymbol{\Gamma}_2$ of second-phase average eigenstrain, determined from the original eigenstrains field. Applying the formula for averages to the fields $\breve{\boldsymbol{\sigma}}^{(\text{a-in})}_{(\alpha)}$ (see (4.32)), the energy $W^{(\text{a-in})}$ can be expressed as follows:

$$W^{(\text{a-in})} \equiv -\frac{1}{2\rho}\left[z_1 \left\langle \breve{\boldsymbol{\sigma}}^{(\text{a-in})}_{(1)} \cdot (-z_2 \Delta\boldsymbol{\Gamma}) \right\rangle_{V_1} + z_2 \left\langle \breve{\boldsymbol{\sigma}}^{(\text{a-in})}_{(2)} z_1 \Delta\boldsymbol{\Gamma} \right\rangle_{V_2} \right]$$
$$= -\frac{1}{2\rho}\left[\left(\boldsymbol{\sigma}^{(\text{a-in})}_{(2)} - \boldsymbol{\sigma}^{(\text{a-in})}_{(1)} \right) \right] z_1 z_2 \Delta\boldsymbol{\Gamma} = \frac{1}{2\rho} z_1 z_2 \Delta\boldsymbol{\Gamma} \mathcal{H}^{(\text{a-in})} \Delta\boldsymbol{\Gamma} \geq 0. \tag{4.34}$$

The energy $W^{(\text{af-in})}$ represents elastic energy generated as a result of mutual interaction of the fields of system "(a-in)" and system "(f-in)."

$$W^{(\text{af-in})} \equiv -\tfrac{1}{2\rho}\left[z_1 \left\langle \breve{\boldsymbol{\sigma}}_{(1)}^{(\text{a-in})} \breve{\boldsymbol{\Gamma}}_{(1)}^{\text{f}}(\boldsymbol{x}) \right\rangle_{V_1} + z_2 \left\langle \breve{\boldsymbol{\sigma}}_{(2)}^{(\text{a-in})} \breve{\boldsymbol{\Gamma}}_{(2)}^{\text{f}}(\boldsymbol{x}) \right\rangle_{V_2} \right] - \tfrac{1}{2\rho}\left[z_1 \left\langle \breve{\boldsymbol{\sigma}}_{(1)}^{(\text{f-in})} (-z_2 \Delta\boldsymbol{\Gamma}) \right\rangle_{V_1} + z_2 \left\langle \breve{\boldsymbol{\sigma}}_{(2)}^{(\text{f-in})} z_1 \Delta\boldsymbol{\Gamma} \right\rangle_{V_2} \right]. \tag{4.35}$$

The pairs $\left(\breve{\boldsymbol{\sigma}}^{(\text{a-in})} \leftrightarrow \breve{\boldsymbol{\varepsilon}}^{(\text{a-in})}\right)$, $\left(\breve{\boldsymbol{\sigma}}^{(\text{f-in})} \leftrightarrow \breve{\boldsymbol{\varepsilon}}^{(\text{f-in})}\right)$ constitute solutions of the elasticity problems with eigenstrains $\boldsymbol{\Gamma}_{(\alpha)}$ and $\breve{\boldsymbol{\Gamma}}_{(\alpha)}^{\text{f}}$, respectively. Substituting in formula (A4.4) derived in Annex 4.1, in place of fields I the fields "(a-in)," and in place of fields II the fields "(f-in)," and taking advantage of the fact that fields "(a-in)" and "(f-in)" are self-equilibrating, residual fields ($\breve{\boldsymbol{t}}^{(\text{b})(\text{a-in})} = \breve{\boldsymbol{t}}^{(\text{b})(\text{f-in})} = \boldsymbol{0}$, see (4.29) and (4.30)) one obtains

$$\left\langle \breve{\boldsymbol{\sigma}}_{(\alpha)}^{(\text{a-in})} \cdot \breve{\boldsymbol{\Gamma}}_{(\alpha)}^{\text{f}} \right\rangle_V = \left\langle \breve{\boldsymbol{\sigma}}_{(\alpha)}^{(\text{f-in})} \cdot \boldsymbol{\Gamma}_{(\alpha)} \right\rangle_V,$$

$$z_1 \left\langle \breve{\boldsymbol{\sigma}}_{(1)}^{(\text{a-in})} \cdot \breve{\boldsymbol{\Gamma}}_{(1)}^{\text{f}}(\boldsymbol{x}) \right\rangle_{V_1} + z_2 \left\langle \breve{\boldsymbol{\sigma}}_{(2)}^{(\text{a-in})} \cdot \breve{\boldsymbol{\Gamma}}_{(2)}^{\text{f}}(\boldsymbol{x}) \right\rangle_{V_2} = z_1 \left\langle \breve{\boldsymbol{\sigma}}_{(1)}^{(\text{f-in})} \cdot (-z_2 \Delta\boldsymbol{\Gamma}) \right\rangle_{V_1} + z_2 \left\langle \breve{\boldsymbol{\sigma}}_{(2)}^{(\text{f-in})} \cdot z_1 \Delta\boldsymbol{\Gamma} \right\rangle_{V_2}. \tag{4.36}$$

During derivation of relations $(4.36)_2$, equality $\left\langle \breve{\boldsymbol{\sigma}}_{(\alpha)}^{(\text{f-in})} \cdot \boldsymbol{\Gamma}_1 \right\rangle_V = \left\langle \breve{\boldsymbol{\sigma}}_{(\alpha)}^{(\text{f-in})} \right\rangle_V \cdot \boldsymbol{\Gamma}_1 = 0$ was used.

After using this expression for the difference of the phase averages for the field "(f-in)" (see $(4.31)_6$), one obtains

$$W^{(\text{af-in})} = 2\left(-\frac{1}{2\rho}\right) \cdot \left[z_1 z_2 \Delta\boldsymbol{\Gamma} \left(-\left\langle \breve{\boldsymbol{\sigma}}_{(1)}^{(\text{f-in})} \right\rangle_{V_1} + \left\langle \left(\breve{\boldsymbol{\sigma}}_{(2)}^{(\text{f-in})} \right) \right\rangle_{V_2} \right) \right] = z_1 z_2 \boldsymbol{\Sigma}^{(\text{f-in})} (\Delta\boldsymbol{\Gamma})/\rho. \tag{4.37}$$

The energy $W^{(\text{af-in})}$ disappears when any of the fields $\Delta\boldsymbol{\Gamma}\left(W^{(\text{a-in})} = 0\right)$ or $\breve{\boldsymbol{\Gamma}}_{(\alpha)}^{\text{f}}\left(W^{(\text{f-in})} = 0\right)$ is a compatible field.

The always nonnegative energy $W^{(\text{f-in})}$ describes the elastic energy generated as a result of geometrical incompatibility of the field of eigenstrains fluctuations.

$$W^{(\text{f-in})} \equiv -\frac{1}{2\rho}\left[z_1 \left\langle \breve{\boldsymbol{\sigma}}^{(\text{f-in})} \cdot \breve{\boldsymbol{\Gamma}}_{(1)}^{\text{f}} \right\rangle_{V_1} + z_2 \left\langle \breve{\boldsymbol{\sigma}}^{(\text{f-in})} \cdot \breve{\boldsymbol{\Gamma}}_{(2)}^{\text{f}} \right\rangle_{V_2} \right] \geq 0 \tag{4.38}$$

Further decomposition of the energy $W^{(\text{f-in})}$ can be done upon remembering that situation studied up to now concerns certain instantaneous "frozen" states appearing during real phase transformation. The progress of phase transformation is described by the value of product phase volume fraction z_2. It is convenient to introduce the concept of "*stored energy of coherency*" ϕ_2^{st}; that is, the energy that would be stored in the SMA macroelement (RVE) after complete phase transformation from pure parent phase into pure product phase $z_2 \to 1$.

$$\phi_2^{st} \equiv \lim_{V_2 \to V} W^{(in)} = \lim_{V_2 \to V} W^{(f\text{-}in)} \equiv -\frac{1}{2\rho} \lim_{V_2 \to V} \left\langle \breve{\sigma}_{(2)}^{(f\text{-}in)} \cdot \breve{\Gamma}_{(2)}^{f} \right\rangle_{V_2} \tag{4.39}$$

This energy naturally does not depend on z_α but it is a function of the final field of eigenstrains fluctuations.

In the case of thermoelastic martensitic transformation, the experimental results indicate that parent phase practically behaves in a purely elastic way. Hence, with good approximation it can be accepted that eigenstrains appearing in this phase during progress of phase transformation can be safely neglected. Usually austenite (parent phase) in an undeformed stress-free state is adopted as a reference configuration. Hence, the field of eigenstrains in the parent phase is assumed to be a zero field $\breve{\boldsymbol{\Gamma}}_{(1)}(\boldsymbol{x}) = \boldsymbol{0} \left(\boldsymbol{\Gamma}_{(1)} = \boldsymbol{0}, \breve{\boldsymbol{\Gamma}}_{(1)}^{f}(\boldsymbol{x}) = \boldsymbol{0} \right)$. A nonzero eigenstrain field $\breve{\boldsymbol{\Gamma}}_{(1)}(\boldsymbol{x}) \neq \boldsymbol{0}$, with a physical origin in the classical plastic slip mechanism or resulting from production of precipitates in the parent phase, can be generated in SMA material submitted to special thermomechanical procedure (training) targeted at inducing two-way memory effect.

The energy $W^{(f\text{-}in)}$ can be expressed at any instant of the phase transformation progress with the following formula:

$$W^{(f\text{-}in)} \equiv -\frac{1}{2\rho} z_2 \left\langle \breve{\boldsymbol{\sigma}}^{(f\text{-}in)} \breve{\boldsymbol{\Gamma}}_{(2)}^{f} \right\rangle_{V_2} = z_2 W^{(f\text{-}in)2} = z_2 \left[\phi_2^{st} + z_1 \overset{*}{W}{}^{(f-in)} \right],$$
$$\overset{*}{W}{}^{(f\text{-}in)}(z_2) \equiv \left(W^{(f\text{-}in)2} - \phi_2^{st} \right) / z_1. \tag{4.40}$$

It is assumed that at $z_2 \to 1$ $(z_1 \to 0)$ the term $\overset{*}{W}{}^{(f\text{-}in)}(z_2) \le C = \text{const}$ is limited, so that then $z_1 \overset{*}{W}{}^{(f\text{-}in)} \to 0$ and $W^{(f\text{-}in)} \to \phi_2^{st}$. The term $\overset{*}{W}{}^{(f\text{-}in)}(z_2)$ does not have to be nonnegative, and it also does not have to be a monotonous function of z_2.

Substituting into (4.19) the expressions (4.20) for $W^{(ex)}(\boldsymbol{\varepsilon}^E)$, (4.33) for $W^{(in)}$, and expressions (4.34), (4.37), and (4.40) for $W^{(a\text{-}in)}$, $W^{(af\text{-}in)}$, and $W^{(f\text{-}in)}$, respectively, one obtains the following formula for elastic energy stored in macroelement $\mathcal{B}$ (RVE):

$$W = \frac{1}{2\rho} \boldsymbol{\varepsilon}^E \mathbf{L}^{ef} \boldsymbol{\varepsilon}^E + W^{(in)}$$
$$= \frac{1}{2\rho} (\boldsymbol{\varepsilon} - \boldsymbol{\varepsilon}^{pt}) \mathbf{L}^{ef} (\boldsymbol{\varepsilon} - \boldsymbol{\varepsilon}^{pt}) + z_1 z_2 \phi_{it} + z_2 \phi_2^{st}. \tag{4.41}$$

The symbol ϕ_{it} denotes specific energy defined with the formula

$$\phi_{it} \equiv \frac{1}{2\rho} \Delta \boldsymbol{\Gamma} \cdot \boldsymbol{\mathcal{H}}^{(a\text{-}in)} \Delta \boldsymbol{\Gamma} + \frac{1}{\rho} \boldsymbol{\Sigma}^{(f\text{-}in)} \cdot \Delta \boldsymbol{\Gamma} + \overset{*}{W}{}^{(f\text{-}in)} \tag{4.42}$$

The energy $\phi_{\rm it}$ does not always have to be nonnegative. It can take negative values under the condition that at the particular instant of phase transformation process the value of stored energy $\phi_2^{\rm st}$ is sufficiently large that energy resulting from incompatibility of the eigenstrain field has a value greater than zero $W^{(\rm in)} = z_2\phi_2^{\rm st} + z_1 z_2 \phi_{\rm it} \geq 0$, in accordance with (4.33).

The mathematical form of formula (4.41) defining elastic energy W is formally identical, with accuracy to a constant, with that defining the sum of energies $\phi^{\rm tm}(\boldsymbol{\varepsilon}, T, z, \boldsymbol{\kappa}(h_K)) + \phi^{\rm coh}(z, T)$ (see (3.3)), in isothermal conditions $T = {\rm const}$ ($\boldsymbol{\alpha} = 0$). In formula (3.3), heuristic assumptions were made regarding the form of functions $\phi_{\rm it}$ and $\phi_2^{\rm st}$ of the type $(3.3)_{5,6}$. Here, with assumptions done only on the mesoscopic level, the formula for elastic energy W of the SMA macroelement (4.41) and (4.42) is exact. The perspective of using these formulas directly in a macroscopic model is limited, as the energy W depends on state parameters measurable on a macroscopic level, such as strain, phase strain, volume fraction of martensitic phase, difference between phase averages of eigenstrain field in both phases $\Delta\boldsymbol{\Gamma}$, and temperature—here treated as a fixed parameter—but it also depends on a number of internal parameters h_K characterizing quantitatively the microstructure of SMA macroelement and its components, such as the elastic properties of individual phases, number, and arrangement of martensitic plates, and so forth.

4.3 Postulate of optimal rearrangement of mesostructure

In this section advantage is taken of the concept of "*optimum rearrangement of mesostructure*" proposed in the work Raniecki and Lexcellent (1998) in order to obtain a more useful form of the formula for elastic energy of SMA macroelement, independent from the set of internal parameters h_K $K = 1,N$ characterizing macroelement (RVE) mesostructure. Hence, it is accepted as a valid modeling postulate that SMA macroelement always remains in thermodynamic mesostructural equilibrium. It is assumed that values of internal variables characterizing mesostructure of the RVE, at prescribed loading conditions on the surface of RVE, minimize the elastic energy $W(\boldsymbol{\varepsilon}^{pt}(h_K), z_\alpha, h_K)$ (see (4.20)), in the general case of nonhomogeneous body or (4.41) in the case of a two-phase body. The RVE mesostructure is then optimally rearranged with respect to actual loading. The equilibrium values of parameters h_K for which W attains extremum can be determined from the equations defining necessary conditions of reaching extremum.

$$\partial W / \partial h_K = 0 \tag{4.43}$$

The following physical simplifying assumptions are accepted before studying the consequences of adopting the postulate of optimal rearrangement of microstructure, expressed by formula (4.43).

The tensor field of elastic compliances modules is uniform and constant in the whole volume of the RVE. This assumption means that tensors of elastic compliance modules of both phases are taken to be the same. The densities of both phases also are taken to be the same.

$$\mathbf{L}^{\mathrm{ef}} = \mathbf{L} = \mathrm{const}, \ \rho_1 = \rho_2 = \rho \tag{4.44}$$

The strains connected with phase transformation are taken to be zero in the case of the parent phase (austenite).

$$\left(\breve{\boldsymbol{\Gamma}}(\boldsymbol{x}) = 0 \ \ \boldsymbol{x} \in V_1\right) \Leftrightarrow \left(\boldsymbol{\Gamma}_1 = 0, \breve{\boldsymbol{\Gamma}}_1^{\mathrm{f}}(\boldsymbol{x}) = 0\right) \tag{4.45}$$

Then, using the formula $(4.14)_2$, it is

$$\begin{aligned} \boldsymbol{\varepsilon}^{\mathrm{pt}}(z_\alpha, h_K) &= \underbrace{\mathbf{M}\left\langle \breve{\boldsymbol{\sigma}}^{(\mathrm{in})} \right\rangle_V}_{=\mathbf{0}} + \underbrace{\boldsymbol{\Gamma}_1}_{=\mathbf{0}} + z_2 \Delta \boldsymbol{\Gamma} = z_2 \boldsymbol{\kappa}(h_K), \\ \boldsymbol{\kappa}(h_K) &= \Delta \boldsymbol{\Gamma} = \boldsymbol{\Gamma}_2 - \underbrace{\boldsymbol{\Gamma}_1}_{=\mathbf{0}} = \frac{1}{V_2} \oint_{V_2} \breve{\boldsymbol{\Gamma}}(\boldsymbol{x}) \mathrm{d}v, \end{aligned} \tag{4.46}$$

where in the last equation $(4.46)_1$, denotation (3.1) has been used. Taking advantage of the relations (4.44), (4.45), (4.46) in (4.19) or $(4.41)_1$, one obtains

$$\begin{aligned} W &= \frac{1}{2\rho}(\boldsymbol{\varepsilon} - z_2 \boldsymbol{\kappa}(h_K)) \cdot \mathbf{L}(\boldsymbol{\varepsilon} - z_2 \boldsymbol{\kappa}(h_K)) + W^{(\mathrm{in})}(z_\alpha, h_K) \\ &= \frac{1}{2\rho} \boldsymbol{\sigma} \cdot \mathbf{M} \boldsymbol{\sigma} + W^{(\mathrm{in})}(z_\alpha, h_K) \end{aligned} \tag{4.47}$$

where the energy of incompatibilities $W^{(\mathrm{in})}$ dependence on z_α and h_K results from its definition (4.22). After application of condition, (4.43) to (4.47) it is obtained that

$$-\partial W / \partial h_K = z_2 \boldsymbol{\sigma} \cdot \partial \boldsymbol{\kappa}(h_K) / \partial h_K - \rho \partial W^{(\mathrm{in})}(z_\alpha, h_K) / \partial h_K = 0 \tag{4.48}$$

The equations (4.48) constitute a set of algebraic equations for unknown equilibrium values of internal parameters h_K^{eq}. As can be easily verified with the use of (4.48), the equilibrium values h_K^{eq} must depend on z_α and $\boldsymbol{\sigma}$. Denoting equilibrium values of the functions $\boldsymbol{\kappa}$ and $W^{(\mathrm{in})}$ by index "eq" one can write

$$\begin{aligned} \boldsymbol{\kappa}^{\mathrm{eq}} &= \boldsymbol{\kappa}^{\mathrm{eq}}(z_\alpha, \boldsymbol{\sigma}) = \boldsymbol{\kappa}(h_K^{\mathrm{eq}}(z_\alpha, \boldsymbol{\sigma})), \\ W^{\mathrm{eq}} &= W^{\mathrm{eq}}(z_\alpha, \boldsymbol{\sigma}) = W^{(\mathrm{in})}(z_\alpha, h_K^{\mathrm{eq}}(z_\alpha, \boldsymbol{\sigma})). \end{aligned} \tag{4.49}$$

Let us multiply N equations (4.48) on both sides by $\partial h_K^{\mathrm{eq}}(z_\alpha, \boldsymbol{\sigma}) / \partial \boldsymbol{\sigma}$, respectively, and sum them up to obtain the following condition of thermodynamic microstructural equilibrium:

$$\begin{aligned} &\left\{ \sum_i \left[z_2 \boldsymbol{\sigma} \cdot \partial \boldsymbol{\kappa}(h_K) / \partial h_K - \rho \partial W^{(\mathrm{in})}(z_\alpha, h_K) / \partial h_K \right] \partial h_K^{\mathrm{eq}}(z_\alpha, \boldsymbol{\sigma}) / \partial \boldsymbol{\sigma} = 0 \right\} \Leftrightarrow \\ &\{ z_2 \boldsymbol{\sigma} \cdot \partial \boldsymbol{\kappa}^{\mathrm{eq}}(z_\alpha, \boldsymbol{\sigma}) / \partial \boldsymbol{\sigma} - \rho \partial W^{\mathrm{eq}}(z_\alpha, \boldsymbol{\sigma}) / \partial \boldsymbol{\sigma} = 0 \} \Leftrightarrow \\ &\{ \partial [z_2 \boldsymbol{\sigma} \cdot \partial \boldsymbol{\kappa}^{\mathrm{eq}}(z_\alpha, \boldsymbol{\sigma}) - \rho W^{\mathrm{eq}}(z_\alpha, \boldsymbol{\sigma})] / \partial \boldsymbol{\sigma} = z_2 \boldsymbol{\kappa}^{\mathrm{eq}} = (\boldsymbol{\varepsilon}^{\mathrm{pt}})^{\mathrm{eq}} \}. \end{aligned} \tag{4.50}$$

The condition $(4.50)_3$ indicates that equilibrium macroscopic phase strain $(\boldsymbol{\varepsilon}^{\mathrm{pt}})^{\mathrm{eq}} = z_2\boldsymbol{\kappa}^{\mathrm{eq}}$ and the tensor of macroscopic phase strains $\boldsymbol{\kappa}^{\mathrm{eq}}$ admits the existence of a potential. Hence, it is hereby postulated that there exists the potential generating $\boldsymbol{\kappa}^{\mathrm{eq}}$

$$\boldsymbol{\kappa}^{\mathrm{eq}}(\boldsymbol{\sigma}) \equiv \rho \frac{\partial g^{\mathrm{ult}}(z_\alpha, \boldsymbol{\sigma})}{\partial \boldsymbol{\sigma}} \tag{4.51}$$

This potential will be called the *ultimate phase eigenstrains potential*. The specific, explicit form of this potential adopted in the $\mathrm{R_L}$ models family takes the form $\rho g^{\mathrm{ult}}(\boldsymbol{\sigma}, T) = \boldsymbol{\kappa}^{\mathrm{eq}} \cdot \boldsymbol{\sigma} = \eta f(y)\sigma_{\mathrm{ef}}$ (see (3.32)).

Combining $(4.50)_3$ and (4.51), the following differential equation is obtained:

$$\frac{\partial}{\partial \boldsymbol{\sigma}}\left[z_2\left(\boldsymbol{\sigma}\frac{\partial g^{\mathrm{ult}}(z_\alpha, \boldsymbol{\sigma})}{\partial \boldsymbol{\sigma}} - g^{\mathrm{ult}}(z_\alpha, \boldsymbol{\sigma})\right) - W^{\mathrm{eq}}(z_\alpha, \boldsymbol{\sigma})\right] = \mathbf{0} \tag{4.52}$$

Integrating the above equation leads to the relation between $W^{\mathrm{eq}}(z_\alpha, \boldsymbol{\sigma})$ and $g^{\mathrm{ult}}(z_\alpha, \boldsymbol{\sigma})$

$$W^{\mathrm{eq}}(z_\alpha, \boldsymbol{\sigma}) = W^{\mathrm{eq}}(z_\alpha, \mathbf{0}) - z_2\left(g^{\mathrm{ult}}(z_\alpha, \boldsymbol{\sigma}) - \boldsymbol{\sigma}\cdot\boldsymbol{\kappa}^{\mathrm{eq}}/\rho\right) \tag{4.53}$$

As is indicated by formula (4.53), when macroscopic potential $g^{\mathrm{ult}}(z_\alpha, \boldsymbol{\sigma})$ is the uniform function of order one with respect to the stress tensor, then the energy of incompatibilities $W^{(\mathrm{in})}$ corresponding to optimum rearrangement of microstructure does not depend on macroscopic stress. Substituting (4.53) to (4.47), one obtains

$$\begin{aligned} W(z_\alpha, \boldsymbol{\sigma}) &= \frac{1}{2\rho}\boldsymbol{\sigma}\cdot\mathbf{M}\boldsymbol{\sigma} + W^{\mathrm{eq}}(z_\alpha, \mathbf{0}) - z_2\left(g^{\mathrm{ult}} - \boldsymbol{\sigma}\cdot\boldsymbol{\kappa}^{eq}/\rho\right), \\ W &= \frac{1}{2\rho}(\boldsymbol{\varepsilon} - z_2\boldsymbol{\kappa}^{\mathrm{eq}})\cdot\mathbf{L}(\boldsymbol{\varepsilon} - z_2\boldsymbol{\kappa}^{\mathrm{eq}}) + W^{\mathrm{eq}}(z_\alpha, \mathbf{0}) - z_2\left(g^{\mathrm{ult}}(\boldsymbol{\sigma}) - \boldsymbol{\sigma}\cdot\boldsymbol{\kappa}^{\mathrm{eq}}/\rho\right), \end{aligned} \tag{4.54}$$

and further, when it is accepted that $\rho g^{\mathrm{ult}}(\boldsymbol{\sigma}) = \boldsymbol{\sigma}\cdot\boldsymbol{\kappa}^{\mathrm{eq}}(\boldsymbol{\sigma})$, it is

$$W = \frac{1}{2\rho}(\boldsymbol{\varepsilon} - z\boldsymbol{\kappa}^{\mathrm{eq}})\cdot\mathbf{L}(\boldsymbol{\varepsilon} - z\boldsymbol{\kappa}^{\mathrm{eq}}) + W^{\mathrm{eq}}(z, \mathbf{0}) = \frac{1}{2\rho}\boldsymbol{\sigma}\cdot\mathbf{M}\boldsymbol{\sigma} + W^{\mathrm{eq}}(z, \mathbf{0}) \tag{4.55}$$

As $W^{\mathrm{eq}}(z, \mathbf{0})$ does not depend on stress, its general form can be reasonably postulated as follows:

$$W^{\mathrm{eq}}(z, \mathbf{0}) = z_1\phi_1^{\mathrm{st}} + z_1 z_2\phi_{12} + z_2\phi_2^{\mathrm{st}}, \tag{4.56}$$

where $\phi_1^{\mathrm{st}}, \phi_{\mathrm{it}}, \phi_2^{\mathrm{st}}$ are treated as material constants. The term $z_1\phi_1^{\mathrm{st}}$ represents the energy of residual mesostresses generated in the parent phase as a result of the existence of incompatible eigenstrains in this phase. In view of the assumption that pure parent phase ($z_1 = 1$) does not have eigenstrains (see (4.45)), no internal mesostresses in pure parent phase are generated ($\breve{\boldsymbol{\sigma}}^{(\mathrm{in})} = \mathbf{0}$) and energy $W^{(\mathrm{eq})}$ must then be equal to zero.

This immediately leads to the conclusion that ϕ_1^{st} must be equal to zero $\left(\phi_1^{st} \equiv \Delta u_1^{st} - T\Delta s_1^{st} = 0\right)$. Special thermomechanical processing, so-called training executed in order to invoke the appearance of two-way shape memory effect, induces the appearance of nonzero energy term $z_1\phi_1^{st}$.

The term $z_2\phi_2^{st}$ represents energy stored as a result of incompatibility of the eigenstrains field after completion of martensitic phase transformation (see (4.39)), while the term $z_1z_2\phi_{it}$ represents the energy of internal interaction between the phases. Upon substitution of (4.56) to (4.55) and taking into account $z_1\phi_1^{st} = 0$, one finally obtains

$$W = \frac{1}{2\rho}(\boldsymbol{\varepsilon} - z_2\boldsymbol{\kappa}^{eq})\cdot\mathbf{L}(\boldsymbol{\varepsilon} - z_2\boldsymbol{\kappa}^{eq}) + z_1z_2\phi_{12} + z_2\phi_2^{st} \tag{4.57}$$

The above formula for elastic energy W takes the same form, with accuracy to a constant value, as the one proposed in the R_L model for Helmholtz free energy function $\phi(\varepsilon, T, z)$ at some fixed constant temperature.

4.4 Summary

The mesomechanical studies performed revealed what terms should appear in the postulated heuristically, phenomenological, macroscopic form of free energy function valid for shape memory alloy materials, in order to enable reliable modeling predictions of their behavior when submitted to various thermomechanical loadings. The direct advantage of executed analysis is that the terms present in the formula defining macroelement elastic energy W (see (4.41)) now have a clear mesomechanical physical interpretation. This allows for an assessment of how specific properties and features of martensitic mesostructures influence the magnitude of specific shape memory effects such as size of hysteresis loop, pseudoelastic flow amplitude, or heat effects. The three principal elements present in W are: energy connected with macroscopic elastic strain Eps^E, which is directly proportional to effective elastic stiffness tensor $\mathbf{L}^{ef}$, interaction energy ϕ_{it}, and stored energy ϕ_2^{st}. An observation of key importance is that energy stored in pure martensitic mesostructure after completion of martensitic phase transition ϕ_2^{st} depends only on a local phase fluctuation field $\breve{\boldsymbol{\Gamma}}^{f}(\boldsymbol{x})$ but not on the average value of phase eigenstrain (see also Figure 4.1). A broad program of combined materials science and continuum mechanics, experimental and theoretical interdisciplinary studies is required to properly understand and exploit the opportunities indicated in this chapter. A step in this direction is the work of Raniecki and Ziółkowski (2015), where it has been shown that the coherence energy ϕ^{coh} and macroscopic phase strain $\boldsymbol{\varepsilon}^{pt}$ of TiNi isotropic alloys are not much influenced by the elastic heterogeneity of the isotropic phases. The actual values of individual energy terms appearing in formula (4.41) can also be estimated by prescribing particular martensitic mesostructures, and solving relevant boundary value problems of elasticity (4.2)–(4.4). Unfortunately, due to the complexity of real martensitic mesostructures, finding solutions to such problems in an explicit form is possible only in a few cases, which forces the researcher to use numerical methods for that purpose; for example, finite element code.

Annex 4.1 Principle of reciprocity in linear theory of elasticity at presence of eigenstrains

Classical reciprocity theorem formulation without eigenstrains can be found, for example, in Chapter 6 of Sadd (2005), together with other useful principles. In order to formulate a *reciprocity theorem with eigenstrains*, let us investigate mutual relations that exist between two mechanical states, denoted with subscripts "I" and "II," that are generated in linear elastic body $\mathcal{B}$ by two different sets of all-round mechanical traction loadings applied at the surface of the body $\breve{\boldsymbol{t}}^{(\mathrm{I})}(\boldsymbol{x})$, $\breve{\boldsymbol{t}}^{(\mathrm{II})}(\boldsymbol{x})$; $\boldsymbol{x} \in \partial V$, and known, symmetric, eigenstrain fields $\breve{\boldsymbol{\Gamma}}^{(\mathrm{I})}(\boldsymbol{x}), \breve{\boldsymbol{\Gamma}}^{(\mathrm{II})}(\boldsymbol{x})$; $\boldsymbol{x} \in V$ prescribed in the volume of body $\mathcal{B}$. In particular, body $\mathcal{B}$ can be RVE. The fields $\breve{\boldsymbol{\Gamma}}^{(\alpha)}(\boldsymbol{x})$ need not satisfy compatibility conditions and also can be discontinuous on a number of internal interfaces $S^{(\mathrm{in})}$ present in the volume of the body $\mathcal{B}$. The elastic properties of body $\mathcal{B}$ are described with a nonhomogeneous field of symmetric and positively determined tensor of elastic compliances $\breve{\mathbf{M}}(\boldsymbol{x})$ $\left(\breve{\mathbf{M}}(\boldsymbol{x}) = \breve{\mathbf{L}}^{-1}(\boldsymbol{x})\right)$.

Denoted by $\left\{\breve{\boldsymbol{\sigma}}^{(\mathrm{I})}(\boldsymbol{x}), \breve{\boldsymbol{\varepsilon}}^{(\mathrm{I})}(\boldsymbol{x}), \breve{\boldsymbol{u}}^{(\mathrm{I})}(\boldsymbol{x})\right\}$ and $\left\{\breve{\boldsymbol{\sigma}}^{(\mathrm{II})}(\boldsymbol{x}), \breve{\boldsymbol{\varepsilon}}^{(\mathrm{II})}(\boldsymbol{x}), \breve{\boldsymbol{u}}^{(\mathrm{II})}(\boldsymbol{x})\right\}$, solution fields of stresses, strains, and displacements constitute a solution of the mechanical equilibrium problem of linear elasticity for two sets of prescribed boundary and eigenstrain conditions. These fields must satisfy mechanical equilibrium equations and respective constitutive relations of linear elasticity in all points of the volume and on the surface of body $\mathcal{B}$.

$$\begin{gathered}\breve{\boldsymbol{\sigma}}^{(\mathrm{I})}(\boldsymbol{x})_{ij,j} = 0, \qquad \breve{\boldsymbol{\sigma}}^{(\mathrm{II})}(\boldsymbol{x})_{ij,j} = 0 \\ \breve{\boldsymbol{\varepsilon}}^{(\mathrm{I})}(\boldsymbol{x}) = \breve{\mathbf{M}}(\boldsymbol{x})\breve{\boldsymbol{\sigma}}^{(\mathrm{I})}(\boldsymbol{x}) + \breve{\boldsymbol{\Gamma}}^{(\mathrm{I})}(\boldsymbol{x}), \quad \breve{\boldsymbol{\varepsilon}}^{(\mathrm{II})}(\boldsymbol{x}) = \breve{\mathbf{M}}(\boldsymbol{x})\breve{\boldsymbol{\sigma}}^{(\mathrm{II})}(\boldsymbol{x}) + \breve{\boldsymbol{\Gamma}}^{(\mathrm{II})}(\boldsymbol{x})\end{gathered} \tag{A4.1}$$

The stresses fields also must fulfill boundary conditions at $\boldsymbol{x} \in \partial V$.

$$\breve{\boldsymbol{t}}^{(\mathrm{I})}(\boldsymbol{x}) = \boldsymbol{n}(\boldsymbol{x}) \cdot \breve{\boldsymbol{\sigma}}^{(\mathrm{I})}(\boldsymbol{x}), \quad \breve{\boldsymbol{t}}^{(\mathrm{II})}(\boldsymbol{x}) = \boldsymbol{n}(\boldsymbol{x}) \cdot \breve{\boldsymbol{\sigma}}^{(\mathrm{II})}(\boldsymbol{x}); \quad \boldsymbol{x} \in \partial V \tag{A4.2}$$

The stress and strain fields may be discontinuous on internal interfaces $S^{(\mathrm{in})}$. However, the surface tractions and displacements must be continuous there in order that static equilibrium and strain compatibility conditions can be satisfied in all points of the interior of body $\mathcal{B}$.

Multiplying $(4.1)_2$ by $\breve{\boldsymbol{\sigma}}^{(\mathrm{II})}(\boldsymbol{x})$ and $(4.1)_3$ by $\breve{\boldsymbol{\sigma}}^{(\mathrm{I})}(\boldsymbol{x})$ and taking advantage of the diagonal symmetry of tensor $\breve{\mathbf{M}}(\boldsymbol{x})$ giving $\breve{\boldsymbol{\sigma}}^{(\mathrm{I})}\breve{\mathbf{M}}(\boldsymbol{x})\breve{\boldsymbol{\sigma}}^{(\mathrm{II})} = \breve{\boldsymbol{\sigma}}^{(\mathrm{II})}\breve{\mathbf{M}}(\boldsymbol{x})\breve{\boldsymbol{\sigma}}^{(\mathrm{I})}$, the following can be obtained:

$$\breve{\boldsymbol{\sigma}}^{(\mathrm{I})}(\boldsymbol{x}) \cdot \left(\breve{\boldsymbol{\varepsilon}}^{(\mathrm{II})}(\boldsymbol{x}) - \breve{\boldsymbol{\Gamma}}^{(\mathrm{II})}(\boldsymbol{x})\right) = \breve{\boldsymbol{\sigma}}^{(\mathrm{II})}(\boldsymbol{x}) \cdot \left(\breve{\boldsymbol{\varepsilon}}^{(\mathrm{I})}(\boldsymbol{x}) - \breve{\boldsymbol{\Gamma}}^{(\mathrm{I})}(\boldsymbol{x})\right) \tag{A4.3}$$

After integrating this relation over the whole volume of the body $\mathcal{B}$ and using the Green-Gauss theorem and static admissibility of fields $\breve{\boldsymbol{\sigma}}^{(\mathrm{I})}, \breve{\boldsymbol{\sigma}}^{(\mathrm{II})}$, the following reciprocal theorem relation is obtained:

$$\oint_{\partial V}\left(\breve{\boldsymbol{t}}^{(\mathrm{I})}\cdot\breve{\boldsymbol{u}}^{(\mathrm{II})}\right)\mathrm{d}s-\oint_{V}\left(\breve{\boldsymbol{\sigma}}^{(\mathrm{I})}\cdot\breve{\boldsymbol{\Gamma}}^{(\mathrm{II})}\right)\mathrm{d}V=\oint_{\partial V}\left(\breve{\boldsymbol{t}}^{(\mathrm{II})}\cdot\breve{\boldsymbol{u}}^{(\mathrm{I})}\right)\mathrm{d}s-\oint_{V}\left(\breve{\boldsymbol{\sigma}}^{(\mathrm{II})}\cdot\breve{\boldsymbol{\Gamma}}^{(\mathrm{I})}\right)\mathrm{d}V \tag{A4.4}$$

In particular,

$$\begin{aligned}
&\oint_{\partial V}\left(\breve{\boldsymbol{t}}^{(\mathrm{I})}\cdot\breve{\boldsymbol{u}}^{(\mathrm{II})}\right)\mathrm{d}s=\oint_{\partial V}\left(\breve{\boldsymbol{t}}^{(\mathrm{II})}\cdot\breve{\boldsymbol{u}}^{(\mathrm{I})}\right)\mathrm{d}s, \text{ when } \breve{\boldsymbol{\Gamma}}^{(\mathrm{II})}(\boldsymbol{x})=\breve{\boldsymbol{\Gamma}}^{(\mathrm{I})}(\boldsymbol{x})=\mathbf{0}\\
&\oint_{V}\left(\breve{\boldsymbol{\sigma}}^{(\mathrm{I})}\cdot\breve{\boldsymbol{\Gamma}}^{(\mathrm{II})}\right)\mathrm{d}v=\oint_{V}\left(\breve{\boldsymbol{\sigma}}^{(\mathrm{II})}\cdot\breve{\boldsymbol{\Gamma}}^{(\mathrm{I})}\right)\mathrm{d}v, \text{ when } \breve{\boldsymbol{t}}^{(\mathrm{I})}=\breve{\boldsymbol{t}}^{(\mathrm{II})}=\mathbf{0}
\end{aligned} \tag{A4.5}$$

Upon writing constitutive relations $(\mathrm{A}4.1)_{2,3}$ in equivalent form,

$$\begin{aligned}
&\breve{\boldsymbol{\sigma}}^{(\mathrm{I})}(\boldsymbol{x})=\breve{\mathbf{L}}(\boldsymbol{x})\breve{\boldsymbol{\varepsilon}}^{(\mathrm{I})}(\boldsymbol{x})-\breve{\mathbf{L}}(\boldsymbol{x})\breve{\boldsymbol{\Gamma}}^{(\mathrm{I})}(\boldsymbol{x}),\\
&\breve{\boldsymbol{\sigma}}^{(\mathrm{II})}(\boldsymbol{x})=\breve{\mathbf{L}}(\boldsymbol{x})\breve{\boldsymbol{\varepsilon}}^{(\mathrm{II})}(\boldsymbol{x})-\breve{\mathbf{L}}(\boldsymbol{x})\breve{\boldsymbol{\Gamma}}^{(\mathrm{II})}(\boldsymbol{x})
\end{aligned} \tag{A4.6}$$

and taking advantage of symmetry of stiffness tensor $\breve{\mathbf{L}}(\boldsymbol{x})$ to get $\breve{\boldsymbol{\varepsilon}}^{(\mathrm{I})}\breve{\mathbf{L}}(\boldsymbol{x})\breve{\boldsymbol{\varepsilon}}^{(\mathrm{II})}=\breve{\boldsymbol{\varepsilon}}^{(\mathrm{II})}\breve{\mathbf{L}}(\boldsymbol{x})\breve{\boldsymbol{\varepsilon}}^{(\mathrm{I})}$, it can be obtained that

$$\breve{\boldsymbol{\varepsilon}}^{(\mathrm{I})}\left[\breve{\boldsymbol{\sigma}}^{(\mathrm{II})}(\boldsymbol{x})+\breve{\mathbf{L}}(\boldsymbol{x})\breve{\boldsymbol{\Gamma}}^{(\mathrm{II})}(\boldsymbol{x})\right]=\breve{\boldsymbol{\varepsilon}}^{(\mathrm{II})}\left[\breve{\boldsymbol{\sigma}}^{(\mathrm{I})}(\boldsymbol{x})+\breve{\mathbf{L}}(\boldsymbol{x})\breve{\boldsymbol{\Gamma}}^{(\mathrm{I})}(\boldsymbol{x})\right] \tag{A4.7}$$

The dual reciprocal theorem relation involving strains is obtained upon repeating analogous operations to the ones leading to relation (A4.4).

$$\oint_{\partial V}\left(\breve{\boldsymbol{t}}^{(\mathrm{I})}\cdot\breve{\boldsymbol{u}}^{(\mathrm{II})}\right)\mathrm{d}s+\oint_{V}\left(\breve{\boldsymbol{\varepsilon}}^{(\mathrm{II})}\cdot\breve{\mathbf{L}}\breve{\boldsymbol{\Gamma}}^{(\mathrm{I})}\right)\mathrm{d}v=\oint_{\partial V}\left(\breve{\boldsymbol{t}}^{(\mathrm{II})}\cdot\breve{\boldsymbol{u}}^{(\mathrm{I})}\right)\mathrm{d}s+\oint_{V}\left(\breve{\boldsymbol{\varepsilon}}^{(\mathrm{I})}\cdot\breve{\mathbf{L}}\breve{\boldsymbol{\Gamma}}^{(\mathrm{II})}\right)\mathrm{d}v. \tag{A4.8}$$

Below, for the text self-sufficiency, Green-Gauss theorem is recalled expressed in several forms

$$\begin{aligned}
&\oint_{V}\partial A(\boldsymbol{x})/\partial x_i \mathrm{d}V=\oint_{\partial V}An_i\mathrm{d}s+\oint_{S^{(in)}}[\![A(\boldsymbol{x})]\!]\nu_i\mathrm{d}s,\ [\![A(\boldsymbol{x})]\!]\equiv A_2(\boldsymbol{x})-A_1(\boldsymbol{x})\\
&\oint_{V}\nabla\cdot\breve{\boldsymbol{A}}(\boldsymbol{x})\mathrm{d}v=\oint_{\partial V}\breve{\boldsymbol{A}}(\boldsymbol{x})\cdot\breve{\boldsymbol{n}}(\boldsymbol{x})\mathrm{d}s+\oint_{S^{(\mathrm{in})}}[\![\breve{\boldsymbol{\sigma}}]\!]\left[\!\!\left[\breve{\boldsymbol{A}}(\boldsymbol{x})\right]\!\!\right]\cdot\boldsymbol{\nu}(\boldsymbol{x})\mathrm{d}s,\ \left[\!\!\left[\breve{\boldsymbol{A}}(\boldsymbol{x})\right]\!\!\right]\equiv\breve{\boldsymbol{A}}_2(\boldsymbol{x})-\breve{\boldsymbol{A}}_1(\boldsymbol{x}),\\
&\oint_{V}\nabla\otimes\breve{\boldsymbol{A}}(\boldsymbol{x})\mathrm{d}v=\oint_{\partial V}\breve{\boldsymbol{A}}(\boldsymbol{x})\otimes\breve{\boldsymbol{n}}(\boldsymbol{x})\mathrm{d}s+\oint_{S^{(\mathrm{in})}}\left[\!\!\left[\breve{\boldsymbol{A}}(\boldsymbol{x})\right]\!\!\right]\otimes\boldsymbol{\nu}(\boldsymbol{x})\mathrm{d}s
\end{aligned} \tag{A4.9}$$

In the case of coherent displacement field $[\![\boldsymbol{u}(\boldsymbol{x})]\!]=0$, and for statically admissible stress field $[\![\breve{\boldsymbol{\sigma}}(\boldsymbol{x})]\!]\cdot\boldsymbol{\nu}(\boldsymbol{x})=0$ for each $\boldsymbol{x}\in V$.

The following useful identities can be obtained upon use of the Green-Gauss theorem (A4.9$_1$) substituting $(A=x_i)$ and $(A=1)$.

$$\mathbf{1}V=\oint_{\partial V}\boldsymbol{x}\otimes\breve{\boldsymbol{n}}\,\mathrm{d}s=\oint_{\partial V}\breve{\boldsymbol{n}}\otimes\boldsymbol{x}\,\mathrm{d}s\Leftrightarrow\oint_{\partial V}\boldsymbol{x}\otimes\breve{\boldsymbol{n}}-\breve{\boldsymbol{n}}\otimes\boldsymbol{x}\,\mathrm{d}s=\mathbf{0},$$

$$\oint_{\partial V}\breve{\boldsymbol{n}}(\boldsymbol{x})\mathrm{d}s=\oint_{\partial V}\mathrm{d}\boldsymbol{s}=0,\quad \mathrm{d}\boldsymbol{s}=\breve{\boldsymbol{n}}(\boldsymbol{x})\mathrm{d}s \tag{A4.10}$$

Introduction of the following definitions of local fields averages proves useful. The average of a tensorial field $\breve{\boldsymbol{A}}(\boldsymbol{x})$ over the total volume V of a body $\mathcal{B}$ is denoted by basic symbol $\boldsymbol{A}$—with no "arc smiley" embellishment and no symbolic coordinate "$(\boldsymbol{x})$"—or with averaging operation symbol $\left\langle\breve{\boldsymbol{A}}(\boldsymbol{x})\right\rangle_V$ $\left(\boldsymbol{A}=\left\langle\breve{\boldsymbol{A}}(\boldsymbol{x})\right\rangle_V\right)$. The fluctuation field of field $\breve{\boldsymbol{A}}(\boldsymbol{x})$ with respect to its average value $\boldsymbol{A}$ is denoted $\breve{\boldsymbol{A}}^{\mathrm{f}}(\boldsymbol{x})$. The following formulas are valid in this respect.

$$\boldsymbol{A}=\left\langle\breve{\boldsymbol{A}}(\boldsymbol{x})\right\rangle_V\equiv\frac{1}{V}\oint_{\mathrm{V}}\breve{\boldsymbol{A}}(\boldsymbol{x})\mathrm{d}v,\ \boldsymbol{x}\in V$$

$$\breve{\boldsymbol{A}}^{\mathrm{f}}(\boldsymbol{x})\equiv\breve{\boldsymbol{A}}(\boldsymbol{x})-\boldsymbol{A},\ \left\langle\breve{\boldsymbol{A}}^{\mathrm{f}}(\boldsymbol{x})\right\rangle_V=0 \tag{A4.11}$$

In the case of a multiphase body, it is purposeful to introduce the concept of phase averages. The phase average of a tensorial field $\breve{\boldsymbol{A}}(\boldsymbol{x})$ over the volume V_α of a body $\mathcal{B}$ occupied by a phase α is denoted by basic symbol $\boldsymbol{A}_{(\alpha)}$ with subscript α in commas—with no "arc smiley" embellishment—or with averaging operation symbol $\left\langle\breve{\boldsymbol{A}}(\boldsymbol{x})\right\rangle_{V_\alpha}$ $\left(\boldsymbol{A}_{(\alpha)}=\left\langle\breve{\boldsymbol{A}}(\boldsymbol{x})\right\rangle_{V_\alpha}\right)$.

The fluctuation field of field $\breve{\boldsymbol{A}}(\boldsymbol{x})$ with respect to its phase average value $\boldsymbol{A}_{(\alpha)}$ is denoted $\breve{\boldsymbol{A}}^{\mathrm{f}}_{(\alpha)}(\boldsymbol{x})$. The following formulas are valid in this respect:

$$\boldsymbol{A}_{(\alpha)}=\left\langle\breve{\boldsymbol{A}}(x)\right\rangle_{V_\alpha}\equiv\frac{1}{V_\alpha}\oint_{V_\alpha}\breve{\boldsymbol{A}}(\boldsymbol{x})\mathrm{d}v,\ (\alpha=1,2,\ldots,n)$$

$$\breve{\boldsymbol{A}}^{\mathrm{f}}_{(\alpha)}(\boldsymbol{x})\equiv\breve{\boldsymbol{A}}(\boldsymbol{x})-\left\langle\breve{\boldsymbol{A}}(\boldsymbol{x})\right\rangle_{V_\alpha}=\breve{\boldsymbol{A}}(\boldsymbol{x})-\boldsymbol{A}_{(\alpha)}\ \boldsymbol{x}\in V_\alpha,\ \left\langle\breve{\boldsymbol{A}}^{\mathrm{f}}_{(\alpha)}(\boldsymbol{x})\right\rangle_{V_\alpha}=0 \tag{A4.12}$$

For a two-phase body $(n=2)$ there exist simple relations between average and fluctuations over the total volume of the body $\mathcal{B}$ and phase averages and phase fluctuations calculated over volumes of the particular phases.

$$\boldsymbol{A}=\left\langle\breve{\boldsymbol{A}}(\boldsymbol{x})\right\rangle_V=z_1\left\langle\breve{\boldsymbol{A}}(\boldsymbol{x})\right\rangle_{V_1}+z_2\left\langle\breve{\boldsymbol{A}}(\boldsymbol{x})\right\rangle_{V_2}=z_1\boldsymbol{A}_{(1)}+z_2\boldsymbol{A}_{(2)},$$

$$\breve{\boldsymbol{A}}^{\mathrm{f}}(\boldsymbol{x})=\begin{cases}\breve{\boldsymbol{A}}^{\mathrm{f}}_{(1)}(\boldsymbol{x})-z_2\Delta\boldsymbol{A} & \boldsymbol{x}\in V_1\\ \breve{\boldsymbol{A}}^{\mathrm{f}}_{(2)}(\boldsymbol{x})+z_1\Delta\boldsymbol{A} & \boldsymbol{x}\in V_2\end{cases},\ \Delta\boldsymbol{A}\equiv\boldsymbol{A}_{(2)}-\boldsymbol{A}_{(1)}$$

$$\left(\boldsymbol{A}_{(1)}=\boldsymbol{A}-z_2\Delta\boldsymbol{A},\ \boldsymbol{A}_{(2)}=\boldsymbol{A}+z_1\Delta\boldsymbol{A}\right) \tag{A4.13}$$

It can be immediately obtained from $(A14.13)_1$ that $\boldsymbol{A}_{(1)} = \boldsymbol{A} - z_2\Delta\boldsymbol{A}, \boldsymbol{A}_{(2)} = \boldsymbol{A} + z_1\Delta\boldsymbol{A}$. Wherefrom when the average over the whole volume of the body $\mathcal{B}$ disappears ($\boldsymbol{A} = \boldsymbol{0}$), it is

$$\boldsymbol{A}_{(1)} = -z_2\Delta\boldsymbol{A},\ \boldsymbol{A}_{(2)} = z_1\Delta\boldsymbol{A} \tag{A4.14}$$

Annex 4.2 The postulate of work compatibility—Hill's postulate, concept of effective properties

In this appendix concepts and assumptions underling transition from mesomechanical description to effective macroscopic description of elastic material behavior are discussed. The equivalence of the descriptions is based on adoption of the *postulate of compatibility of mechanical work*. Before this postulate can be mathematically formulated an introduction of preliminaries is required.

Let's assume that it is a known distribution of surface tractions $\breve{\boldsymbol{t}}^{(\mathrm{b})}(\boldsymbol{x}) \equiv \breve{\boldsymbol{t}}(\boldsymbol{x})|_{x\in\partial V}$, and surface displacements $\breve{\boldsymbol{u}}^{(\mathrm{b})}(\boldsymbol{x}) = \breve{\boldsymbol{u}}(\boldsymbol{x})|_{x\in\partial V}$, constant in time, operating on the outer surface of a body $\mathcal{B}$ treated as material macroelement (RVE).

It is convenient to decompose boundary displacements on the surface of body $\mathcal{B}$ into linear boundary displacements (l.b.d.) $\tilde{\boldsymbol{u}}^{(u)}$, and fluctuant boundary displacements $\breve{\boldsymbol{u}}^{f}$

$$\begin{aligned}
&\breve{\boldsymbol{u}}^{(\mathrm{b})}(\boldsymbol{x}) = \tilde{\boldsymbol{u}}^{(\mathrm{u})}(\boldsymbol{x}) + \breve{\boldsymbol{u}}^{\mathrm{f}}(\boldsymbol{x}),\ \boldsymbol{x} \in \partial V\\
&\tilde{\boldsymbol{u}}^{(\mathrm{u})}(\boldsymbol{x}) = \boldsymbol{h}\boldsymbol{x} = (\boldsymbol{\varepsilon} + \boldsymbol{\omega})\boldsymbol{x},\ \breve{\boldsymbol{u}}^{\mathrm{f}}(\boldsymbol{x}) \equiv \breve{\boldsymbol{u}}^{(\mathrm{b})}(\boldsymbol{x}) - \tilde{\boldsymbol{u}}^{(\mathrm{u})}(\boldsymbol{x}),
\end{aligned} \tag{A4.15}$$

where $\breve{\boldsymbol{h}}(\boldsymbol{x}) = \partial\breve{\boldsymbol{u}}(\boldsymbol{x})/\partial\boldsymbol{x}$ denotes displacements gradient tensor.

Using a definition of small strains $\left(\breve{\boldsymbol{\varepsilon}}(\boldsymbol{x}) = 0.5\left[(\nabla\otimes\breve{\boldsymbol{u}}) + \nabla\otimes\breve{\boldsymbol{u}})^T\right]\right)$, taking advantage of the Green-Gauss theorem $(A4.9)_3$ and continuity of the displacement field $\breve{\boldsymbol{u}}(\boldsymbol{x})$ in the whole body $\mathcal{B}$, it can be shown that the average (macroscopic) total strain $\boldsymbol{\varepsilon}$ is completely determined by the surface values of the displacement field $\breve{\boldsymbol{u}}^{(\mathrm{b})}(\boldsymbol{x})$

$$\begin{aligned}
&\boldsymbol{\varepsilon} \equiv \langle\breve{\boldsymbol{\varepsilon}}(\boldsymbol{x})\rangle_V = \frac{1}{V}\oint_V \breve{\boldsymbol{\varepsilon}}(\boldsymbol{x})\mathrm{d}v = \frac{1}{V}\oint_V 0.5\left[\nabla\otimes\breve{\boldsymbol{u}} + (\nabla\otimes\breve{\boldsymbol{u}})^{\mathrm{T}}\right]\mathrm{d}v\\
&= \frac{1}{V}\oint_{\partial V} 0.5\left[\breve{\boldsymbol{n}}\otimes\breve{\boldsymbol{u}} + (\breve{\boldsymbol{n}}\otimes\breve{\boldsymbol{u}})^{\mathrm{T}}\right]\mathrm{d}s = \frac{1}{2V}\oint_{\partial V}\left[\breve{\boldsymbol{n}}\otimes\breve{\boldsymbol{u}}^{(b)} + \breve{\boldsymbol{u}}^{(b)}\otimes\breve{\boldsymbol{n}}\right]\mathrm{d}s,\\
&\boldsymbol{\omega} \equiv \langle\breve{\boldsymbol{\omega}}\rangle_V = \frac{1}{V}\oint_V \breve{\boldsymbol{\omega}}(\boldsymbol{x})\mathrm{d}v\oint = \frac{1}{V}\oint_{\partial V} 0.5\left[\breve{\boldsymbol{n}}\otimes\breve{\boldsymbol{u}}^{(b)} - \breve{\boldsymbol{u}}^{(b)}\otimes\breve{\boldsymbol{n}}\right]\mathrm{d}s,\\
&\boldsymbol{h} = \frac{1}{V}\oint_{\partial V} \breve{\boldsymbol{u}}\otimes\breve{\boldsymbol{n}}\mathrm{d}s.
\end{aligned} \tag{A4.16}$$

The formulas for average value of spin $\boldsymbol{\omega}$ and displacement gradient $\boldsymbol{h}$ in terms of boundary displacements are obtained in an analogous way to $\boldsymbol{\varepsilon}$, using their defining expressions; that is, $\breve{\boldsymbol{\omega}}(\boldsymbol{x}) = 0.5\left[(\nabla\otimes\breve{\boldsymbol{u}}) - \nabla\otimes\breve{\boldsymbol{u}})^T\right]$, $\breve{\boldsymbol{h}}(\boldsymbol{x}) = \partial\breve{\boldsymbol{u}}(\boldsymbol{x})/\partial\boldsymbol{x}$.

The following relations are valid:

$$\oint_{\partial V} \breve{\boldsymbol{u}}^{\mathrm{f}} \otimes \breve{\boldsymbol{n}} \mathrm{d}s = \mathbf{0}, \ \left\langle \breve{\boldsymbol{\varepsilon}}^{\mathrm{f}} \right\rangle_V = \frac{1}{2V} \oint_{\partial V} \left[\breve{\boldsymbol{n}} \otimes \breve{\boldsymbol{u}}^{\mathrm{f}} + \breve{\boldsymbol{u}}^{\mathrm{f}} \otimes \breve{\boldsymbol{n}} \right] \mathrm{d}s = \mathbf{0} \tag{A4.17}$$

Relation $(\mathrm{A4.17})_1$ is obtained by multiplying identity $(\mathrm{A4.10})_1$ with $\boldsymbol{h}$ and subtracting it from $(\mathrm{A4.17})_1$. Relation $(\mathrm{A4.17})_2$ is immediately obtained upon the substitution of displacement fluctuation field $\breve{\boldsymbol{u}}^{\mathrm{f}}$ into $(\mathrm{A4.16})_1$ and taking advantage of $(\mathrm{A4.17})_1$.

In order that the body $\mathcal{B}$ will be in a static mechanical equilibrium resultant force and resultant moment acting on the body must be equal to zero. Thus, surface tractions acting on the body must fulfill the relations

$$\mathcal{F} = \oint_{\partial \mathrm{V}} \breve{\boldsymbol{t}}^{(\mathrm{b})} \mathrm{d}s = \mathbf{0}, \ \mathcal{M} = \oint_{\partial \mathrm{V}} \boldsymbol{x} \times \breve{\boldsymbol{t}}^{(\mathrm{b})} \mathrm{d}s = \mathbf{0}, \ \boldsymbol{x} \in \partial V \tag{A4.18}$$

When the second relation $(\mathrm{A4.18})_2$, written in index form, is multiplied by permutation symbol $\mathscr{E}_{ijk}$ and identity $\varepsilon_{ijk}\varepsilon_{imn} = \delta_{jm}\delta_{kn} - \delta_{jn}\delta_{km}$ is used, then it is obtained that

$$\oint_{\partial V} \breve{\boldsymbol{t}}^{(\mathrm{b})} \otimes \boldsymbol{x} - \boldsymbol{x} \otimes \breve{\boldsymbol{t}}^{(\mathrm{b})} \mathrm{d}s = 0, \ \boldsymbol{x} \in \partial V, \tag{A4.19}$$

which assures symmetry of stress tensor.

It is also convenient to decompose boundary tractions into uniform boundary tractions (u.b.t.) $\tilde{\boldsymbol{t}}^{(u)}$, and fluctuation boundary tractions $\breve{\boldsymbol{t}}^{f}$

$$\begin{aligned} &\breve{\boldsymbol{t}}^{(\mathrm{b})}(\boldsymbol{x}) \equiv \tilde{\boldsymbol{t}}^{(\mathrm{u})}(\boldsymbol{x}) + \breve{\boldsymbol{t}}^{\mathrm{f}}(\boldsymbol{x}), \ \boldsymbol{x} \in \partial V \\ &\tilde{\boldsymbol{t}}^{(\mathrm{u})}(\boldsymbol{x}) = \breve{\boldsymbol{n}}(\boldsymbol{x})\boldsymbol{\sigma}, \ \breve{\boldsymbol{t}}^{\mathrm{f}}(\boldsymbol{x}) = \breve{\boldsymbol{t}}^{(\mathrm{b})}(\boldsymbol{x}) - \tilde{\boldsymbol{t}}^{(\mathrm{u})}(\boldsymbol{x}) \end{aligned} \tag{A4.20}$$

Taking advantage of the property that the mesoscopic stress field $\breve{\boldsymbol{\sigma}}(\boldsymbol{x})$ fulfills conditions of mechanical equilibrium $(\nabla \cdot \breve{\boldsymbol{\sigma}} = 0)$ and static conditions of compatibility $[\![\breve{\boldsymbol{\sigma}}]\!]_{\boldsymbol{x} \in \mathcal{S}^{(\mathrm{in})}} \cdot \boldsymbol{\nu}(\boldsymbol{x}) = \mathbf{0}$, it can be shown that the value of average (macroscopic) stress $\boldsymbol{\sigma}$ is completely determined by the surface tractions field $\breve{\boldsymbol{t}}^{(\mathrm{b})}(\boldsymbol{x})$

$$\begin{aligned} \boldsymbol{\sigma} \equiv \langle \breve{\boldsymbol{\sigma}}(\boldsymbol{x}) \rangle_V &= \frac{1}{V} \oint_{\mathrm{V}} \breve{\boldsymbol{\sigma}}(\boldsymbol{x}) \mathrm{d}v = \frac{1}{V} \oint_V \left[\nabla \left(\breve{\boldsymbol{\sigma}}(\boldsymbol{x}) \otimes \boldsymbol{x} \right) \right]^T \mathrm{d}v \\ &= \frac{1}{V} \oint_{\partial V} \left[\breve{\boldsymbol{n}} \cdot \left(\breve{\boldsymbol{\sigma}} \otimes \boldsymbol{x} \right) \right]^T \mathrm{d}s + \frac{1}{V} \oint_{S_i} \left[\underbrace{\nu(x) \cdot \left([\![\breve{\sigma}]\!] \otimes \boldsymbol{x} \right)}_{=0} \right]^T \mathrm{d}s \\ &= \frac{1}{V} \oint_{\partial V} \boldsymbol{x} \otimes \breve{\boldsymbol{t}}^{(\mathrm{b})}(\boldsymbol{x}) \mathrm{d}s \end{aligned}$$

$$\boldsymbol{\sigma} = \oint_{\partial V} \boldsymbol{x} \otimes \breve{\boldsymbol{t}}^{(\mathrm{b})}(\boldsymbol{x}) \mathrm{d}s \overset{(A4.19)}{\widehat{=}} \frac{1}{2V} \oint_{\partial V} \boldsymbol{x} \otimes \breve{\boldsymbol{t}}^{(\mathrm{b})}(\boldsymbol{x}) + \breve{\boldsymbol{t}}^{(\mathrm{b})}(\boldsymbol{x}) \otimes \boldsymbol{x} \mathrm{d}s, \tag{A4.21}$$

The following identities have been used in the above derivation $\breve{\boldsymbol{\sigma}}(\boldsymbol{x}) = [(\nabla\cdot\breve{\boldsymbol{\sigma}})\otimes\boldsymbol{x}) + \breve{\boldsymbol{\sigma}}^{\mathrm{T}}\cdot\mathbf{1}^{(2)}]^{\mathrm{T}} = [\nabla\cdot(\breve{\boldsymbol{\sigma}}\otimes\boldsymbol{x})]^{\mathrm{T}}$, $\nabla \equiv \boldsymbol{e}_i\partial/\partial x_i$, $\mathbf{1}^{(2)} = \nabla\otimes\boldsymbol{x}$.

When the body $\mathcal{B}$ is in static equilibrium under surface tractions $\breve{\boldsymbol{t}}^{(\mathrm{b})}$, then it is also in static equilibrium under uniform boundary tractions $\tilde{\boldsymbol{t}}^{(\mathrm{u})}$ and surface tractions fluctuations $\breve{\boldsymbol{t}}^{(\mathrm{f})}$

$$\oint_{\partial V}\tilde{\boldsymbol{t}}^{(\mathrm{u})}(\boldsymbol{x})\mathrm{d}s = \mathbf{0}, \quad \oint_{\partial V}\breve{\boldsymbol{t}}^{\mathrm{f}}(\boldsymbol{x})\mathrm{d}s = \mathbf{0},$$

$$\left\langle\breve{\boldsymbol{\sigma}}^{\mathrm{f}}(\boldsymbol{x})\right\rangle_V = \frac{1}{V}\oint_{\partial V}\boldsymbol{x}\otimes\breve{\boldsymbol{t}}^{\mathrm{f}}(\boldsymbol{x})\mathrm{d}s = \mathbf{0}. \tag{A4.22}$$

The last relation $(\mathrm{A4.22})_3$ results from the observation that average stress of surface tractions and uniform tractions is the same and equal to $\boldsymbol{\sigma}$. Substituting $\breve{\boldsymbol{t}}^{(\mathrm{b})}$ and $\tilde{\boldsymbol{t}}^{(\mathrm{u})}$ to $(\mathrm{A4.22})_2$ and deducting the left and right sides of obtained relations upon $(\mathrm{A4.20})_3$, gives $\mathbf{0} = \frac{1}{2V}\oint_{\partial V}\left[\boldsymbol{x}\otimes\left(\breve{\boldsymbol{t}}^{(\mathrm{b})} - \tilde{\boldsymbol{t}}^{(\mathrm{u})}\right) + \left(\breve{\boldsymbol{t}}^{(\mathrm{b})} - \tilde{\boldsymbol{t}}^{(\mathrm{u})}\right)\otimes\boldsymbol{x}\right]\mathrm{d}s = \left\langle\breve{\boldsymbol{\sigma}}^{\mathrm{f}}(\boldsymbol{x})\right\rangle_V$.

The total work $W^{(\mathrm{Tot})}$ spent on carrying tractions $\breve{\boldsymbol{t}}^{(\mathrm{b})}$ throughout displacements $\breve{\boldsymbol{u}}^{(\mathrm{b})}$, constant in time, operating on the outer surface of a body $\mathcal{B}$ treated as material macroelement (RVE) can be expressed as follows:

$$\begin{aligned}\rho W^{(\mathrm{Tot})} &\equiv \frac{1}{V}\oint_{\partial V}\breve{\boldsymbol{t}}^{(\mathrm{b})}\cdot\breve{\boldsymbol{u}}^{(\mathrm{b})}\mathrm{d}s = \frac{1}{V}\oint_{\partial V}\breve{\boldsymbol{n}}\breve{\boldsymbol{\sigma}}\cdot\breve{\boldsymbol{u}}^{(\mathrm{b})}\mathrm{d}s = \frac{1}{V}\oint_V\nabla\cdot(\breve{\boldsymbol{\sigma}}\breve{\boldsymbol{u}})\mathrm{d}v\\ &= \frac{1}{V}\oint_V(\breve{\boldsymbol{\sigma}}\cdot\breve{\boldsymbol{\varepsilon}})\mathrm{d}v = \langle\breve{\boldsymbol{\sigma}}\cdot\breve{\boldsymbol{\varepsilon}}\rangle_V.\end{aligned} \tag{A4.23}$$

The above formula is obtained by use of the Green-Gauss theorem $(\mathrm{A4.9})_2$ and the identity $\breve{\boldsymbol{\sigma}}\cdot\breve{\boldsymbol{\varepsilon}} = \breve{\boldsymbol{\sigma}}\cdot(\breve{\boldsymbol{\varepsilon}} + \breve{\boldsymbol{\omega}}) = \breve{\boldsymbol{\sigma}}\cdot(\nabla\otimes\breve{\boldsymbol{u}}) = \nabla\cdot(\breve{\boldsymbol{\sigma}}\breve{\boldsymbol{u}}) - (\nabla\cdot\breve{\boldsymbol{\sigma}})\breve{\boldsymbol{u}} = \nabla\cdot(\breve{\boldsymbol{\sigma}}\breve{\boldsymbol{u}})$, where it has been used as the property that local stress field fulfills conditions of static mechanical equilibrium $\nabla\cdot\breve{\boldsymbol{\sigma}}(\boldsymbol{x}) = 0$.

The total work can also be expressed in an equivalent form taking advantage of the decompositions (A4.15), (A4.20)

$$\begin{aligned}\rho W^{(\mathrm{Tot})} &= \frac{1}{V}\oint_{\partial V}\left(\tilde{\boldsymbol{t}}^{(\mathrm{u})} + \breve{\boldsymbol{t}}^{\mathrm{f}}\right)\cdot\left(\tilde{\boldsymbol{u}}^{(\mathrm{u})} + \breve{\boldsymbol{u}}^{\mathrm{f}}\right)\mathrm{d}s\\ &= \frac{1}{V}\oint_{\partial V}\tilde{\boldsymbol{t}}^{(\mathrm{u})}\cdot\tilde{\boldsymbol{u}}^{(\mathrm{u})}\mathrm{d}s + \frac{1}{V}\oint_{\partial V}\tilde{\boldsymbol{t}}^{(\mathrm{u})}\cdot\breve{\boldsymbol{u}}^{\mathrm{f}}\mathrm{d}s\\ &\quad + \frac{1}{V}\oint_{\partial V}\breve{\boldsymbol{t}}^{\mathrm{f}}\cdot\underline{\tilde{\boldsymbol{u}}}^{(\mathrm{u})}\mathrm{d}s + \frac{1}{V}\oint_{\partial V}\breve{\boldsymbol{t}}^{\mathrm{f}}\cdot\breve{\boldsymbol{u}}^{\mathrm{f}}\mathrm{d}s = \langle\breve{\boldsymbol{\sigma}}\rangle_V\langle\breve{\boldsymbol{\varepsilon}}\rangle_V + \rho W^{\mathrm{f}}\\ \rho W^{\mathrm{f}} &\equiv \frac{1}{V}\oint_{\partial V}\breve{\boldsymbol{t}}^{\mathrm{f}}(\boldsymbol{x})\cdot\breve{\boldsymbol{u}}^{\mathrm{f}}(\boldsymbol{x})\mathrm{d}s.\end{aligned} \tag{A4.24}$$

When deriving formula (A4.24), the following properties have been used

$$\oint_{\partial V} \breve{t}^{(u)} \cdot \breve{u}^{(u)} ds = \oint_{\partial V} (\breve{n}\sigma)\cdot((\varepsilon+\omega)x)ds$$
$$= \oint_{\partial V} (\breve{n}\otimes x)ds[\sigma\cdot(\varepsilon+\omega)] = V\langle\breve{\sigma}\rangle_V\langle\breve{\varepsilon}\rangle_V,$$
$$\oint_{\partial V} (\breve{n}(x)\otimes x)ds = \oint_V \nabla\otimes x dv = \oint_V 1^{(2)}dv = V1^{(2)},$$
$$\oint_{\partial V} \breve{t}^{(u)} \cdot \breve{u}^{f} ds = \oint_{\partial V} \breve{n}\sigma\cdot\breve{u}^{f} ds = \oint_{\partial V} \breve{n}\otimes\breve{u}^{f} ds\sigma = \mathbf{0} \quad (\text{see}(A4.17)_1)$$
$$\oint_{\partial V} \breve{t}^{f} \cdot \breve{u}^{(u)} ds = \oint_{\partial V} \breve{t}^{f} (\varepsilon+\omega)x ds = \oint_{\partial V} \breve{t}^{f} \otimes x ds\cdot(\varepsilon+\omega) = \mathbf{0} \quad (\text{see } (A4.22)_3)$$
(A4.25)

Relations (A4.23) and (A4.24) are valid regardless of material properties; the advantage has been taken of static admissibility of field $\breve{\sigma}(x)$ and kinematical admissibility of field $\breve{\varepsilon}(x)$, only. Formula (A4.24) reveals that total work $\rho W^{(Tot)}$ spent on deformation of body $\mathcal{B}$ depends not only on the product of average (macroscopic) values of stress and strain fields but also on work of fluctuation fields of tractions and displacements operating on the surface of a macroelement ρW^{f}.

Such a situation disables meaningful transition from mesoscale to macroscale description of material behavior; that is, introduction of the concept of so-called effective macroscopic properties of a material. No meaningful relation can be delivered for uniform, effective material properties to substitute modules of mesoscopic fields of elastic compliance or stiffness, as no meaningful functional relation can be delivered for total work $\rho W^{(Tot)}$ in terms of effective material properties and macroscopic state variables σ (ε) solely.

In order to surmount this difficulty, Hill (1963) proposed to impose a constraint, so-called *Hill's postulate* or *postulate of mechanical work compatibility*, requiring that work density of boundary fluctuation fields operating on macroelement surface vanishes.

$$\rho W^{f} = \frac{1}{V}\oint_{\partial V} \breve{t}^{f} \cdot \breve{u}^{f} ds = \mathbf{0} \Rightarrow \rho W^{(Tot)} = \langle\breve{\sigma}\cdot\breve{\varepsilon}\rangle_V = \langle\breve{\sigma}\rangle_V\langle\breve{\varepsilon}\rangle_V \tag{A4.26}$$

Hill's postulate is fulfilled in a trivial way by two especially simple cases of boundary traction and boundary displacement fields: (i) so-called uniform boundary tractions, and (ii) so-called linear boundary displacements. In the case of uniform boundary tractions (u.b.t.), fluctuations of surface tractions are identically equal to zero. In the case of linear boundary displacements (l.b.d.), fluctuations of surface displacements are identically equal to zero

$$\breve{t}^{(u)}(x) = n(x)\sigma, \ \rightarrow \breve{t}^{f}(x) = \mathbf{0}, \ x\in\partial V \ - \ (\text{u.b.t.}),$$
$$\breve{u}^{(u)}(x) = \varepsilon x, \ \rightarrow \breve{u}^{f}(x) = \mathbf{0}, \ x\in\partial V \ - \ (\text{l.b.d}). \tag{A4.27}$$

One should note that tractions and displacements fields on the surface of a macroelement are a combination of controlled information in the form of prescribed boundary conditions and a not controlled "response" of the macroelement-dependent on its mesostructure. Surface tractions and surface displacements cannot be simultaneously controlled on any portion of a macroelement surface. As mesostructure of a macroelement is usually assumed to be fixed, the postulate of work compatibility $(A4.26)_1$ in fact imposes constraints on prescribed boundary conditions that are prudent for investigation in the mechanical equilibrium problem of elastic mesomechanics in the process leading to determination of effective (macroscopic) elastic properties of the macroelement. The surface fields, making the solution of mesomechanics, fulfilling Hill's postulate $(A4.26)_1$, enable embedding, or packing, all the most essential information coming from mesoscale of observation in effective elastic properties of the material without the necessity for any further referring to mesoscale information. Thus, the postulate of work compatibility delivers an enabling tool for meaningful realization of the notion of effective (macroscopic) properties of the material.

Let's assume that all-round surface tractions $\breve{\boldsymbol{t}}^{(b)}(\boldsymbol{x})$ boundary conditions have been prescribed on the surface of the body $\mathcal{B}$ in the mechanical equilibrium problem of mesomechanics. Then, it can be shown that when mechanical equilibrium solution fields of stress and displacement lead to fulfillment of Hill's postulate, there must exist a functional relation between surface tractions fluctuations field $\breve{\boldsymbol{t}}^{f}(\boldsymbol{x})$ and average value of stress field $\boldsymbol{\sigma}\left(\breve{\boldsymbol{t}}^{f} = \breve{\boldsymbol{t}}^{f}(\boldsymbol{x};\, \boldsymbol{\sigma})\right)$. When additionally the body $\mathcal{B}$ is linearly elastic, then the actual solution stress field $\breve{\boldsymbol{\sigma}}(\boldsymbol{x})$ can be expressed as a linear function of average (macroscopic) stress with the aid of the stress concentration tensor $\boldsymbol{\mathcal{B}}(\boldsymbol{x})$ $\left(\breve{\boldsymbol{\sigma}}(\boldsymbol{x}) = \boldsymbol{\sigma} + \breve{\boldsymbol{\sigma}}^{f}(\boldsymbol{x}) = \boldsymbol{\mathcal{B}}(\boldsymbol{x})\boldsymbol{\sigma},\ \Rightarrow \boldsymbol{t}^{f} = \boldsymbol{n}(\boldsymbol{x})\boldsymbol{\mathcal{B}}(\boldsymbol{x})\boldsymbol{\sigma}\right)$. For linearly elastic body $\mathcal{B}$, the mechanical work $\rho W^{(\mathrm{Tot})}$ is simply the doubling of the elastic energy stored in the body $\left(W^{(\mathrm{Tot})} = 2W = \left(\frac{1}{\rho}\right)\langle \breve{\boldsymbol{\sigma}} \cdot \breve{\boldsymbol{\varepsilon}}_{(e)} \rangle_V\right)$ due to surface loading; see, fo example, Sadd (2005). It is worth it to indicate that the postulate of work compatibility does not lead to unique effective properties.

Functional dependence of fluctuation fields on average stress (strain) enables transition to a macroscopic description of material behavior, introduction of the concept of effective elastic properties, and pursuit of the following symbolic scheme: $\left\{\breve{\boldsymbol{t}}^{(b)}\right\} \to \left\{\boldsymbol{\sigma}, \breve{\boldsymbol{t}}^{f}(\boldsymbol{x};\, \boldsymbol{\sigma}), \boldsymbol{\mathcal{B}}^{t}(\boldsymbol{x})\right\} \to W\left(\boldsymbol{\sigma}, \boldsymbol{\varepsilon}\left(\mathbf{M}^{ef}\boldsymbol{\sigma}\right)\right)$. Please note that upon fixed mesostructure of the body $\mathcal{B}$, different all-round prescribed traction boundary conditions leading to fulfillment of Hill's postulate and giving the same value of average stress $\boldsymbol{\sigma}$ result in different fluctuations fields of surface traction. Hence, they lead to different values of elastic energy stored in material macroelement. This, in turn, leads to different values of effective modules. Usefulness of Hill's postulate is not limited to prescribed all-round surface tractions boundary conditions. It finds applicability to mixed and all-round displacement boundary conditions. In the case of prescribed all-round displacement boundary conditions on the surface of body $\mathcal{B}$ leading to fulfillment of Hill's postulate giving the same value of average strain $\boldsymbol{\varepsilon}$, results in different surface displacement fluctuation fields $\breve{\boldsymbol{u}}^{f} = \breve{\boldsymbol{u}}^{f}(\boldsymbol{x};\, \boldsymbol{\varepsilon})$ parametrically depending on average strain $\boldsymbol{\varepsilon}$. Similar to the case of traction boundary conditions, the following symbolic scheme leads to macroscopic description of the material behavior: $\left\{\breve{\boldsymbol{u}}^{(b)}\right\} \to \left\{\boldsymbol{\varepsilon}, \breve{\boldsymbol{u}}^{f}(\boldsymbol{x};\, \boldsymbol{\varepsilon}), \boldsymbol{\mathcal{B}}^{u}(\boldsymbol{x})\right\} \to W\left(\boldsymbol{\varepsilon}, \boldsymbol{\sigma}\left(\mathbf{L}^{ef}\boldsymbol{\varepsilon}\right)\right)$.

It can be shown that among all prescribed boundary conditions leading to fulfillment of Hill's postulate and to the same average value of stress $\boldsymbol{\sigma}$, prescribed uniform boundary tractions boundary conditions $(A4.27)_1$ give the largest value of elastic energy of linearly elastic body, keeping the microstructure fixed; that is, a norm of tensor of *effective modules of elastic compliance* $\mathbf{M}_{\text{ubt}}^{\text{ef}}$ then takes the largest value. Linear boundary displacements boundary conditions $(A4.27)_2$ give the smallest value of elastic energy of linearly elastic body, keeping the microstructure fixed; that is, the norm of tensor of effective compliance modules $\mathbf{M}_{\text{lbd}}^{\text{ef}}$ takes the smallest value (see, e.g., Nemat-Naser and Hori (1993)). Hence, these two boundary conditions deliver limit assessments of the value of elastic energy, which can be stored in linearly elastic body $\mathcal{B}$, and consequently estimates on values of effective modules $\left(\left|\mathbf{M}_{\text{lbd}}^{\text{ef}}\right| \leq \left|\mathbf{M}^{\text{ef}}\right| \leq \left|\mathbf{M}_{\text{ubt}}^{\text{ef}}\right|\right)$. For some special microstructures these assessments can reach the value of the well-known Reuss estimate—when stress field uniform in the whole volume of the body $\mathcal{B}$ constitutes a solution of the problem of mesomechanics (4.2)–(4.4), or the Voigt estimate—when strain field uniform in the whole volume of the body $\mathcal{B}$ constitutes a solution of the problem of micromechanics (4.2)–(4.4).

The boundary conditions (A4.27) are very useful in performing evaluation whether the size and shape of the domain of body $\mathcal{B}$ treated as macroelement RVE is properly selected; that is, whether it fulfills the postulate of so-called statistical representativeness. Fulfillment of this postulate is one of the requirements necessary for the body $\mathcal{B}$ that it can be treated as the *macroelement* (RVE). In order to actually perform the assessment, one should solve the problem of mechanical balance with uniform boundary tractions $\breve{\boldsymbol{t}}(\boldsymbol{x}) = \boldsymbol{n}(\boldsymbol{x})\boldsymbol{\sigma}$ of the type $(A4.27)_1$—leading to average (macroscopic) stress $\boldsymbol{\sigma}$. Solution of this problem gives average (macroscopic) strain $\boldsymbol{\varepsilon}^{\boldsymbol{\sigma}} = \langle \breve{\boldsymbol{\varepsilon}}(\boldsymbol{x}, \boldsymbol{\sigma}) \rangle_V$ and leads to elastic energy $W^{\text{U}} = \frac{1}{2}\boldsymbol{\sigma} \cdot \boldsymbol{\varepsilon}^{\boldsymbol{\sigma}}$. Next the mechanical equilibrium problem should be solved, for the same mesostructure, with linear displacements boundary conditions $\breve{\boldsymbol{u}}(\boldsymbol{x}) = \boldsymbol{\varepsilon}^{\boldsymbol{\sigma}}\boldsymbol{x}$ of the type $(A4.27)_2$—leading to average (macroscopic) strain $\boldsymbol{\varepsilon}^{\boldsymbol{\sigma}}$. Solution of this second problem gives average (macroscopic) stress $\boldsymbol{\sigma}^{\boldsymbol{\sigma}} = \langle \breve{\boldsymbol{\sigma}}(\boldsymbol{x}, \boldsymbol{\varepsilon}^{\boldsymbol{\sigma}}) \rangle_V$ and leads to elastic energy $W^L = \frac{1}{2}\boldsymbol{\sigma}^{\boldsymbol{\sigma}} \cdot \boldsymbol{\varepsilon}^{\boldsymbol{\sigma}}$. When energies $W^U \cong W^L$, and hence stresses $\boldsymbol{\sigma}^{\boldsymbol{\sigma}} \cong \boldsymbol{\sigma}$ are nearly equal, then body $\mathcal{B}$ can be treated as statistically representative and thus proves to be correct macroelement (RVE); see also Nemat-Naser and Hori (1993).

In order to illustrate the process of determination of effective elastic properties, let us investigate the problem of mechanical balance of a body $\mathcal{B}_0$ made of linearly elastic material without eigenstrains—$\breve{\boldsymbol{\varepsilon}}_0(\boldsymbol{x}) = \breve{\mathbf{M}}(\boldsymbol{x})\breve{\boldsymbol{\sigma}}_0(\boldsymbol{x})$, with prescribed on its surface force boundary conditions fulfilling Hill's postulate $\breve{\boldsymbol{t}}_0^{(\text{b})}(\boldsymbol{x}) = \breve{\boldsymbol{n}}(\boldsymbol{x})\boldsymbol{\sigma}_0 + \breve{\boldsymbol{t}}^{\text{f}}(\boldsymbol{x})$. Fulfillment of Hill's postulate by the stress vector $\breve{\boldsymbol{t}}_0^{(\text{b})}$ field ensures that fields of strains and displacements $\breve{\boldsymbol{\varepsilon}}_0(\boldsymbol{x}, \boldsymbol{\sigma}_0)$, $\breve{\boldsymbol{u}}_0(\boldsymbol{x}, \boldsymbol{\sigma}_0)$, constituting the solution of this problem, linearly depend only on the mean value of the stress tensor field also belonging to the solution of the posed problem $\boldsymbol{\sigma}_0 = \langle \boldsymbol{\mathcal{B}}^0(\boldsymbol{x})\boldsymbol{\sigma}_0 \rangle_V$; $\breve{\boldsymbol{\sigma}}_0(\boldsymbol{x}) = \boldsymbol{\mathcal{B}}^0(\boldsymbol{x})\boldsymbol{\sigma}_0$. Effective elastic properties can then be determined in a unique way by computation of the average value of the strain field $\breve{\boldsymbol{\varepsilon}}_0(\boldsymbol{x})$; that is, $\boldsymbol{\varepsilon}_0 = \langle \breve{\boldsymbol{\varepsilon}}_0(\boldsymbol{x}) \rangle_V = \left\langle \breve{\mathbf{M}}(\boldsymbol{x})\boldsymbol{\mathcal{B}}^0(\boldsymbol{x})\boldsymbol{\sigma}_0 \right\rangle_V = \mathbf{M}^{\text{ef}}\boldsymbol{\sigma}_0$. Elastic energy can be expressed with the formula $\rho W = \frac{1}{2}\boldsymbol{\sigma}_0 \cdot \boldsymbol{\varepsilon}_0 = \frac{1}{2}\boldsymbol{\sigma}_0 \cdot \mathbf{M}^{\text{ef}}\boldsymbol{\sigma}_0$.

Experimental validation of R_L model assumptions for NiTi alloy

5

5.1 Introduction

The results of a broad program of experimental studies on the behavior of $Ni_{51at\%}$-Ti alloy in the range of its pseudoelastic behavior submitted to multiaxial stress loadings are discussed in this chapter. The program consisted of isothermal, stress-controlled, mechanical tests, in which a tubular sample made of NiTi alloy was submitted to proportional in stress space loading paths. Raw data from thermomechanical tests has been elaborated to a form convenient for analytical studies with the aid of advanced software for data processing (DADISP). The experimental data analysis enabled verification of three fundamental theoretical assumptions made in the R_L model family in the case of a monoclinic martensitic transformation taking place in an investigated polycrystalline NiTi alloy: specifically, the assumption on elastic isotropy, the assumption on admissibility of the mathematical representation of pseudoelastic flow surface with the use of the concept of shape function $f(y)$, and the assumption on inelastic isotropy—admissibility of the representation of the macroscopic tensor of phase transition connected with the monoclinic transformation with the aid of the isotropic tensor function of zero order with respect to the stress tensor. An effective procedure for identification of the material functions and constants of the R_L model is discussed. This procedure has actually been used for identification of modeling data of the investigated $Ni_{51at\%}$-Ti alloy. The assessment of the quality of predictions of the R_L model has been done by comparison of time runs of total strain tensor components signals actually registered during mechanical tests with those computed with the aid of the R_L model.

The $Ni_{51at\%}$-Ti alloy and tubular specimen used in the multiaxial experimental program was prepared from elemental materials at the University of Tsukuba. In Japan, fundamental, thermal (DSC), and mechanical (uniaxial tension) tests of the designed and manufactured NiTi alloy material were also performed (Tanaka et al., 1999). The multiaxial tests program of isothermal tests, executed at various temperatures on the tubular specimen, has been performed in the Strength of Materials Laboratory in the Institute of Fundamental Technological Research, Polish Academy of Science, Poland. The experimental methodology employed in this research work is explained in the paper by Raniecki et al. (1999a). The experimental and theoretical results obtained have been published in a number of papers: Raniecki et al. (1999a,b), Socha et al. (2000), Raniecki et al. (2001), and Ziółkowski et al. (2004). Here, the results are collected and discussed together with new information concerning experimental verification of theoretical assumptions of the R_L model. This is done by analysis of strain effects appearing in the NiTi alloy undergoing thermoelastic martensitic

Pseudoelasticity of Shape Memory Alloys. http://dx.doi.org/10.1016/B978-0-12-801697-8.00005-3

phase transformations in the pseudoelastic range of its behavior. A complete set of material constants and functions of the R_L model for the investigated $Ni_{51at\%}$-Ti alloy is specified.

Metallurgical details regarding the $Ni_{51at\%}$-Ti alloy and the manufacturing specimen for mechanical tests, methodology concerning characterizing the SMA material, and its fundamental characteristics are discussed in the work by Tanaka et al. (1999). Here, the most essential data are recalled. In view of the cost of the material and samples manufacturing, it was desirable to design and prepare an alloy that would have high stability of its shape memory effects behavior. The desired repeatability of the SME material response has been reached by professor Miyazaki through proper selection of the chemical composition of the polycrystalline alloy, that is, $Ni_{51at\%}$-Ti and expert finishing thermal treatment, where the last step consisted of aging at a temperature of 673 K for 3600 s and subsequent cooling in a furnace. The grain size after thermal treatment was around 30 μm. From such a prepared alloy, thin-walled tubular samples were manufactured with an external diameter of 19 mm, a wall thickness in the gauge section of 1.5 mm, a maximum external diameter in the grip section of 26 mm, and a total length of 122 mm. Despite complicated multiaxial and high stress loadings of the sample reaching 800 MPa, the planned experimental program of around 250 tests was carried out using only one tubular NiTi sample before it underwent damage.

Differential scanning calorimetry (DSC) tests revealed that in the NiTi alloy after SME thermomechanical treatment, two-phase transformations of martensitic type take place. The NiTi sample remaining initially, at high temperature, in austenitic state undergoes phase transformation into the rhombohedral martensitic phase $\mathcal{A} \rightarrow \mathcal{R}$ at the lowering of temperature, in a sequence called *R-phase transformation*. Further decreasing of temperature leads to the next phase transformation of the rhombohedral martensite into monoclinic martensite $\mathcal{R} \rightarrow \mathcal{M}$, in a sequence called *monoclinic transformation*. During heating, reverse phase transformation $\mathcal{M} \rightarrow \mathcal{A}$ and/or $\mathcal{R} \rightarrow \mathcal{A}$ takes place. *Characteristic temperatures of phase transitions* were determined with the aid of the *DSC method*; see Figure 2 in Tanaka et al. (1999). These temperatures are as follows:

- R-phase transformation:

$$R_s^0 = 306\left(R_s^0 = 312\right), \quad R_f^0 = 293\left(R_f^0 = 298\right),$$

$$RA_s^0 = 305\left(RA_s^0 = 303\right), \quad RA_f^0 = 316\left(RA_f^0 = 319\right)(\mathrm{K}).$$

- Monoclinic transformation:

$$M_s^0 = 253\left(M_s^0 = 250\right), \quad M_f^0 = 208\left(M_f^0 = 193\right),$$

$$A_s^0 = 269\left(A_s^0 = 268\right), \quad A_f^0 = 306\left(A_f^0 = 306\right).$$

Subscripts "s" and "f" denote the start and finish temperatures of phase transition. Superscript "0" indicates that the specified characteristic temperatures of p.t. concern transformations taking place at zero macroscopic stress. The present author executed DSC measurement on the same $Ni_{51at\%}$-Ti alloy 14 years later. The thermograph of this measurement is shown in Figure 5.1 and the values of characteristic

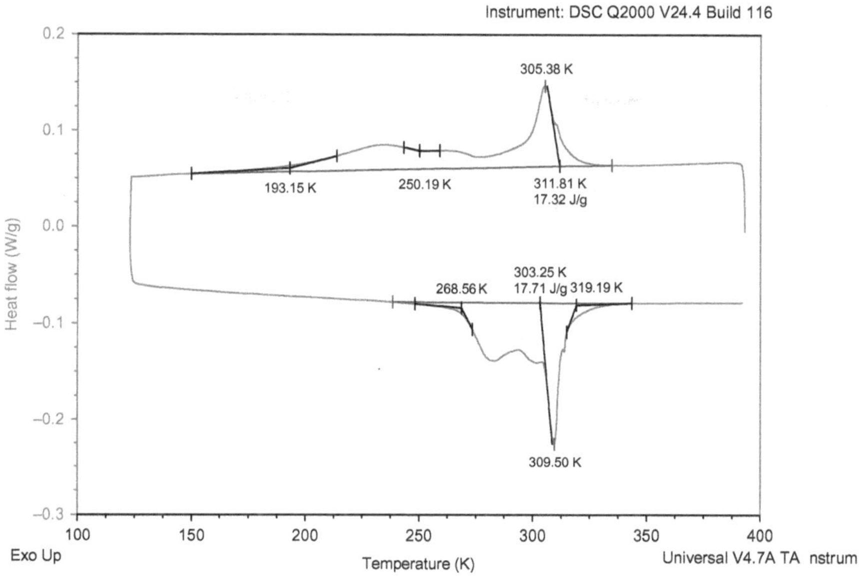

Figure 5.1 DSC thermograph of $Ni_{51at\%}$-Ti alloy; TMT 400 °C, 1 h.

temperatures obtained are shown in commas above. The very small differences between corresponding values of characteristic temperatures, some of which can surely be attributed to the very smeared character of monoclinic martensitic transition, proves the excellent time stability of the investigated NiTi alloy material properties. Latent heat of phase transformation has been measured to be $\Delta Q = 17{,}500$ (J/kg).

The DSC tests allowed the temperature range to be determined in which multiaxial, mechanical tests for testing pseudoelastic behavior of the NiTi alloy should be performed. Besides DSC tests, Tanaka et al. (1999) executed a series of isothermal, quasi-static, uniaxial tension tests in the range of temperatures from 212 to 363 K with strain control and loading strain amplitude from 0% to 8%. The results of these tests has been collected together, and overlaid as shown in Figure 3.4 in Chapter 3. Tanaka et al. (1999) also executed a series of iso-stress tests; that is, tests in which temperature was varied while assisting loading stress was kept constant. The assisting stresses values were in the range from 30 to 500 MPa. During the iso-stress tests, the sample was cooled and subsequently heated in a temperature range from 400 to 200 K with a constant rate of 4 K/min. Results of these last tests will not be discussed here as they go beyond the pseudoelastic range of the alloy behavior. It is worth mentioning that strain effect accompanying R-phase transformation amounts to around 0.5% and is an order of magnitude smaller than the strain effect accompanying monoclinic transformation, which amounts to around 6%. The stress level required to induce R-phase transformation is considerably lower than the stress level required to induce monoclinic transformation at the same isothermal test. Taking into account the risk of damaging the specimen, the uniaxial test data indicated that at first the program of tests for R-phase investigation should be preformed, and only later the tests for monoclinic phase transformation investigation. Such a sequence of events has actually been adopted.

It is also worth mentioning that the level of pseudoelastic flow plateau obtained in the isothermal, uniaxial mechanical tests of Tanaka et al. (1999) differ by about 25% from such a level obtained in analogical multiaxial tests performed on tubular specimen tested in Poland. Similarly, considerable differences appeared in the size of the "hardening" effect. The reasons for these differences remain unknown to the author. One of the reasons might be the method of loading control (strain control versus stress control); see also Figure 12 in Raniecki et al. (2001).

5.2 Methodology of experimental studies on NiTi alloy submitted to multiaxial stress states loadings

Methodology of thermomechanical experimental studies on $Ni_{51at\%}$-Ti alloy tubular specimen submitted to biaxial stress states has been described in detail in Raniecki et al. (1999a). Here, the key features of the approach are recalled.

All thermomechanical tests have been executed on a single tubular specimen. After each single test, the specimen has been "revitalized" by executing the reset procedure, which consisted of increasing its temperature to $T = 350$ K (about 50 K above austenite finish temperature—$A_f^0 = 306$K—and keeping it at this temperature for 10 min. This is done to assure that the sample is transformed into a pure austenitic state before each new test, and to remove any possible residual phase transition strain connected with untransformed martensitic phase. Excellent reversal to the initial dimensions of the specimen after such procedure and repeatability of observed results allows judging that consecutive loading cycles did not introduce substantial microstructural changes into the specimen. The testing program schedule has been planned in such a way as to first investigate NiTi alloy behavior at low stresses and strains (in the range of R-phase transformation) and next execute tests involving relatively high strains and stresses (to the range of monoclinic martensitic transformation).

Isothermal mechanical tests have been preformed at temperatures 250, 260, 280, 300, 310, 315, and 322.5 K. The range of temperatures of mechanical testing in relation to characteristic temperatures of phase transformations taking place in the investigated NiTi alloy determined with DSC is shown schematically in Figure 5.2a. Mechanical loadings employed in the testing program are shown schematically in Figure 5.2b. The tests' loading cycles have been executed with stress control, during which stress was increased linearly in time up to a certain maximum value $\sigma_{ij}(t) = \left(\sigma_{ij}^{\mathrm{Max}}/t_{\mathrm{Max}}\right)t\left(\sigma_{ij}^{\mathrm{Max}},\ t_{\mathrm{Max}} = \mathrm{const}\right)$, and next it was decreased to zero value. The rate of effective stress during the test was kept constant and amounted to $\dot{\sigma}_{\mathrm{ef}} = 1\,\mathrm{MPa/s}$. The following five biaxial tests with different proportional loading paths for each temperature have been executed:

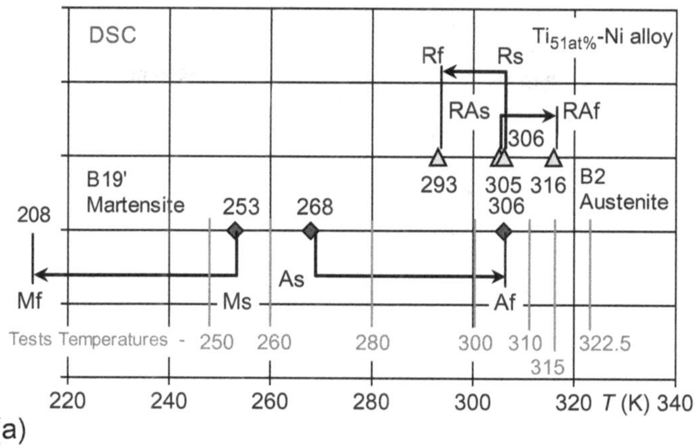

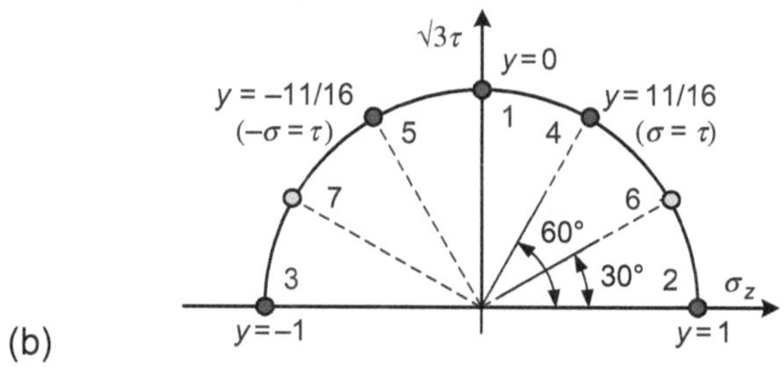

Figure 5.2 (a) Temperatures of isothermal mechanical tests in reference to characteristic temperatures of phase transformations—DSC. (b) Scheme of mechanical loadings program.

Path No. 1, $\tau \neq 0, \sigma_z = 0, y = 0$	Torsion
Path No. 2, $\tau = 0, \sigma_z \geq 0, y = 1$	Tension
Path No. 3, $\tau = 0, \sigma_z < 0, y = -1$	Compression
Path No. 4, $\tau = \sigma_z, \sigma_z > 0, y = 11/16$	Torsion and tension
Path No. 5, $\tau = -\sigma_z, \sigma_z < 0, y = -11/16$	Torsion and compression

where σ_z denotes the axial component of the stress tensor, and τ is the shear component.

Several occasional tests with paths 6 and 7 have been executed.

The equivalent stress σ_{ef} $\left(\sigma_{ef} = \sqrt{\frac{3}{2}s_{ij}s_{ij}}\right)$ and loading "direction" parameter y were computed with the aid of the formulas (see also (3.37))

$$\sigma_{ef} = \sqrt{\sigma_z^2 + 3\cdot\tau^2}, \quad y = \frac{\sigma_z\left(2\sigma_z^2 + 9\tau^2\right)}{2\left(\sigma_z^2 + 3\tau^2\right)^{3/2}} = \text{sign}(\sigma_z)\frac{(2+9m^2)}{2(1+3m^2)^{3/2}}, \quad m = \tau/\sigma_z. \tag{5.1}$$

As the testing machine on which the tests were actually performed was not equipped with the unit enabling loading in the form of internal pressure applied to tubular sample, the loadings were limited to plane stress state type of loadings. This technical limitation has the following limitations regarding the mechanical tests results obtained (Raniecki et al., 1999b):

(i) It is not possible to directly confirm that the investigated material is isotropic, as the tension-torsion (compression-torsion) tests do not allow generating the set of equivalent paths; i.e., such paths that $\boldsymbol{\sigma}^{\mathrm{R}}(t)=\mathbf{R}^{\mathrm{T}}\boldsymbol{\sigma}(t)\mathbf{R}$; $\mathbf{R}^{\mathrm{T}}\mathbf{R}=\mathbf{I}$, $\det(\mathbf{R})=1$.
(ii) It is not possible to directly investigate the influence of hydrostatic pressure on material behavior, as it is not possible to keep constant the value of pressure during the test with the loading scheme specified above.
(iii) It is not possible to induce the states of pure tension ($\sigma_1>0, \sigma_2>0, \sigma_3=0$) in the sample, nor the states of pure compression ($\sigma_1<0, \sigma_2<0, \sigma_3=0$), where σ_1 and σ_2 denote the principal values of stress tensor ($\sigma_3\equiv 0$), as in the tests of compression (tension) with torsion the principal values of stress tensor σ_1, σ_2 have opposite signs, when $\sigma_\theta\equiv 0$.

Despite the limitations listed above, an effort has been undertaken to verify, in an indirect manner, whether the investigated polycrystalline NiTi alloy can be treated as isotropic material, that is modeling assumption made in the $\mathrm{R_L}$ models family.

Ten channels of data collected through the testing machine control unit were registered during each test. These channels were the following signals: time t, displacement u, rotation φ, force F, torque M, axial strain ε_z, shear strain ε_θ, hoop stain $\varepsilon_{z\theta}$, and two temperatures measured at both ends of the gauge section of tubular sample T_1, T_2. The component strains were measured with strain gauges, and temperatures were measured with thermocouples. Care has been taken that uniform, stabile thermal conditions of the sample are reached before starting the loading process of the sample in each test. This has been assured by waiting until the temperature gradient on the axial gauge length of the sample reached a small difference, on the order of 0.1 K.

A typical time graph of registered data $\sigma_z(t), \tau(t), \varepsilon_z(t), \varepsilon_\theta(t), \varepsilon_{z\theta}(t), T(t)$, obtained during one isothermal mechanical test, is shown in Figure 5.3. The values of axial and shear components of stress in the test were calculated from registered digital signals of the values of force and torque with the use of the following formulas:

$$\sigma_z(t)=\frac{4F(t)}{\pi(D^2-d^2)},\quad \tau(t)=\frac{16M(t)}{\pi(D^4-d^4)},\quad D=19(\mathrm{mm}), d=16(\mathrm{mm}). \tag{5.2}$$

When it is accepted that, with good approximation, investigated polycrystalline NiTi alloy behaves in an isotropic manner in the elastic and inelastic ranges of deformation, then it can be expected that plane state of stress applied during sample loading will generate material response in the form of the following plane state of strain:

$$\begin{bmatrix} \sigma_z & \tau & 0 \\ \tau & 0 & 0 \\ 0 & 0 & 0 \end{bmatrix} \rightarrow \begin{bmatrix} \varepsilon_z & \varepsilon_{z\theta} & 0 \\ \varepsilon_{z\theta} & \varepsilon_\theta & 0 \\ 0 & 0 & \varepsilon_r \end{bmatrix}, \tag{5.3}$$

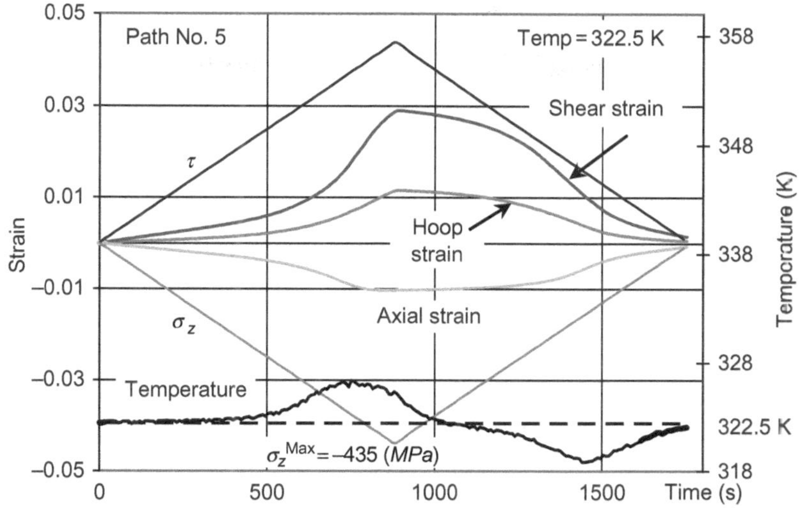

Figure 5.3 Illustrative example of charts of signals registered during isothermal tests. Data obtained in test of simultaneous compression and torsion, path No. 5 (Torsion and Compression), performed in temperature $T = 322.5°K$.

where $\varepsilon_z = \varepsilon_{11}$ denotes axial strain, $\varepsilon_{z\theta} = \varepsilon_{12}$ is shear strain, $\varepsilon_\theta = \varepsilon_{22}$ is hoop strain, and $\varepsilon_r = \varepsilon_{33}$ denotes radial strain.

As already indicated, there were measured and registered components $\varepsilon_z(t)$, $\varepsilon_{z\theta}(t)$, $\varepsilon_\theta(t)$ of total strain tensor during the tests but due to technical difficulties, the radial component of total strain tensor $\varepsilon_r(t)$ has not been physically measured. The lack of experimental values of radial component of strain creates difficulties with calculation of equivalent strain defined with the formula $\varepsilon_{eq} = \sqrt{\frac{2}{3}\bar{\varepsilon}_{ij}\bar{\varepsilon}_{ij}}$ ($\bar{\varepsilon}_{ij}$ denotes deviatoric components of strain tensor). In the elastic range of the material behavior, this difficulty has been overcome by adopting an assumption about isotropic elastically behavior of the investigated material; that is, radial component has been determined indirectly using the formula $\varepsilon_r(t) = -\nu\sigma_z(t)/E = -\nu\varepsilon_z(t)$. In order to calculate effective strain in the inelastic range of the material behavior, the inelastic radial component of strain ε_r^{pt} has been determined with the use of an assumption on the incompressibility of macroscopic tensor of phase strains $\mathrm{tr}(\boldsymbol{\varepsilon}^{pt}) \equiv 0$ (see also (3.1) in Chapter 3). This last assumption has not been verified experimentally in the testing program discussed here. However, the literature indicates that volumetric changes during thermoelastic martensitic phase transformation responsible for shape memory effects usually does not exceed 0.2% (Otsuka and Ren, 2005). As the R_L model has been developed within the small deformations formalism, the additive decomposition of total strain tensor into elastic and inelastic (phase transformation) parts is valid. Taking advantage of the assumption on elastic isotropy and tracelessness of phase transformation tensor, the components of the latter can be calculated with the aid of the following formulas:

$$\boldsymbol{\varepsilon}^{\mathrm{pt}} \equiv \boldsymbol{\varepsilon} - \mathbf{M}\boldsymbol{\sigma}, \quad \mathrm{tr}(\boldsymbol{\varepsilon}^{\mathrm{pt}}) \equiv 0,$$
$$\varepsilon_z^{\mathrm{pt}} = \varepsilon_z - \frac{\sigma_z}{E}, \quad \varepsilon_{z\theta}^{\mathrm{pt}} = \varepsilon_{z\theta} - \frac{\tau}{2\mu},$$
$$\varepsilon_\theta^{\mathrm{pt}} = \varepsilon_\theta - \varepsilon_\theta^{\mathrm{e}} = \varepsilon_\theta - \nu\varepsilon_z^{\mathrm{pt}} = \varepsilon_\theta + \left(\frac{1}{2\mu} - \frac{1}{E}\right)\sigma_z, \quad \varepsilon_{\mathrm{r}}^{\mathrm{pt}} = -\left(\varepsilon_\theta^{\mathrm{pt}} + \varepsilon_z^{\mathrm{pt}}\right). \tag{5.4}$$

Effective phase strain takes the following form:

$$\varepsilon_{\mathrm{eq}}^{\mathrm{pt}} \equiv \left[(2/3)\overline{\boldsymbol{\varepsilon}}^{\mathrm{pt}} \cdot \overline{\boldsymbol{\varepsilon}}^{\mathrm{pt}}\right]^{1/2} = \frac{2\sqrt{3}}{3}\sqrt{\left(\varepsilon_z^{\mathrm{pt}} - \varepsilon_\theta^{\mathrm{pt}}\right)^2 + 3\varepsilon_z^{\mathrm{pt}}\varepsilon_\theta^{\mathrm{pt}} + \left(\varepsilon_{z\theta}^{\mathrm{pt}}\right)^2}, \quad \mathrm{tr}(\boldsymbol{\varepsilon}^{\mathrm{pt}}) \equiv 0. \tag{5.5}$$

5.3 Experimental data processing for validation of theoretical assumptions of the R_L model

Two sessions of tests on NiTi tubular specimen were executed, one with "low" loadings and one with "high" loadings. In the first session with "low" loadings, limited to about 300 MPa, executed in the broad temperature range of 260 ÷ 315 K, only oriented R-phase was induced, and no monoclinic transformation took place. In the second session with "high" loadings, up to 850 MPa, in the temperature range of 310 ÷ 322.5 K (a range of pseudoelastic behavior of investigated SMA material), two-phase transformations took place: R-phase and monoclinic. In Figure 5.4a, effective stress-equivalent total strain curves are shown, plotted for isothermal tests results of loading path No. 5 (torsion-compression) performed in the temperature range

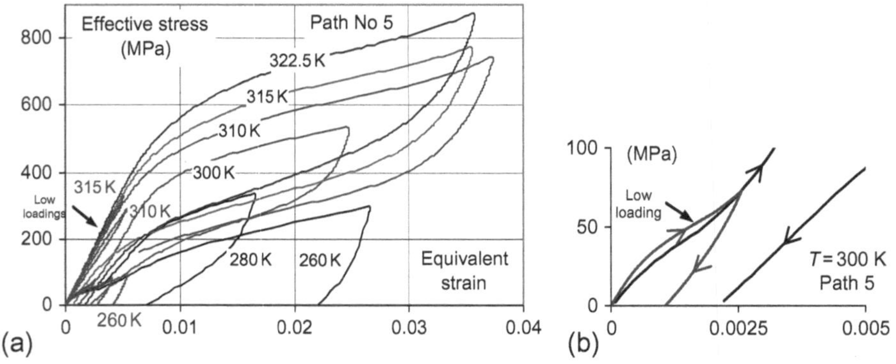

Figure 5.4 (a) Charts of effective stress-equivalent strain obtained for NiTi alloy submitted to isothermal tests of simultaneous compression and torsion, path No. 5, in temperature range $T = 260 \div 322.5$ K. Charts with strains not exceeding 0.005 represent data from tests with low loadings. (b) Overlaid charts of stress-strain obtained at low and high loading cycle, path No. 5, $T = 300$ K.

$260 \div 322.5$ K. The graphs prepared using data from tests with "low" loadings are contained within area (0, 0.005, 0, 400)—, and remaining charts were prepared using data from series of tests with "high" loadings. Figure 5.4b shows "zoomed" stress-strain plots of test results obtained in "low" and "high" loading cycle tests executed for the same path No. 5 at temperature 300 K. It is worth noting that the difference lies within the acceptable limits of engineering accuracy of about 10%. Taking into account that in between any two "low" loading and corresponding "high" loading tests, plotted overlaid in Figure 5.4a, several dozens of other tests involving multiaxial stress states were preformed, it can be stated that repeatability (stability) of the investigated NiTi material response is very good.

Experimental data obtained in the first session of tests, involving low loadings, concern behavior of NiTi alloy undergoing R-phase transformation in temperature ranges of one-way memory effect and pseudoelasticity. They are elaborated and discussed on theoretical grounds in Raniecki et al. (1999b). This information is not discussed in this book. Here, processing and theoretical elaboration of the data obtained in a "high" loadings session involving pseudoelastic behavior of investigated NiTi alloy due to monoclinic martensitic transformation is discussed. In this process, R-phase transition phenomena are treated as disturbances instead. Original results of such processing have been already published in several works; for example, Raniecki et al. (2001), Ziółkowski et al. (2004), and Ziółkowski (2006). Here, they are reworked and updated in view of new information that became available due to progress in research on thermoelastic martensitic transformations.

First, it is checked whether loading paths of the specimen were, indeed, proportional paths in the space of stress tensor components. For that purpose, graphs of loading in $\sigma_z \leftrightarrow \tau$ coordinates are prepared. The set of these graphs for five paths of loading at temperature 315 K is presented in Figure 5.5a. Based on this figure it can be stated that the quality of control loading during the tests was perfect. In the case of tests performed at temperatures of 310 and 322.5 K, the quality of loading control was the same.

Figure 5.5b shows strain response of the material to induced force loading for five loading paths (i.e., graphs $\varepsilon_z \leftrightarrow \varepsilon_{z\theta}$). Inspection of graphs in this figure indicates that material response can be divided into two subranges. This is illustrated symbolically with two straight lines for loading path No. 5. Concerning absolute values of strain components from 0% to about 0.5% in the first range, the material behaves elastically. Concerning absolute values of strain components from about 0.5% to a maximum 4% in the second range, the material flows pseudoelastically as a result of progress of monoclinic martensitic phase transformation. The force loadings and material response to those loadings obtained in low loadings series of tests and in high loadings series of tests of the specimen are plotted together in Figure 5.5c and d, respectively. The material response obtained in the session of tests with low loadings is marked with symbol "L," and that obtained in the session of tests with high loadings with symbol "H." The comparison of these responses allows stating that, with good approximation, it can be adopted that the history of the NiTi specimen has negligible influence on the behavior of polycrystalline NiTi alloy. The biggest discrepancy in the response of the sample appeared for path No. 2. Note that in between any two tests corresponding to

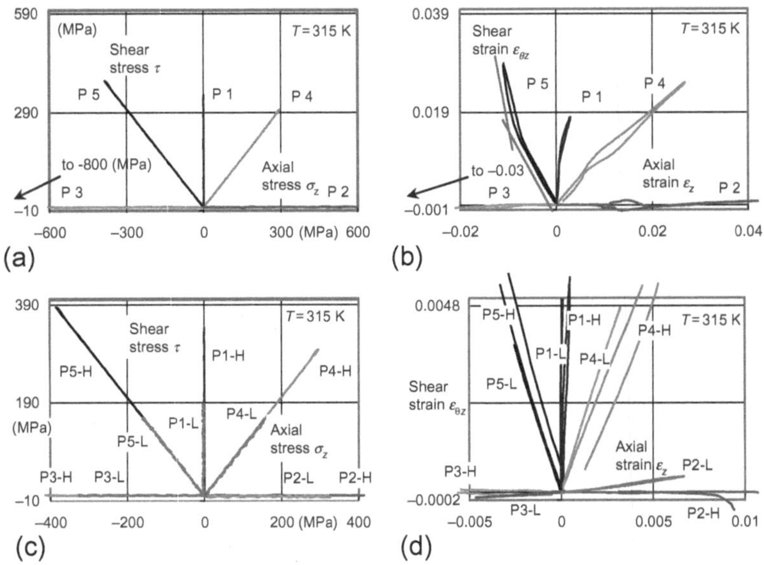

Figure 5.5 Actual force loading paths and NiTi alloy strain response in the case of tests performed at temperature 315 K. (a) Force loadings of the sample. (b) Induced by force loadings strain responses obtained in series of tests with high loadings. (c and d) Comparison of force loadings of the sample and obtained strain responses of the material in a series of tests with "low" and "high" loadings. The symbol "L" denotes low and the symbol "H" high loading.

each other for some prescribed loading path (e.g., torsion "P1-L" and "P1-H"), about fifty other tests with different loading schemes took place; hence, the rather involved history of the loading of the sample. The results obtained for other temperatures have the same character and quality as the one discussed in detail here for a temperature of 315 K. The experimental graphs plotted in Figure 5.5d indicate that it can be accepted with very good approximation that the ratio of strain components $\varepsilon_{z\theta}/\varepsilon_z$ remains constant in the elastic range of polycrystalline NiTi alloy behavior. This constant ratio of components of strain, in the case of loading paths No. 4 and No. 5, takes place at a preserving constant ratio of τ/σ_z. This indicates a constant value of Poisson coefficient of NiTi alloy in the elastic range of its behavior. The graphs plotted in Figure 5.5d show that the application of torsional stress (path No. 1) does not induce axial strains (graphs P1-L and P1-H). Similarly, the application of tension or compression stress (paths No. 2 and No. 3) does not induce shear strain (graphs P3-L, P3-H, and P2-H; a small shear strain of the sample appeared in the case of graph P2-L). Isotropic theory of elasticity predicts lack of coupling between axial and shear deformation effects.

In the case of tests with loading paths No. 1, 2, and 3, excellent lack of coupling between axial and shear strain components indicates the acceptability of the R_L model assumption of elastic isotropy for the case of investigated shape memory alloy material.

In the subsequent step of experimental data processing, the stress-strain graphs of mutually corresponding components of stress and strain tensors $\sigma_z \leftrightarrow \varepsilon_z, \tau \leftrightarrow \varepsilon_{z\theta}$ are prepared for five loading paths and three temperatures $T = 310, 315, 322.5$K. Next, there are prepared graphs of effective stress as the function of equivalent strain $\sigma_{ef} \leftrightarrow \varepsilon_{eq}$. In Figure 5.6a, a set of such graphs is plotted for five loading paths of tests performed at an obtained temperature of 322.5 K (the elastic range of material behavior is marked with an ellipse). The graphs indicate the linearly elastic behavior of NiTi alloy. Hence, experimental confirmation finds assumption adopted in family of R_L models on linearity of elastic behavior of shape memory alloy materials.

There are determined Young's modules E and shear modules 2μ for five tests with different loading paths at three testing temperatures. The elastic modules are defined here as slopes of tangent to the stress-strain curves of axial and shear components, respectively. They were determined with the use of the mean square method with strain range for average tangent determination limited to $\varepsilon \in \langle 0, 0.001 \rangle$. The values of elastic constants obtained are listed in Table 5.1. A survey of data contained in Table 5.1 shows that the maximum difference between values of Young's modulus

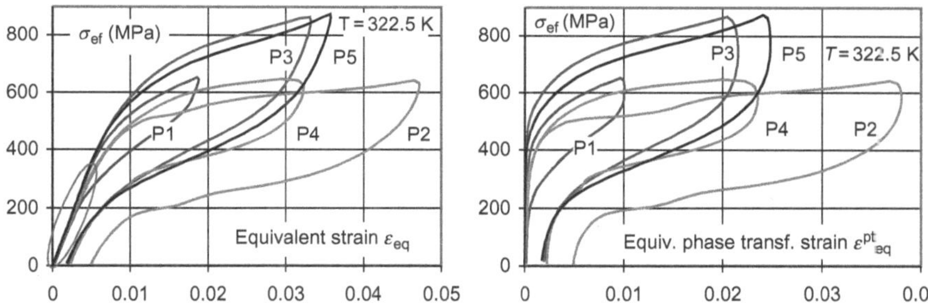

Figure 5.6 Graphs of five proportional loading paths at temperature $T = 322.5$ K. (a) Effective stress ↔ Equivalent strain. (b) Effective stress ↔ Equivalent phase transformation strain.

Table 5.1 Elastic properties of $Ni_{51at\%}$-Ti alloy (offset 0.1%)

Path	No. 1	No. 2	No. 3	No. 4	No. 5
***T* = 310 (K)**					
2 μ (GPa)	42.7	-	-	43.1	42.5
E (GPa)	-	61.2	62.3	53.5	62.3
***T* = 315 (K)**					
2 μ (GPa)	45.5	-	-	45.6	45.0
E (GPa)	-	62.9	67.3	57.2	66.9
***T* = 322.5 (K)**					
2 μ (GPa)	47.2	-	-	47.5	47.1
E (GPa)	-	66.9	68.3	58.8	70.5

does not exceed more than 9% from the average value for specific test temperatures and different loading paths. The values of shear modulus practically do not exhibit any dependence on the direction of loading. Hence, it can be concluded that experimentally determined values of elasticity constants, similar to the data contained in Figure 5.5d, indicate that it is not necessary to take into account possible anisotropic elasticity of the investigated NiTi material. The methodologically strict confirmation of the correctness of the assumption on elastic isotropy and size of possible deviations from this assumption will be possible only upon performing proportional tests on tubular samples with loading including internal pressure, as only then it is possible to generate equivalent loading paths; see text below formula (5.1).

Taking advantage of the knowledge of the values of elasticity constants, the components of macroscopic phase strain tensor are determined using formulas (5.4) and next the equivalent phase transformation strain using formula (5.5). The assumption on the additivity of elastic and inelastic strains is used and lack of volumetric changes of phase transformation strain tensor (see $(5.4)_2$). The effective stress-equivalent phase transformation strain $\sigma_{\mathrm{ef}} \leftrightarrow \varepsilon_{\mathrm{eq}}^{\mathrm{pt}}$ graphs are subsequently prepared. The graphs obtained for five loading paths at temperature $T=322.5\mathrm{K}$ are plotted in Figure 5.6b. Using these graphs, critical effective stresses of pseudoelastic flow start σ^{AM} connected with forward monoclinic phase transformation, critical effective stresses of partial reverse phase transformation start σ_{ef}^{*}, and corresponding phase transformation equivalent strain $\varepsilon_{\mathrm{eq}}^{\mathrm{pt*}}$ are determined. From these, the data width of partial hysteresis loops $\Delta\sigma_{\mathrm{m}} = \sigma_{\mathrm{ef}}^{\max} - \sigma_{\mathrm{ef}}^{*}$ are determined. When due to preceding loading complete forward p.t. $(z=1)$ would have taken place, then σ_{ef}^{*} would be equal to critical stress σ^{MA}. In such a case, a stress-strain graph would have revealed so-called full hysteresis loop. In the experimental studies discussed here, obtaining such a full hysteresis loop has not been dared due to the serious risk of specimen damage. For that reason, the data for partial (internal) hysteresis loops only were obtained.

The effective critical stresses of pseudoelastic flow were determined using the fixed offset method of equivalent phase transformation strain, which was accepted on the level of 0.002. The operational process of critical stresses of pseudoelastic flow determination connected with forward and reverse phase transformation is shown schematically in Figure 5.7. The values of effective critical stresses of pseudoelastic flow for forward phase transition $\sigma^{\mathrm{AM}}(y,T)$ in dependence of temperature and loading direction (parameter y) are listed in Table 5.2.

Similar to the aid of fixed offset method (0.002), critical values of components of stress tensor $\sigma_z^{\mathrm{AM}}, \tau^{\mathrm{AM}}$ are determined at which the process of pseudoelastic flow starts for five loading paths and three temperatures. Such experimental points of critical stresses of austenite to oriented martensite transformation $(\mathcal{A} \rightarrow \mathcal{M})$ are obtained and marked with symbols in the chart shown in Figure 5.8 plotted in $\sigma - \sqrt{3}\tau$ coordinates.

By definition, these points determine the experimental surface of pseudoelastic flow start in a space of stress tensor components. For comparison purposes, in Figure 5.8 is plotted a hypothetical surface corresponding to the Huber-Mises model ($J_2{}'$ theory), which includes experimental point $\sigma^{\mathrm{AM}}(y=0, T=315\mathrm{K})$—path No. 1 (torsion). The experimental points of critical pseudoelastic flow stresses clearly show that they cannot be well accepted to be located on a hypothetical Huber-Mises surface.

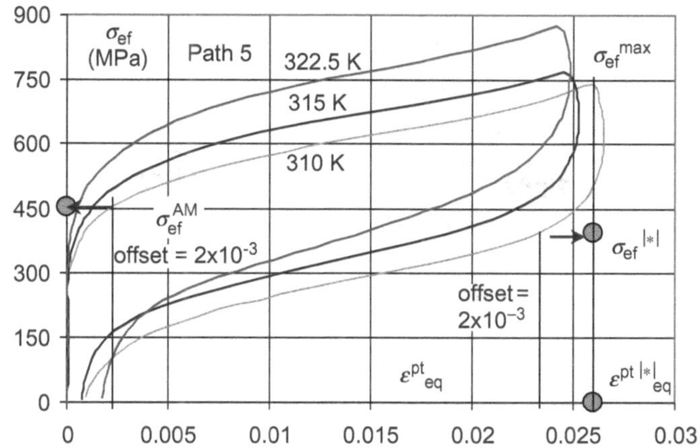

Figure 5.7 Schematic illustration of operational determination of critical stresses of pseudoelastic flow start σ^{AM}; that is, here start of forward phase transformation and critical stresses of start of partial reverse transformation σ_m.

Table 5.2 Experimental and modeling effective critical stresses of pseudoelastic flow $\sigma^{AM}(y,T)$ MPa, offset $\varepsilon_{eq}^{pt}=0.2\%$

Path	No. 3	No. 5	No. 1	No. 4	No. 2
Experiment $T=310$ (K)	510	462	407	380	329
Model	506	460	405	378	371
Experiment $T=315$ (K)	570	500	440	410	419
Model	550	500	440	411	403
$T=322.5$ (K)	604	558	491	459	458
Model	617	561	494	461	452

Modeling $\sigma^{AM}(y,T)=\rho\Delta s_0^2(T-M_s^0)/(\gamma f(y))$.

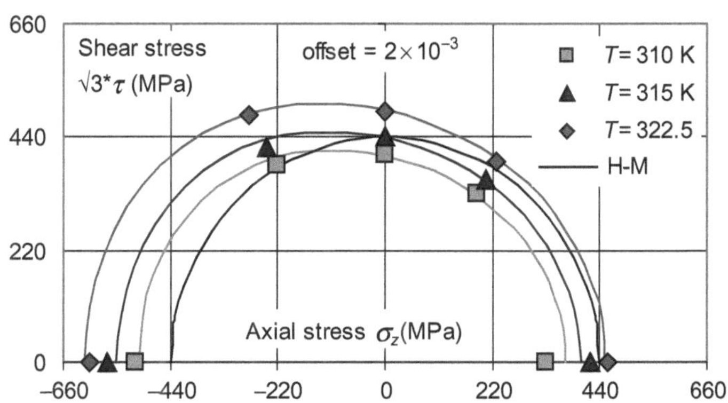

Figure 5.8 Values of critical stresses of pseudoelastic flow $A\rightarrow M$. Experimental values are marked with symbols; lines represent predicted by R_L model flow surfaces obtained upon the use of identified for the investigated NiTi alloy and monoclinic p.t. shape function $f(y)$. The symbol H-M denotes hypothetical Huber-Mises surface.

This outcome confirms earlier experimental results specified in the SMA literature regarding the appearance of considerable deviations of critical stresses of pseudoelastic flow from the surface predicted by J_2' theory. Orgeas and Favier (1996) reported that NiTi shape memory alloy investigated by them exhibited different critical stress of pseudoelastic flow in tension and in compression; the difference reaching about 40%. The appearance of such considerable deviation from the Huber-Mises condition has also been found by Sittner et al. (1995) and Rogueda et al. (1996) in the case of Cu-based alloys; see Section 3.2 in Chapter 3. These experimental findings indicate that in order to substantially improve the accuracy of modeling behavior of shape memory alloy materials it is desirable to take into account the dependence of the pseudoelastic flow limit in the constitutive model, not only on the second invariant of stress deviator but also on the third stress deviator; that is, stress tensor direction parameter y.

When elaborating the constitutive model properly predicting the strength differential (SD) effect described above, it is desirable to remain within the isotropic theory in view of the benefit of relative simplicity of the model. The experimental finding that volumetric effects connected with thermoelastic martensitic phase transformation are small indicates that the SD effect is not induced by pressure, at least until reliable information is available describing the extent of such an influence. Further steps to enable modeling of the SD effect are clearly indicated by the general theory of isotropic functions. In accordance with the theorems of isotropic scalar function—here limit surface of pseudoelastic flow; of tensorial argument—here stress tensor, independent from trace of tensorial argument—here pressure, can be a function of only second and third invariant of tensorial argument deviator—stress deviator.

When it is accepted that the limit surface can be determined by a function of second and third invariant of stress tensor deviator with the aid of constraint relation in the form

$$\begin{aligned} &F\left(J_2'(\overline{\boldsymbol{\sigma}}),J_3'(\overline{\boldsymbol{\sigma}}),T\right)=0,\\ &J_2'\equiv \mathrm{tr}\left(\overline{\boldsymbol{\sigma}}^2\right),\;\; J_3'\equiv \det\left(\overline{\boldsymbol{\sigma}}\right)=\tfrac{1}{3}\mathrm{tr}\left(\overline{\boldsymbol{\sigma}}^3\right),\;\; \overline{\boldsymbol{\sigma}}=\mathrm{dev}(\boldsymbol{\sigma}), \end{aligned} \tag{5.6}$$

that is, when it is accepted that the most general geometric form of pseudoelastic flow surface has the shape of a "star", then its equation can be written in the form

$$F^{\mathrm{AM}}=G_0^*(\overline{\boldsymbol{\sigma}},T)-1=0, \tag{5.7}$$

where the function G_0^* is homogeneous of the first order with respect to stress tensor; that is, it fulfills the condition $G_0^*(\lambda\overline{\boldsymbol{\sigma}},T)=\lambda G_0^*(\overline{\boldsymbol{\sigma}},T)$. The draft of the proof of the above statement consists of the observation that when limit surface has the shape of a star then the location of a point lying on this surface can be parametrically described with the aid of a product of a certain function describing the direction of "radius tensor" in the space of stress tensor components and scalar describing length of the radius tensor. Notice next that the function of the direction of the radius tensor is a function homogeneous of the zero order, and the length of the radius tensor is a function homogeneous of the first order with respect to the stress tensor. Thus, the conclusion is reached specified above. Coming back to the issue of limit surface of pseudoelastic flow. The stress deviator can be parametrically described as

$$\overline{\boldsymbol{\sigma}} = \sigma' \boldsymbol{n}^{\sigma} = \sqrt{\tfrac{2}{3}}\sigma_{\mathrm{ef}}\boldsymbol{n}^{\sigma}, \quad \boldsymbol{n}^{\sigma} = \overline{\boldsymbol{\sigma}}/\sigma', \quad \sigma' = \sqrt{(\overline{\boldsymbol{\sigma}}\cdot\overline{\boldsymbol{\sigma}})} = \sqrt{\tfrac{2}{3}}\sigma_{\mathrm{ef}},$$
$$\mathrm{tr}(\boldsymbol{n}^{\sigma}) = 0, \quad \mathrm{tr}(\boldsymbol{n}^{\sigma}\boldsymbol{n}^{\sigma}) = 1, \quad \sigma_{\mathrm{ef}} = \sqrt{\tfrac{3}{2}J_2'}, \tag{5.8}$$

where $\boldsymbol{n}^{\sigma}$ denotes versor of direction of stress tensor deviator. Classical Huber-Mises condition of inelastic flow, which is independent of the direction of stress deviator, can be expressed as

$$F^{\mathrm{H-M}}\left(J_2'(\overline{\boldsymbol{\sigma}})\right) = \sigma_{\mathrm{ef}}/\sqrt{3}k_{\mathrm{HM}} - 1 = 0,$$

where k_{HM} denotes limit of inelastic flow at pure shear, which is usually treated as a material parameter and it is determined from the experiment. The above form of Huber-Mises condition delivers premises in what form the rule of inelastic flow should be proposed, dependent on direction of stress deviator $\boldsymbol{n}^{\sigma}$

$$F^{\mathrm{AM}}\left(J_2', J_3'\right) = \sigma_{\mathrm{ef}}/\sigma^{\mathrm{AM}}(\boldsymbol{n}^{\sigma}, T) - 1 = \left(\sigma_{\mathrm{ef}}/\sqrt{3}k(T)\right)f^*(\boldsymbol{n}^{\sigma}, T) - 1 = 0,$$
$$f^*(\boldsymbol{n}^{\sigma}, T) = \sqrt{3}k(T)/\sigma^{\mathrm{AM}}(\boldsymbol{n}^{\sigma}, T), \tag{5.9}$$

where $k(T)$ denotes limit of pseudoelastic flow at pure shear possibly depending on temperature, and $\sigma^{\mathrm{AM}}(\boldsymbol{n}^{\sigma}, T)$ is effective critical stress of pseudoelastic flow for direction of proportional loading defined with direction versor $\boldsymbol{n}^{\sigma}$. Instead of using tensorial parameter of stress deviator direction $\boldsymbol{n}^{\sigma}$, it is more convenient to use scalar parameter of direction $y(\overline{\boldsymbol{\sigma}}) \equiv 3\sqrt{6}J_3'/\left(J_2'\right)^{3/2}$ (see (3.37)). Then the condition of pseudoelastic flow $F^{\mathrm{AM}} = 0$ can be expressed in the form

$$F^{\mathrm{AM}} = \sigma_{\mathrm{ef}} f(y, T)/\sqrt{3}k(T) - 1 = 0. \tag{5.10}$$

The described modeling approach, targeted at taking into account the dependence of flow limit on the direction of stress deviator (the so-called strength differential (SD) effect) was proposed in the 1998 version of the R_L model (Raniecki and Lexcellent, 1998). The *shape function* of pseudoelastic flow surface is defined with formula

$$f(y, T) \equiv \frac{\sigma^{\mathrm{AM}}(y=0, T)}{\sigma^{\mathrm{AM}}(y, T)}, \quad f(0, T) = 1, \tag{5.11}$$

where $\sigma^{\mathrm{AM}}(y=0, T) = \sqrt{3}k(T)$ is effective stress of pseudoelastic flow in pure shear—path No. 1 (see also (3.34)).

The experimental determination of pseudoelastic flow surface, which is equivalent to determination of a number of values of shape function—sufficient for the needed accuracy—require knowledge of the effective stress of pseudoelastic flow at pure shear and at least several other critical stresses for different loading paths directions characterized by parameter y. The critical effective stresses of pseudoelastic flow for

Figure 5.9 Predictions of identified shape function $f(y)$ for investigated $Ni_{51at\%}$-Ti alloy undergoing monoclinic phase transition—marked with continuous line. The symbols denote experimental data.

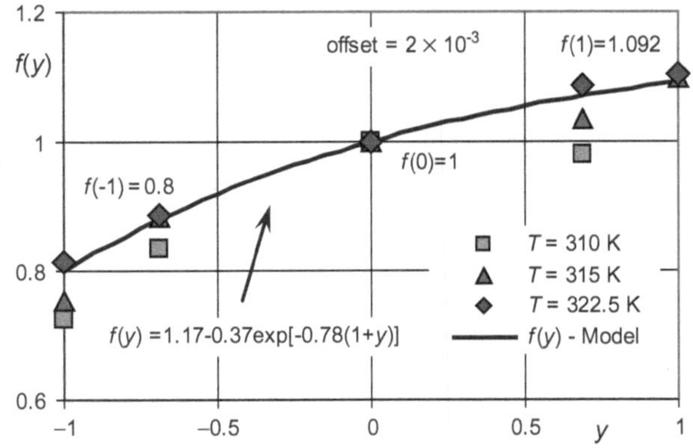

five paths and three temperatures determined for investigated NiTi alloy undergoing monoclinic martensitic transformation are collected in Table 5.2. These values served for computation of experimental values of function $f(y, T)$ in accordance with formula (5.11). They are plotted in Figure 5.9. Examination of experimental data contained in Figure 5.9 allowed the following explicit parametric mathematical form of shape function for NiTi alloy undergoing monoclinic transformation to be proposed, accepted to be independent from temperature (Raniecki et al., 2001):

$$f(y) = 1.17 - 0.37\exp[-0.78(y+1)]. \tag{5.12}$$

The predictions of pseudoelastic flow surface shape in accordance with shape function (5.12) are plotted in Figure 5.9 with continuous line. Further extensive experimental tests program is necessary to more precisely verify the viability of the isotropic shape function concept, with at least a dozen or so experimental points lying on the pseudoelastic flow surface. Although the results shown here are encouraging for the expectation that it is enough to employ isotropic shape function to describe pseudoelastic flow surfaces and obtain good conformity of modeling and experimental results for SMA materials, without the necessity to refer to the anisotropic one, the issue of finding an optimum, parametric, mathematical form of the shape function $f(y)$ for shape memory alloys and a number of parameters appearing in it constitutes an open scientific problem.

The start of pseudoelastic flow surface from pure austenitic phase predicted on theoretical grounds by the R_L model is determined by the condition of zeroing of thermodynamic driving force of phase transition π^{1-2} at $z = 0$; that is, by the formula (3.15). After substituting the expression for potential g^{ult} (formula (3.32)) to this formula, modeling condition of pseudoelastic flow in the form below is obtained (see also $(3.36)_1$)

$$F^{\mathrm{AM}} = \sigma_{\mathrm{ef}} f(y) - (\rho/\gamma)\left(\Delta s_2^0 - s_{12}\right)\left(T - M_S^0\right) = 0, \quad \sigma^{\mathrm{AM}}(y, T) = f(y)\sigma_{\mathrm{ef}}. \tag{5.13}$$

Comparing the above with formula (5.10), one obtains

$$\sqrt{3}k(T) = (\rho/\gamma)(\Delta s_2^0 - s_{12})(T - M_s^0) = \sigma^{AM}(y=0,T). \tag{5.14}$$

Hence, the R_L model predicts that critical effective stress of pseudoelastic flow in pure shear linearly depends on temperature. This relation constitutes the counterpart of the well-known *Clausius-Clapeyron relation* (CC relation) from fluid mechanics. The R_L models family actually predicts linear dependence of pseudoelastic flow stress for forward and reverse phase transition for any proportional stress loading path y in accordance with formulas (3.36) in Chapter 3; that is, infinitely many Clausius-Clapeyron lines. Taking advantage of formulas (3.36), the following expressions are obtained for slope values of Clausius-Clapeyron lines expressed in terms of R_L model material constants

$$\frac{d\sigma^{AM}(y,T)}{dT} = \frac{\rho(\Delta s_2^0 - s_{12})}{\gamma f(y)}, \quad \frac{d\sigma^{MA}(y,T)}{dT} = \frac{\rho(\Delta s_2^0 + s_{12})}{\gamma f(y)}. \tag{5.15}$$

Knowledge of experimental values of Clausius-Clapeyron lines slopes provides a good basis for identification of material constants of the R_L model.

The values of critical effective stresses of pseudoelastic flow listed in Table 5.2 are marked with symbols in the graph $\sigma_{ef} - T$ shown in Figure 5.10. These points make elements of "experimental" Clausius-Clapeyron curves for specific proportional loading path with direction y. An attempt to fit Clausius-Clapeyron lines for each proportional path direction, y, on the basis of only three available experimental points from temperatures 310, 315, and 322.5 K with the aid of least squares method, leads to miserable results. However, as hinted by common sense and by formula (5.13), when the theoretical predictions are extrapolated to zero stress value $\sigma_{ef} = 0$ then all CC lines should converge for different proportional loading paths in temperature M_s^0. Using this premise, an attempt to interpolate experimental CC lines was undertaken, using the least squares method, with the constraint that they converge in temperature $T = 253\text{K} = M_s^0$ for $\sigma_{ef} = 0$. The results of this operation are shown with continuous lines in bigger perspective in Figure 5.10a, and in zoom in Figure 5.10b. Notice that experimental data points locate very well on the CC lines, except perhaps for path No. 2, which indicates that the modeling assumption on linear dependence with temperature and convergence of Clausius-Clapeyron p.t. start lines (σ^{AM}) at temperature (M_s^0) at zero stress is well conceived. Thus, in the sequel below it is regarded as well justified that martensitic phase transformation start temperature determined by extrapolation of Clausius-Clapeyron lines (σ^{AM}) from pseudoelastic range of SMA material behavior to stress-free state is identical with the M_s^0 temperature determined in stress-free state with the aid of DSC.

The smallest slope of Clausius-Claperon line appears for loading in tension and amounts to 6.5 MPa/K, while the largest value appears for loading in compression and amounts to 8.9 MPa/K. In fact, with the aid of shape function $f(y)$, all the plethora of CC lines for different loading paths directions can be reduced to a single

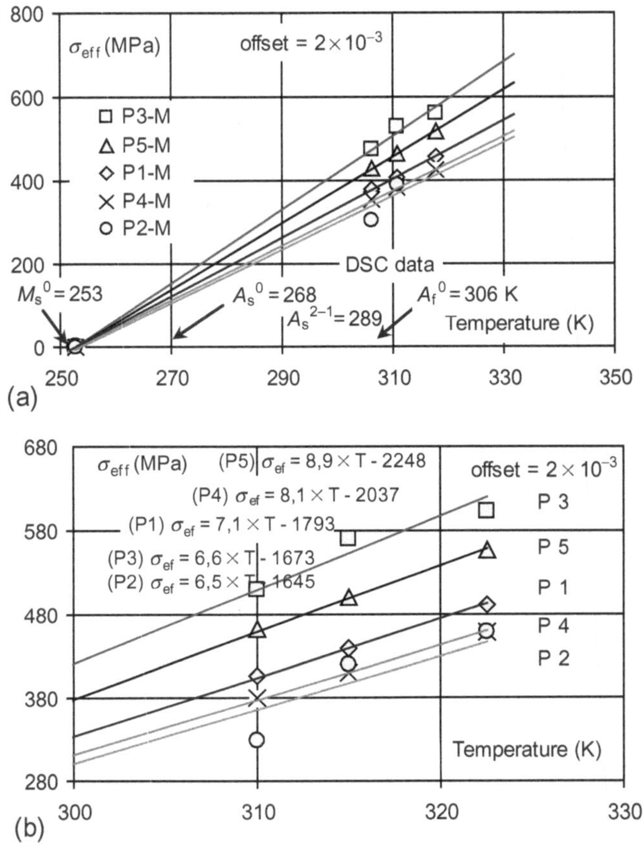

Figure 5.10 (a) Clausius-Clapeyron relations—CC, for $A \rightarrow M$ phase transition in the range of pseudoelastic behavior of NiTi alloy and monoclinic transformation. The symbols denote experimental data. Continuous lines are interpolations determined with the aid of the least squares method with assumed extrapolation to the point ($\sigma_{ef} = 0, T = 253\text{K}$); that is, to the point of martensitic phase transformation start temperature in stress-free state. (b) Location of experimental points of Clausius-Clapeyron relation in "zoom view" and numerical, experimental values of slopes of CC lines determined by interpolation with the use of the least squares method.

Clausius-Claperon line valid for all proportional loading paths directions. This observation lays the foundation for the introduction of ultimate phase potential g^{ult} in the shape (3.32), which in view of the above gains the physical interpretation that the value of ultimate phase potential is proportional to critical stress of forward pseudoelastic flow and pseudoelastic flow amplitude is the coefficient of proportionality $\rho g^{\text{ult}} = \gamma \sigma^{\text{AM}}$. This is an extremely valuable and interesting theoretical result.

Upon comparing $(5.15)_{1,2}$, it is easy to notice that constant s_{12} characterizes the difference in slopes between the Clausius-Clapeyron lines $\sigma^{\text{AM}}(y,T)$ and $\sigma^{\text{MA}}(y,T)$ or, in other words, the parallelism of these lines at fixed y. The experimental data show that, for NiTi alloys, these lines can be well accepted as parallel; see, for example, the data in Figure 46, p. 588, in a paper by Otsuka and Ren (2005) or data shown in Figure 3.4b in Chapter 3 of this book concerning the $Ni_{51at\%}$-Ti alloy investigated here. This leads to the conclusion that it is reasonable to accept a zero value for the material constant $s_{12}(s_{12}=0)$.

Taking advantage of the data obtained by Tanaka et al. (1999) in uniaxial tension tests there has been identified for the investigated $Ni_{51at\%}$-Ti alloy a value of pseudoelastic flow amplitude in simple tension $\gamma_{tn}=0.059$ (see also Figure 3.4a) and related text in Chapter 3. Substituting this value and the value of identified earlier function $F(y)$ ($f(y)$) for simple tension direction $(y=1)$, the value of pseudoelastic flow amplitude in pure shear has been obtained $\gamma=\gamma_{tn}/F(y)=\gamma_{tn}/f(y=1)=0.059/1.092=0.054$ (see $(3.40)_2$).

Upon substitution of the values of the CC line slope for pure shear (path No. 1) $\mathrm{d}\sigma^{AM}(y=0)/\mathrm{d}T=7.1\,\mathrm{MPa/K}$ identified in Figure 5.10b, value of known from literature density of NiTi alloy $\rho=6500\,(\mathrm{kg/m^3})$, value of shape function for pure shear $f(y=0)=1$, material constant $s_{12}=0$, and amplitude $\gamma=0.054$ into simply transformed formula $(5.15)_1$, the following value of entropy of formation of oriented martensite is obtained:

$$\Delta s_2^0=\frac{\mathrm{d}\sigma^{AM}(y=0)}{\mathrm{d}T}\cdot\gamma/\rho,\quad s_{12}=0,\quad \rho=6500\left(\mathrm{kg/m^3}\right),\quad \gamma=0.054.$$
$$\Delta s_2^0=59\,\mathrm{J/kgK},\quad \rho\Delta s_2^0=0.3834\,\mathrm{MPa/K}. \tag{5.16}$$

Please note that the value of material constant Δs_2^0 can be determined in a similar manner using data from simple tension (path No. 2) or simple compression (path No. 5) tests without the necessity for explicit knowledge of shape function $f(y)$. For that purpose, instead of using the individual elements of product $\gamma\cdot f(y)$ in formula $(5.15)_1$, directly *pseudoelastic flow amplitude* in simple tension $\gamma_{tn}=\bar{\gamma}(y=1)=\gamma f(y=1)$ or simple compression $\gamma_{comp}=\bar{\gamma}(y=-1)=\gamma f(y=-1)$, respectively, can be used (see (3.40)). The values γ_{tn} or γ_{comp} can be directly obtained experimentally upon executing the respective test.

The heat of complete $(z=0\rightarrow z=1)$ phase transition from pure austenite to pure self-accommodating martensite $l_f^{sa}=\Delta u_3^0=\Delta Q=17{,}500\,\mathrm{J/kg}$, measured in DSC test characteristic transformation temperatures $M_s^0=253\,\mathrm{K}$, $A_s^0=268\,\mathrm{K}$, (see thermograph in Figure 3.1 and formula $(3.22)_8$). Characteristic temperature $A_s^{2-1}=289\,\mathrm{K}$ and entropy of oriented martensite formation $\Delta s_2^0=59\,\mathrm{J/kgK}$ were determined with the aid of data shown in Figure 3.4 (see $(5.16)_5$). Together, they enable the determination of further R_L model material constants. Simple transformations of (3.16) lead to the following expressions:

$$\Delta u_2^0=\tfrac{1}{2}\left(A_s^{2-1}+M_s^0\right)\cdot\Delta s_2^0,\quad \phi_{12}=\tfrac{1}{2}\left(A_s^{2-1}-M_s^0\right)\cdot\Delta s_2^0,$$
$$\Delta u_2^0=15{,}990\,\mathrm{J/kg},\quad \phi_{12}=1062\,\mathrm{J/kg},\quad (s_{12}=0),$$
$$\Delta u_3^0=\tfrac{1}{2}\left(A_s^0+M_s^0\right)\cdot\Delta s_3^0,\quad \phi_{13}=\tfrac{1}{2}\left(A_s^0-M_s^0\right)\cdot\Delta s_3^0,$$
$$\Delta s_3^0=67.2\,\mathrm{J/kgK},\quad \phi_{13}=504\,\mathrm{J/kg},\quad (s_{13}=0). \tag{5.17}$$

In the above, the Δu_2^0, ϕ_{12} are output values computed from M_s^0, A_s^{2-1}, and Δs_2^0 treated as input data. It is worth pointing out that in accordance with formula (3.22)$_5$, Δu_2^0 is equal to the heat of complete p.t. from pure oriented martensite to pure austenite $l_f(\boldsymbol{\sigma}=0)=\Delta u_2^0$, which can and actually has been experimentally measured in a properly designed DSC procedure; see Figure 3.5 and related text. Experimental data plotted in thermographs of Figure 3.5b also show that $16.5\,\mathrm{J/g}=\Delta u_3^0>\Delta u_2^0=15\,\mathrm{J/g}$. Such a relation can be acquired on theoretical grounds, used here under the constraint that $M_s^{1-2}=M_s^0$, only when $\Delta s_3^0>\Delta s_2^0\ (A_s^{2-1}>A_s^0)$. In the case of self-accommodating martensite, the Δs_3^0, ϕ_{13} values are computed from M_s^0, A_s^0, and Δu_3^0 treated as input data.

The volume fraction of martensitic phase z has not been directly experimentally measured in the isothermal tests program of NiTi alloy discussed here. However, the volume fraction of martensitic phase can be determined indirectly using experimental values of effective phase transformation strains. Upon substitution of experimental maximum values of effective phase transformation strains $\varepsilon_{\mathrm{eq}}^{\mathrm{pt}*}$ (see Figure 5.7) into formula $z=\varepsilon_{\mathrm{eq}}^{\mathrm{pt}}/\gamma F(y)$ (see (3.40) and (5.5)), semi-empirical values $z_{\max}$ of maximum volume fractions of martensitic phase achieved during individual mechanical tests can be computed. The obtained values are shown in Figure 5.11. They deliver another opportunity for verification of the $\mathrm{R_L}$ model predictions upon the condition that volume fraction of martensitic phase is measured by, for example, measurement of variation of electrical resistance of the NiTi specimen.

In the $\mathrm{R_L}$ model it is assumed that for macroscopic phase transformation strain tensor there exists inelastic potential g^{ult} defined explicitly with formula (3.32). This form of potential leads to formula (3.39) for ultimate phase eigenstrains tensor $\boldsymbol{\kappa}$. After substituting (3.39) to formula (3.38)$_3$, the following expression for macroscopic phase strain tensor $\boldsymbol{\varepsilon}^{\mathrm{pt}}$ is obtained:

$$\begin{aligned}&\boldsymbol{\varepsilon}^{\mathrm{pt}}=z\partial\left(\rho g^{\mathrm{ult}}(\boldsymbol{\sigma},T)\right)/\partial\boldsymbol{\sigma}=\Lambda_0\mathbf{G}^*(\boldsymbol{\sigma}),\\&\mathbf{G}^*(\boldsymbol{\sigma})\equiv f(y)\frac{3}{2}\left\{\boldsymbol{n}_{\mathrm{ef}}^{\sigma}+3x\left[3\left(\boldsymbol{n}_{\mathrm{ef}}^{\sigma}\right)^2-\frac{2}{3}\boldsymbol{I}-y\boldsymbol{n}_{\mathrm{ef}}^{\sigma}\right]\right\},\\&x\equiv(1/f)\mathrm{d}f(y)/\mathrm{d}y,\quad \Lambda_0\equiv\gamma z,\quad \boldsymbol{n}_{\mathrm{ef}}^{\sigma}=\sqrt{\frac{2}{3}}\boldsymbol{n}^{\sigma},\quad \boldsymbol{n}^{\sigma}=\overline{\boldsymbol{\sigma}}/\sigma',\quad \mathrm{tr}(\boldsymbol{\varepsilon}^{\mathrm{pt}})=0.\end{aligned}\tag{5.18}$$

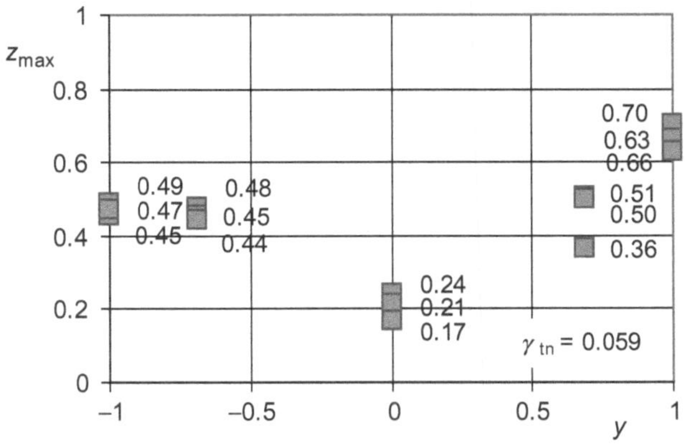

Figure 5.11 Semi-empirical values of maximum advancement of phase transition attained during proportional, isothermal tests.

By direct calculation, it is easy to show that tensor function $\mathbf{G}^*$ is isotropic, homogeneous of order zero with respect to stress tensor $\mathbf{G}^*(\lambda\boldsymbol{\sigma}) = \mathbf{G}^*(\boldsymbol{\sigma})$, where λ denotes here scalar parameter. Hence, in accordance with the R_L model conjectures also that macroscopic phase strains tensor $\boldsymbol{\varepsilon}^{\mathrm{pt}}$ is isotropic, zero order function of stress tensor. The form of function $\mathbf{G}^*$ $(5.18)_2$ does not constitute essential limitation on the range of applicability of a developed model within the framework of isotropic theory, as this form presents the most general representation of isotropic tensor function of stress deviator under the condition that x and y are independent scalar functions of the second and third invariant of stress deviator. Formula $(5.18)_2$ is special only to this extent: that there is an imposed relation $(5.18)_3$ between scalars x and y, ensuring conformity of the R_L model with the physical observations, and there are neglected effects connected with hydrostatic pressure. In the general case, postulate $(5.18)_1$ leads to rather involved pseudoelastic flow rule

$$\dot{\boldsymbol{\varepsilon}}^{\mathrm{pt}} = \dot{\Lambda}_0 \mathbf{G}^* + \Lambda_0 \dot{\mathbf{G}}^*, \quad \dot{\Lambda}_0 \equiv \gamma \dot{z}. \tag{5.19}$$

The phase transformation kinetics relations (3.26) proposed in the R_L model are insensitive to change of the time scale. The rate of phase transformation itself is proposed as a function homogeneous of order one with respect to inducing it factors $\dot{z}(\dot{\boldsymbol{\sigma}}, \dot{T}) = (\partial \dot{z}/\dot{\boldsymbol{\sigma}}) \cdot \dot{\boldsymbol{\sigma}} + (\partial \dot{z}/\dot{T})\dot{T}$, $\lambda \dot{z} = \dot{z}(\lambda \dot{\boldsymbol{\sigma}}, \lambda \dot{T})$. The postulate on the independence of phase transformation kinetics rule from the time scale (i.e., its independence from the rate of inducing it factors), results from property, commonly accepted in materials science community, that thermoelastic martensitic transformation is athermal; see, for example, Otsuka and Wayman (1999). The function $\dot{\mathbf{G}}^*$ is a function homogeneous of order one with respect to the time rate of stress tensor $\dot{\mathbf{G}}^* = \left[\partial \dot{\mathbf{G}}^*/\partial \dot{\boldsymbol{\sigma}}\right] \dot{\boldsymbol{\sigma}}$ $\lambda \dot{\mathbf{G}}^*(\dot{\boldsymbol{\sigma}}) = \dot{\mathbf{G}}^*(\lambda \dot{\boldsymbol{\sigma}})$. In view of the properties of $\dot{z}$ and $\dot{\mathbf{G}}^*$ mentioned above, the pseudoelastic flow rule (5.19) is insensitive to change of the time scale. All tests performed on the $Ni_{51at\%}$-Ti specimen discussed here are performed with a constant rate of loading; hence, its results do not allow verifying the admissibility of the postulate on independence of flow rule (5.19) from the time scale.

In the case of proportional loading paths, that is, such paths that $\boldsymbol{\sigma}(t) = \boldsymbol{\sigma}^{\mathrm{Max}} \cdot (t/t_{\mathrm{Max}})$, the term $\dot{\mathbf{G}}^*(\boldsymbol{\sigma}) = \mathbf{0}$, as then $\dot{\boldsymbol{n}}^{\sigma} = 0$ and pseudoelastic flow rule (5.19) for this class of loadings simplifies to the form

$$\begin{gathered} \dot{\boldsymbol{\varepsilon}}^{\mathrm{pt}} = \dot{\Lambda}_0 \mathbf{G}^*(\boldsymbol{\sigma}), \quad \dot{\Lambda}_0 \equiv \gamma \dot{z}, \quad \boldsymbol{\sigma}(t) = \boldsymbol{\sigma}^{\mathrm{Max}} \cdot (t/t_{\mathrm{Max}}), \\ y, f(y), \frac{df(y)}{dy}, x, = \mathrm{const}, \quad \boldsymbol{n}^{\sigma} = \overline{\boldsymbol{\sigma}}^{\mathrm{Max}} / \sqrt{\overline{\boldsymbol{\sigma}}^{\mathrm{Max}} \overline{\boldsymbol{\sigma}}^{\mathrm{Max}}}, \quad \overline{\boldsymbol{\sigma}}^{\mathrm{Max}}, t_{\mathrm{Max}} = \mathrm{const}. \end{gathered} \tag{5.20}$$

The flow rule (5.20) predicts that proportionality of loading path in stress components space results in proportionality of the material response path in the space of inelastic strains connected with phase transformation $\left(\dot{\varepsilon}^{\mathrm{pt}}_{\mathrm{KL}} \equiv C_{\mathrm{KLMN}} \dot{\varepsilon}^{\mathrm{pt}}_{\mathrm{MN}};\ C_{\mathrm{KLMN}} = \mathrm{const}\right)$. Investigation of the graphs of ratios of rates $C^{\exp}_{\mathrm{KLMN}} = \dot{\varepsilon}^{\mathrm{pt}}_{\mathrm{KL}} / \dot{\varepsilon}^{\mathrm{pt}}_{\mathrm{MN}}$ determined on the basis of experimental data originating from proportional tests allows for verification of the admissibility of mathematical form proposed in the R_L model flow rule.

In research work targeted at the verification of R-phase reorientation law, Raniecki et al. (1999b) proposed an original procedure for verification of postulated isotropy of reorientation flow rule in the case when there are available results of tests with proportional loading paths. Here, a similar procedure is used for experimental verification of special mathematical form of pseudoelastic flow rule (5.19), which reduces to the formula (5.20) at proportional loading paths to the case of monoclinic martensitic transformation.

The validity of relation (5.20) (see also $(5.18)_2$), requires that at any time instant three scalar values $\dot{\Lambda}_0$, y, and x determine five components of macroscopic phase strain deviator $\boldsymbol{\varepsilon}^{\mathrm{pt}}$ expressed in terms of—treated as parameters—components of constant loading tensor $\boldsymbol{\sigma}^{\mathrm{Max}} = \mathrm{const}$. Such mathematical structure allows concluding that in the general case two constraint relations must exist linking components of tensor $\dot{\boldsymbol{\varepsilon}}^{\mathrm{pt}}$ with components of tensor $\boldsymbol{\sigma}^{\mathrm{Max}}$, valid for any values of scalars $\left(\dot{\Lambda}_0, y, x\right)$ mentioned earlier. In the case of tests without internal pressure loadings and at an assumed isotropy of flow rule (5.20), only one of these relations is nontrivial. Taking advantage of (5.20) and (5.18), the following relation can be derived between ratios of components of tensor $\dot{\boldsymbol{\varepsilon}}^{\mathrm{pt}}$ and components of tensor $\boldsymbol{\sigma}^{\mathrm{Max}}$ (see (5.3))

$$R_1 = \frac{\dot{\varepsilon}_{z\theta}^{\mathrm{pt}}}{\dot{\varepsilon}_z^{\mathrm{pt}}} = \frac{3}{2}\left[1 - \frac{3\left(\tau^{\mathrm{Max}}\right)^2 x}{M}\right]\frac{\tau^{\mathrm{Max}}}{\sigma_z^{\mathrm{Max}}}, \quad R_2 = \frac{\dot{\varepsilon}_{\theta}^{\mathrm{pt}}}{\dot{\varepsilon}_z^{\mathrm{pt}}} = -\frac{1}{2}\left[1 - 3\frac{3\left(\tau^{\mathrm{Max}}\right)^2 x}{M}\right],$$
$$M = x\left[2\left(\sigma_z^{\mathrm{Max}}\right)^2 + 3\left(\tau^{\mathrm{Max}}\right)^2 - 2y\sigma_z^{\mathrm{Max}}\sigma_{\mathrm{ef}}^{\mathrm{Max}}\right] + \frac{2}{3}\sigma_z^{\mathrm{Max}}\sigma_{\mathrm{ef}}^{\mathrm{Max}}. \tag{5.21}$$

Upon elimination of term $3(\tau^{\mathrm{Max}})^2 x/M$ from formulas $(5.21)_{1,2}$ one obtains searched constraint relation

$$\sigma_z^{\mathrm{Max}}\dot{\varepsilon}_{z\theta}^{\mathrm{pt}} + \tau^{\mathrm{Max}}\left(\dot{\varepsilon}_{\theta}^{\mathrm{pt}} - \dot{\varepsilon}_z^{\mathrm{pt}}\right) = 0, \tag{5.22}$$

Introducing the definition of function Y as below and using the definition of parameter $m = \tau^{\mathrm{Max}}/\sigma_z^{\mathrm{Max}}$ (see $(5.1)_3$), the constraint relation (5.22) takes the form

$$Y = m, \quad Y \equiv \frac{\dot{\varepsilon}_{z\theta}^{\mathrm{pt}}/\dot{\varepsilon}_z^{\mathrm{pt}}}{1 - \dot{\varepsilon}_{\theta}^{\mathrm{pt}}/\dot{\varepsilon}_z^{\mathrm{pt}}}. \tag{5.23}$$

Constraint relation (5.23) indicates that if the isotropic flow rule (5.20) for investigated material is valid, then the value of Y calculated upon experimental data should be equal to the value of the ratio of components of loading stress tensor m. Relation (5.23) is valid for any values of scalar parameters x, y and $\dot{\Lambda}_0$. It delivers a tool enabling, in an indirect way, the experimental verification of the assumption on inelastic isotropy of the flow rule (5.20), and thus (5.18) for the case of proportional loading paths. Deviation of experimental data from theoretical relation (5.23) indicates the dependence of material properties of investigated material on direction of loading; that is, the existence of an inelastic anisotropy component in the material behavior.

In order to experimentally validate the admissibility of made in $\mathrm{R_L}$ models family postulate of inelastic isotropy expressed by relation (5.18), and defining flow rule (5.20), two ratios are calculated: the ratio of shear phase strain rate to the rate of axial phase strain $R_1 = \dot{\varepsilon}_{z\theta}^{\mathrm{pt}}/\dot{\varepsilon}_z^{\mathrm{pt}}$, and the ratio of hoop phase strain rate to the rate of axial phase strain $R_2 = \dot{\varepsilon}_\theta^{pt}/\dot{\varepsilon}_z^{pt}$ for paths Nos. 2, 3, 4, and 5 and three temperatures $T = 310, 315, 322.5$ K. In the case of torsion loading (path No. 1), the following ratios are calculated: $\dot{\varepsilon}_z^{pt}/\dot{\varepsilon}_{\theta z}^{pt}$ and $\dot{\varepsilon}_\theta^{pt}/\dot{\varepsilon}_{\theta z}^{pt}$.

Figure 5.12 presents plotted ratios R_1 and R_2 in dependence of effective stress σ_{ef} for a few selected paths and temperatures: for path No. 3 at temperature 310°K, for

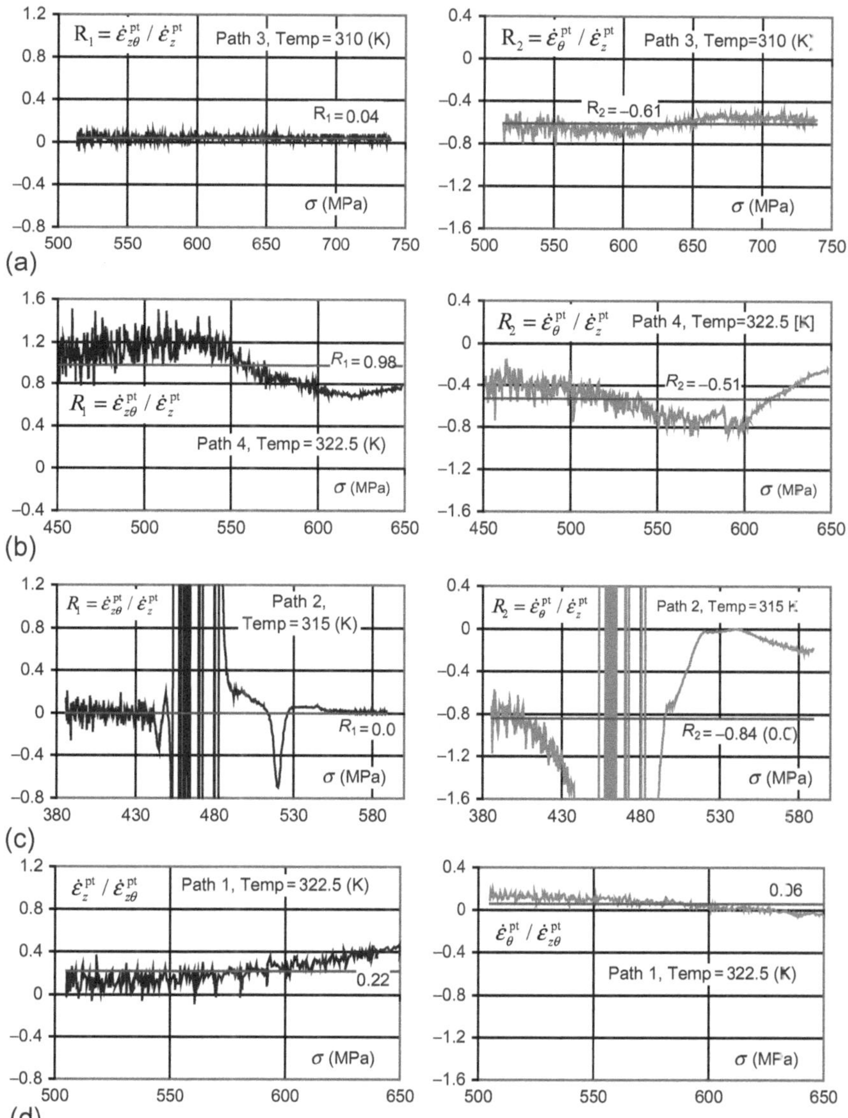

Figure 5.12 The evolution of ratios of velocity of phase transformation shear strain and velocity of phase transformation axial strain $R_1 = \dot{\varepsilon}_{z\theta}^{pt}/\dot{\varepsilon}_z^{pt}$, and ratios of velocity of phase transformation hoop strain and velocity of phase transformation axial strain $R_2 = \dot{\varepsilon}_\theta^{pt}/\dot{\varepsilon}_z^{pt}$. (a) Path No. 3, $T = 310$ K. (b) Path No. 4, $T = 322.5$ K. (c) Path No. 2, $T = 315$ K. (d) Ratios $\dot{\varepsilon}_z^{pt}/\dot{\varepsilon}_{z\theta}^{pt}$ and $\dot{\varepsilon}_\theta^{pt}/\dot{\varepsilon}_{z\theta}^{pt}$ Path No. 1, $T = 322.5$ K.

path No. 4 at temperature $T = 322.5°K$, and for path No. 2 at temperature 315°K. There are also plotted ratios $\dot{\varepsilon}_z^{pt}/\dot{\varepsilon}_{z\theta}^{pt}$ and $\dot{\varepsilon}_\theta^{pt}/\dot{\varepsilon}_{z\theta}^{pt}$ for path No. 1 at temperature 322.5 K. The "noise" visible in the graphs plotted in Figure 5.12 results from point wise digital registration of data. In order not to "distort" the picture no numerical "smoothing" of raw registered signals $\varepsilon_{ij}^{pt}(t)$ has been done before calculating numerical "derivates" of $\dot{\varepsilon}_{ij}^{pt}(t)$ signals to subsequently compute ratios R_1,R_2. The graphs contained in Figure 5.12 representatively illustrate certain characteristic situations encountered.

In the case of path No. 3, nearly ideal lines of constant values of ratio R_1 for values of effective stress remaining within the plateau of pseudoelastic flow for all temperatures were obtained. The value of ratio R_2 remained constant up to half of the pseudoelastic plateau, and then after a short period of drifting it reached constant value of -0.5. The averaged values calculated for the range of the whole pseudoelastic plateau are listed in Table 5.3. The graphs of ratios R_1 and R_2 for path No. 3 at a temperature of 310 K are plotted in Figure 5.12a.

In the case of path No. 4 the invariability of values of ratios R_1,R_2 for all temperatures is worse than the ones obtained for path No. 3. The characteristic picture is well illustrated by plots of ratios R_1 and R_2 for path No. 4 at temperature 322.5 K shown in Figure 5.12b. The ratios R_1 and R_2 preserve constant value in some quite extensive ranges of pseudoelastic flow plateau but in the range of the whole plateau they undergo quite considerable variations. For example ratio R_1 amounts initially to about 1.1 in the range of effective stresses $\langle 450, 540 \rangle$ MPa, to stabilize at the value of about 0.8 in the range of stresses $\langle 580, 650 \rangle$ MPa. The average value of R_1 over the whole plateau has the value of 0.98. The average ratio R_2 over the pseudoelastic flow plateau range amounts to -0.51 with variation of the order ± 0.2.

In the case of path No. 2, for temperatures of 315 and 322.5 K an interesting situation emerged. The ratios R_1 and R_2 preserve constant value in the initial stage of phase transformation; that is, in the range of effective stresses $\langle 390, 440 \rangle$ MPa for temperature $T = 315°K$ and $\langle 450, 500 \rangle$ MPa for temperature $T = 322.5°K$, respectively. These values are listed in Table 5.4 as the first numbers without commas. Next, a stage of very strong oscillations takes place in the range of effective stresses $\langle 440, 500 \rangle$ MPa for temperature $T = 315°K$ and $\langle 500, 560 \rangle$ MPa for temperature $T = 322.5°K$, respectively. This indicates the presence of some kind of instability in the process of phase transformation. In the third stage, in the range of effective stresses $\langle 500, 590 \rangle$ MPa for temperature $T = 315°K$ and $\langle 560, 640 \rangle$ MPa for temperature $T = 322.5°K$, the values of ratios R_1 and R_2 again stabilize. However, while ratio R_1 again returns to zero value similar to the first stage of pseudoelastic flow, the ratio R_2 takes different a value than those in the initial stage of phase transformation. These values are listed in commas in Table 5.4. It may be suspected that the effects described above constitute an expression of multiaxial instability resulting from a play between hoop and radial components of phase transformation strain taking over a momentary, dominant role in the deformation mechanism. The instability described above does not appear for path No. 2 at temperature 310 K.

In the case of path No. 5, ratios R_1, R_2 remain constant at all temperatures to only about half of pseudoelastic flow plateau and in the second half they rapidly diminish in

Table 5.3 **Comparison of determined on the base of experimental data and predicted theoretically ratios of velocities of components of strain tensor and values of function $Y(m)$ for temperatures 310, 315, and 322.5 K**

	No. 3	No. 5	No. 4	No. 2	No. 1
Path	**Compression**	**Compression +torsion**	**Tension +torsion**	**Tension**	**Torsion**
$R_1 = \dot{\varepsilon}^{pt}_{z\theta}/\dot{\varepsilon}^{pt}_{z}$					$\dot{\varepsilon}^{pt}_{z}/\dot{\varepsilon}^{pt}_{\theta z}$
Exp. $T=310$	0.04	−2.0	1.17	0.00	0.32
Exp. $T=315$	0.05	−2.0	1.11	0.00	0.15
Exp. $T=322.5$	0.09	−2.2	0.98	0.00	0.22
Theory J_3'	0.00	−3.03	1.28	0.00	0.00
Theory J_2'	0.00	−1.50	1.50	0.00	0.00
$R_2 = \dot{\varepsilon}^{pt}_{\theta}/\dot{\varepsilon}^{pt}_{z}$					$\dot{\varepsilon}^{pt}_{\theta}/\dot{\varepsilon}^{pt}_{\theta z}$
Exp. $T=310$	−0.61	−1.00	−0.41	−0.26	−0.09
Exp. $T=315$	−0.63	−1.00	−0.48	−0.84 (−0.02)	0.11
Exp. $T=322.5$	−0.7	−1.00	−0.51	−0.78 (−0.05)	0.06
Theory J_3'	−2.94	−2.03	−0.28	−0.13	0.00
Theory J_2'	−0.50	−0.50	−0.50	−0.50	0.00
Path	3	5	4	2	1
Auxiliary data common for all temperatures					
$m=\tau/\sigma_z$	0	−1	1	0	$(1/m=0)$
y	−1.00	−0.69	0.69	1.00	0
$f(y)$	0.800	0.880	1.071	1.092	1.0
$df(y)/dy$	0.289	0.226	0.077	0.061	0.132
$F(y)$	0.800	1.009	1.084	1.092	1.076
$\gamma(y)$	0.0432	0.0545	0.0585	0.0590	0.0581
$x \equiv \frac{1}{f}\frac{df(y)}{dy}$	0.361	0.257	0.072	0.056	0.132
$3x\tau^2/M$	−1.62	−1.02	0.14	0.25	1

value reaching even -15.0. For example, in the case of path No. 5, $T=315$ K ratios R_1, R_2 remain practically constant for range $\sigma_{ef} \in \langle 480, 600\rangle$ MPa taking the values $R_1 = -2.0$, $R_2 = -1.0$, then in the range $\sigma_{ef} \in \langle 600, 720\rangle$ MPa R_1 diminishes from value -2 to -11, and R_2 drops in value from -1 to -5.

In the case of path No. 1, the ratios $\dot{\varepsilon}^{pt}_{z}/\dot{\varepsilon}^{pt}_{\theta z}$ and $\dot{\varepsilon}^{pt}_{\theta}/\dot{\varepsilon}^{pt}_{\theta z}$ have been computed and plotted. The plots of these ratios for temperature 322.5 K are shown in

Table 5.4 Comparison of determined from experimental data and predicted theoretically values of function *Y*(*m*) for temperatures 310, 315, and 322.5°K

Path	No. 3	No. 5	No. 1	No. 4	No. 2
	$Y(m)$				
Exp. $T=310$	0.02	−1.00	-	0.83	0.00
Exp. $T=315$	0.03	−1.00	-	0.75	0.00
Exp. $T=322.5$	0.05	−1.10	-	0.65	0.00
Theory J_3'	0.00	−1.00	-	1.00	0.00
Theory J_2'	0.00	−1.00	-	1.00	0.00

Figure 5.12d. The ratio $\dot{\varepsilon}_z^{\mathrm{pt}}/\dot{\varepsilon}_{\theta z}^{\mathrm{pt}}$ preserves a pretty constant value of about 0.12 up to about half of the pseudoelastic flow plateau and then starts to drift to reach the value of 0.35 at maximum stress load. The ratio $\dot{\varepsilon}_\theta^{\mathrm{pt}}/\dot{\varepsilon}_{\theta z}^{\mathrm{pt}}$ drifts regularly from an initial value of 0.13 to 0.0. The average values of these ratios in the range of effective stress corresponding to pseudoelastic flow are listed in Table 5.3.

The experimental data discussed above indicate that the constant value of ratios of components of inelastic strain R_1, R_2 predicted by the $\mathrm{R_L}$ model find quite reasonable confirmation in experimental results obtained for NiTi alloy. However, there exist substantial deviations from this regularity; for example, strong instabilities in phase transformation progress in the case of path No. 2 at temperatures of 315 and 322.5 K (strong instability does not appear at temperature $T=310$ K). The effect of a considerable drop of the values of these ratios close to the loading maximum also appears in the case of path No. 5.

The acceptability of an assumed specific, isotropic inelasticity expressed by the adopted explicit form of function **G*** (see $(5.18)_2$) and the correctness of the adoption of a specific form of shape function $f(y)$ (see (5.12)) can be checked by comparison of the computed from experimental data ratios $R_{1\_\mathrm{exp}}=\dot{\varepsilon}_{z\theta}^{\mathrm{pt}}/\dot{\varepsilon}_z^{\mathrm{pt}}$, $R_{2\_\mathrm{exp}}=\dot{\varepsilon}_\theta^{\mathrm{pt}}/\dot{\varepsilon}_z^{\mathrm{pt}}$ with the theoretical ratios predicted by the $\mathrm{R_L}$ model $R_{1\_\mathrm{Mod}}, R_{2\_\mathrm{Mod}}$; that is, calculated from the formulas (5.21). The average values of ratios $R_{1\_\mathrm{exp}}, R_{2\_\mathrm{exp}}$ calculated on the basis of experimental data and ratios $R_{1\_\mathrm{Mod}}, R_{2\_\mathrm{Mod}}$ calculated on the theoretical basis for J_3' theory ($f(y)\neq\mathrm{const}$) and J_2' theory ($f(y)=\mathrm{const}$; $x=0$) are listed together in Table 5.3. In Table 5.3, values of auxiliary expressions are also specified, which have been used for calculating modeling values of ratios of rates. These values are temperature independent; hence, they are common for all tests with the same loading path.

The experimental values of ratio $\dot{\varepsilon}_{z\theta}^{\mathrm{pt}}/\dot{\varepsilon}_z^{\mathrm{pt}}$ for paths No. 2 and 3 indicate that in the tension and compression mode of loading, no coupling between axial and shear strain components occurs. In the shear mode of deformation (path No. 1), ratio $\dot{\varepsilon}_z^{\mathrm{pt}}/\dot{\varepsilon}_{z\theta}^{\mathrm{pt}}$ reaches a value of 0.2, indicating small coupling between axial and shear strain. Perfect lack of coupling between axial and shear strain components for paths No. 2 and 3 and small coupling in the case of path No. 1 argues for acceptability of the assumption on inelastic isotropy made in the R_L models family and specific form of function **G*** $(5.18)_2$.

Comparing modeling and experimental values of velocities ratios listed in Table 5.3, good agreement is observed for both J_2' and J_3' theories for paths No. 1, 2, and 3, suggesting admissibility of conjecture on inelastic isotropy, but considerable discrepancies can be noticed for J_2' theory and multiaxial paths No. 4 and 5. This indicates that not very good modeling accuracy can be expected when using J_2' theory for description of SMA materials behavior within the framework of inelastic isotropy. Introduction of another variable into the modeling (i.e., J_3' theory) improves the situation but considerable discrepancies between modeling and experimental values are still present for compression loadings (path No. 5). However, the modeling values of velocity ratios are very sensitive to the value of parameter x; that is, the assumed specific mathematical form of shape function $f(y)$. When within J_3' theory the value of parameter x is taken to be $x=0.08$ for path No. 5, then the values of respective modeling ratios become $R_{1\_\mathrm{Mod}}(5)=-1.98$ and $R_{2\_\mathrm{Mod}}(5)=-0.98$ and excellent compatibility between modeling predictions and experimental results is obtained. The above discussion indicates the necessity for further research targeted at a more precise determination of function $f(y)$.

A convenient indicator on acceptability of assumption on isotropy of inelastic flow, not burdened with the assumption on specific mathematical form of the shape function $f(y)$, is the function $Y(m)$ defined with formula (5.23). The experimental values and modeling prediction of the values of this function calculated for different proportional loading directions m (paths No. 2, 3, 4, and 5), are listed in Table 5.4. These data again indicate that behavior of studied $Ni_{51at\%}$-Ti alloy undergoing monoclinic martensitic transformation can be modeled with good accuracy upon accepting the assumption of inelastic isotropy; that is, upon assumption that macroscopic tensor of phase strains can be expressed with the aid of isotropic function of stress tensor.

For torsion loadings (path No. 1) constraint relation (5.22) leads to the constraint relation $\sigma_z^{\mathrm{Max}}/\tau^{\mathrm{Max}}=\dot{\varepsilon}_z^{\mathrm{pt}}/\dot{\varepsilon}_{z\theta}^{\mathrm{pt}}-\dot{\varepsilon}_\theta^{\mathrm{pt}}/\dot{\varepsilon}_{z\theta}^{\mathrm{pt}}=0$. Experimental values contained in Table 5.3 show that this constraint relation is well fulfilled.

5.4 Procedure for identification of the R_L model material parameters. Comparison of modeling predictions with experimental evidence for NiTi alloy

In the previous section, attention was focused on the experimental validation of selected fundamental theoretical assumptions, on which development of the family of constitutive R_L model of SMA materials is based. The actual user of the R_L model will also be interested in the most rapid and the most efficient way of putting the model to work, so that the behavior of an interesting SMA alloy can be predicted. In this section, a pragmatic procedure for the efficient identification of material constants and functions of the R_L model and the types and numbers of experimental tests required are discussed.

The complete set of material data required for modeling behavior of isotropic SMA material in pseudoelastic range of its behavior is given in Chapter 3 (see (3.44)). These data can be subdivided into four groups,

(i) physical and mechanical (elastic) properties $\rho, (E, \nu \rightarrow \mathbf{L})$,
(ii) thermal properties α_0, c_p, λ_c,
(iii) material constants and function characterizing phase transformation $\Delta u_2^0, \Delta s_2^0, \phi_{12}, s_{12}, \gamma, f(y)$,
(iv) material constants characterizing phase transformation kinetics $m_1, m_2, r_1, r_2, p_1, p_2$.

5.4.1 Data of group (i)—physical and mechanical properties

Density ρ, Young's modulus E, and Poisson ratio ν are the data delivered in a standard way by the companies manufacturing shape memory alloys. The elastic properties can be determined in a simple uniaxial tension or compression test, for inspection purposes. The value of Poisson coefficient ν can be conveniently determined in a torsion (shear) test, which must be performed anyway in order to determine the critical pseudoelastic flow stress in shear needed for identification of parameters of shape function $f(y)$.

5.4.2 Data of group (ii)—thermal properties

Thermal expansion coefficient α_0, heat capacity c_p, and coefficient of heat conduction λ_c are data delivered in a standard way by the companies manufacturing shape memory alloys.

5.4.3 Data of group (iii)—shape memory effects static properties

These data are specific for constitutive model of SMA material to be used for modeling of its behavior. Some data are common for all exiting SMA models (e.g., pseudoelastic flow amplitude); the others are generic for a particular model. Due to that, these data do not constitute standard data delivered by SMA materials manufacturers. They must be determined by the user of the R_L model.

The scope of material data identification work depends on actual modeling needs and accuracy of modeling predictions. It is different for uniaxial loading cases and different for multiaxial loadings.

In the case of the very popular *uniaxial loadings*, for example, simple tension or compression loadings at different temperatures, the required static SME properties $\Delta u_2^0, \Delta s_2^0, \phi_{12}, s_{12}, \gamma, f(y)$ (see (3.44)) can be effectively determined upon adopting one of two approaches.

The first approach requires minimum experimental effort of:

- executing one DSC measurement to determine characteristic temperatures of self-accommodating martensite, austenite p.t. characteristic temperatures $M_f^0, M_s^0, A_s^0, A_f^0$ (alternatively, such data can be obtained from the SMA material manufacturer).
- executing one special DSC measurement with SMA material specimen being initially in purely oriented martensite state,

- executing one isothermal mechanical test involving outermost hysteresis loop in pseudoelastic range of SMA material behavior $(T > A_f^{2-1})$, such as shown in Figure 3.3a,
- making one assumption that Clausius-Clapeyron lines for forward and reverse p.t. are parallel, which is equivalent to adopting constant $s_{12} = 0$ (see formulas (5.15) and Figures 3.3b and 3.4b).

In the first DSC, characteristic temperatures determined allow estimating at which temperature an isothermal mechanical test should be executed to obtain closed pseudoelastic hysteresis loop, and they allow correctly choosing a temperature at which the SMA specimen can remain in a purely oriented martensite state upon mechanical unloading to zero stress.

Upon the execution of an isothermal mechanical test, pseudoelastic flow amplitude can be determined in tension γ_{tn} or compression γ_{comp}, respectively. It is not necessary to identify shape function $f(y)$ for modeling of uniaxial loadings.

The second DSC measurement requires special preparation of an SMA material specimen. In order to convert SMA material into a purely oriented martensite state, an SMA sample (e.g., in wire form) must be loaded mechanically at a temperature below M_f^0 with mechanical loading high enough to obtain pure oriented martensite. Upon mechanical unloading, a small DSC sample must be cut and placed in a DSC apparatus taking care that no reverse phase transition takes place beforehand. Next, a DSC measurement must be executed starting with a heating segment from low temperature and one or two complete heating-cooling cycles during which pure austenitic and pure martensitic phases are obtained; see also text discussing results shown in Figure 3.5b. In such a DSC procedure start temperature of oriented martensite to austenite phase transition, self-accommodating martensite start temperature, and latent heat of complete phase transition from oriented martensite to austenite $A_s^{2-1}, M_s^0, Q_{2-1}$ are determined. Taking advantage of formula (3.16) and DSC measurement data, the material model constants $\Delta u_2^0, \Delta s_2^0, \phi_{12}$ can be determined

$$\Delta u_2^0 = Q_{2-1}, \quad \Delta s_2^0 = 2\Delta u_2^0 / \left(M_s^0 + A_s^{2-1}\right), \quad \phi_{12} = \tfrac{1}{2}\left(A_s^{2-1} - M_s^0\right)\cdot \Delta s_2^0, \quad s_{12} = 0 \tag{5.24}$$

Naturally, the approach described above is not always practically feasible, especially when M_f^0 temperature is very low.

The second approach requires minimum experimental effort of:

- executing one DSC measurement to determine the characteristic temperatures of self-accommodating martensite, austenite p.t. characteristic temperatures $M_f^0, M_s^0, A_s^0, A_f^0$
- executing two isothermal mechanical tests involving the outermost hysteresis loops in the pseudoelastic range of SMA material behavior at two temperatures $(T_1, T_2 > A_f^{2-1} > A_f^0)$ (see, e.g., Figure 3.3a), similar to the first approach adopting constant $s_{12} = 0$.

Pseudoelastic flow amplitude in tension γ_{tn} or compression γ_{comp} can be determined from any of the respective isothermal mechanical tests executed in the pseudoelastic range of SMA material behavior—$\overline{\gamma}(y = 1) = \gamma_{tn}$, $\overline{\gamma}(y = -1) = \gamma_{comp}$.

In order to identify other R_L model material constants, critical stresses of forward pseudoelastic flow $\sigma^{AM}(T_1), \sigma^{AM}(T_2)$ must be determined from plots of two isothermal mechanical tests (see, e.g., Figure 5.7). Then, constant Δs_2^0 can be determined from the formula

$$\Delta s_2^0 = \frac{[\sigma^{AM}(T_2, y) - \sigma^{AM}(T_1, y)]}{T_2 - T_1} \frac{\overline{\gamma}(y)}{\rho}, \quad s_{12} = 0 \tag{5.25}$$

Next, critical stresses of reverse pseudoelastic flow $\sigma^{MA}(T_1), \sigma^{MA}(T_2)$ must be determined from plots of two isothermal mechanical tests involving outer hysteresis loops (see, e.g., Figure 3.6b). Through two experimental points $(T_1, \sigma^{MA}(T_1))$, $(T_2, \sigma^{MA}(T_2))$ an experimental Clausius-Clapeyron line $\sigma^{MA}(T) = \Delta s_2^0 (T - A_s^{2-1})$ can be drawn. Extrapolation of this line to zero macroscopic stress $\sigma^{AM} = 0$ determines characteristic temperature A_s^{2-1} (see also Figure 3.4b). Simple transformations of (3.16) and substitution of already known $M_s^0, \Delta s_2^0, A_s^{2-1}$ allows for determination of $\Delta u_2^0, \phi_{12}$ as follows:

$$\Delta u_2^0 = \frac{1}{2}\left(A_s^{2-1} + M_s^0\right) \cdot \Delta s_2^0, \quad \phi_{12} = \frac{1}{2}\left(A_s^{2-1} - M_s^0\right) \cdot \Delta s_2^0, \quad s_{12} = 0 \tag{5.26}$$

When conjecture on the parallelism of forward and reverse Clausius-Clapeyron lines cannot be accepted for specific SMA material, then reverse CC lines must be experimentally determined in order to determine value of $s_{12} = 0$ prior to executing the identification procedures delineated above. Nonzero value of constant $s_{12} \neq 0$ does not create any serious difficulties for the material constants identification procedure or the slightly more complicated mathematical expressions.

In a case when modeling of SMA material submitted to *multiaxial loadings* is required, then the only additional material datum that is needed besides the material data necessary for uniaxial loadings is identification of the shape function $f(y)$. The minimum experimental effort required in this process is the execution of at least three isothermal mechanical tests with three different mechanical loading directions, for example, simple tension $(y = 1)$, simple compression $(y = -1)$, and pure shear $(y = 0)$. From plots of these tests, critical pseudoelastic flow stresses $(\mathcal{A} \rightarrow \mathcal{M})$ must be determined $\sigma^{AM}(y = -1), \sigma^{AM}(y = 0), \sigma^{AM}(y = 1)$. Then, using definition of shape function (3.34), experimental values of shape function values for specific SMA material can be computed for $y = -1$ and $y = 1$. Next, using three known values of shape function $\{f(y = -1), f(y = 0) = 1, f(y = 1)\}$, allows identifying three values of constants appearing in the formula $f(y) = a_1 - a_2 \cdot \exp[-a_3 \cdot (y + 1)]$ (see (5.12)). It is advisable to perform a checking procedure whenever values of shape function parameters identified for specific SMA material do not lead to nonconvex pseudoelastic flow surface. The problem is as follows. While function (5.12) is a convex function—in accordance with the mathematical definition of a function convexity—it represents only one-sixth of a trace of the pseudoelastic flow surface on an octahedral plane. Hence, convexity of shape function itself does not guarantee convexity of the pseudoelastic flow surface figure. The situation is analogous to several flower petals that are convex by themselves but together make a nonconvex flower face shape. More information on this problem can be found in a paper by Raniecki and Mróz (2008).

5.4.4 Data of group (iv)—phase transformation kinetics constants

No additional experimental work is necessary to determine constants of forward and reverse martensitic phase transition kinetics taking place in SMA material besides that already required for identification of static SME constants; that is, data from group (iii). Results of only one or more isothermal, mechanical tests involving the outermost hysteresis loop in the pseudoelastic range of SMA material behavior are necessary for this. In order to identify kinetics constants for forward (m_1, r_1, p_1) and reverse (m_2, r_2, p_2) phase transitions, the registered experimental time series of stress and strain components must be elaborated to prepare two plots at the minimum. It is convenient to prepare plots of volume fraction of martensitic phase as a function of normalized thermodynamic driving force of phase transformation—$z(D)$. The semi-experimental volume fraction of martensitic phase z can be computed from experimental strain signals with the aid of formulas (3.40). The normalized thermodynamic driving force D can be computed using stress components signals, semi-experimental volume fraction signal, and temperature value of isothermal test taking advantage of the formulas (3.35) and $(3.25)_5$ with normalizing factor (e.g., $\pi_N^{1-2} = \phi_{12}$). The volume fraction of martensitic phase must always remain in the range $z \in (0, 1)$, while the range of values of normalized thermodynamic driving force cannot be known in advance for SMA material as yet not identified. It is more convenient to place volume fraction of martensitic phase z on ordinate axis and track the phase transitions horizontally rather than vertically, but naturally the reciprocal curves $D(z)$ can equally well serve kinetics constants identification purposes. In the case of test results of pseudoelastic curves embracing the outermost hysteresis loops, respective forward and reverse pseudoelastic parts of the curves will represent complete, monotonic forward phase transition ($z=0 \to z=1$; $z_f(D \geq 0), dz > 0$) and complete, monotonic, reverse phase transition ($z=1 \to z=0$; $z_r(D \leq 0), dz < 0$). The experimental curves $z(D)$ or ($D(z)$) must be parametrically matched by varying kinetics constants values, as best as possible with the R_L model curves resulting from relations $(3.25)_{1,2}$ valid for monotonic, complete transitions

$$
\begin{aligned}
&D(z) = \left[-r_1 \ln(1-z) + p_1 z\right]^{1/m_1}, \quad -D(z) = \left[-r_2 \ln(z) + p_2(1-z)\right]^{1/m_2} \\
&D_s^{(1)}(z^* = 0) = 0, D_s^{(2)}(z^* = 1) = 0.
\end{aligned}
\tag{5.27}
$$

The parametric identification procedure can be done in many ways; that is, with the aid of some curve fitting software or by trial and error to get required or possible accuracy. Similarly, as in the case of parameters of shape function, some combinations of values of p.t. kinetic parameters may lead to features not observed in reality; for example, weakening and not hardening of stress-strain curve during forward or reverse pseudoelastic flow. For that reason also, in the case of p.t. kinetics parameters it is advisable to perform a checking procedure to eliminate possible unrealistic modeling effects.

The actual identification of R_L model kinetics constants of $Ni_{51at\%}$-Ti alloy investigated in this chapter on a dozen or so isothermal curves is presented in detail and discussed in Chapter 6. Input data for identification process are different than outlined

Table 5.5 Complete set of material constants of R_L model for $Ni_{51at\%}$-Ti alloy (monoclinic transformation)

Constant	Value	Constant	Value
ρ	6500 (kg/m^3)	Δu_0^2	15,990 (J/kg)
E	64 (GPa) (at $T=315$ (K))	Δs_0^2	59.0 (J/kg K)
ν	0.41 (at $T=315$ (K))	ϕ_{12}	1062 (J/kg)
α_0	11e−6 (1/K) A	s_{12}, s_{13}	0.0 (J/kg K)
α_0	11e−6 (6e−6^M) M	γ	0.054
c_p	490 (J/kg K)	Δu_0^3	17,500 (J/kg)
λ_c	18 (W/mK)	Δs_0^3	67.2 (J/kg K)
$f(y)$	$a_1=1.17, a_2=0.37, a_3=0.78$	ϕ_{13}	504 (J/kg)
m_1	1.9	m_2	1.9
r_1	6.8	r_2	2.6
p_1	0.0	p_2	0.0

here due to the unavailability of isothermal curves involving outermost hysteresis loops. Possibly the best fit between all investigated experimental and modeling prediction curves has been obtained for parameter values $m_1=m_2=1.9$, $r_1=6.8, r_2=2.6, p_1=p_2=0.0$. Identified values of kinetics constants of the R_L model kinetics—(3.25)—are listed in Table 5.5.

The procedure of identification of material constants and functions of the R_L model described above takes minimum work to attain the target. Obtaining additional experimental information, at the tradeoff of bearing some extra cost, provides increased accuracy of the identified material data and, as a result, increased precision of modeling predictions. Table 5.5 contains a complete set of R_L model material constants for the $Ni_{51at\%}$-Ti alloy investigated here. The values of $\rho, \alpha_0, c_p, \lambda_c$ have been lent from a data sheet of commercial Flexwire AMPS-100 material with nominal chemical composition $Ni_{56.0wt\%}$-Ti$=$$Ni_{50.95at\%}$-Ti produced by Belgian company AMT.

The identified material data listed in Table 5.5 has been used to predict, with the R_L model, NiTi material response submitted to uniform, proportional isothermal loadings applied in an actual experimental program of NiTi material. The modeling time plots of total strain tensor components $\varepsilon_z(t), \varepsilon_\theta(t), \varepsilon_{z\theta}(t)$ and plots of experimental signals registered in the actual test, during which the specimen has been submitted to simultaneous compression and torsion (path No. 5) at $T=315$ K, are plotted in Figure 5.13a. The modeling and experimental plots of effective stress-total equivalent strain $(\sigma_{ef}-\varepsilon_{eq})$ for the same test are shown in Figure 5.13b. In Figure 5.13c and d respective plots are shown for tension loading at a temperature of 310 K. In numerical computations, the requirement that maximum total effective strain ε_{eq}^{max} in modeling simulations be kept the same as that actually reached in the experimental test has been imposed. The thicker lines mark actual data obtained in experimental tests, while the thinner ones mark modeling predictions.

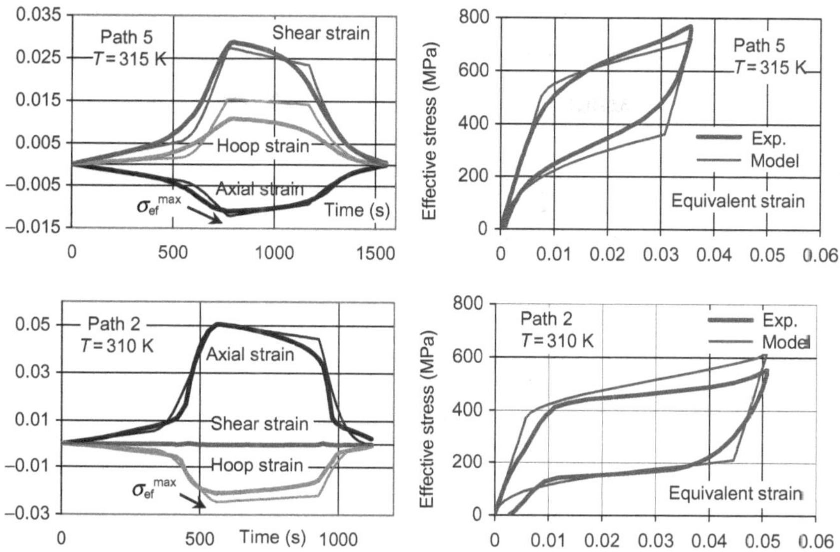

Figure 5.13 Comparison of actual variation in time of total strain tensor components $\varepsilon_z(t), \varepsilon_\theta(t), \varepsilon_{z\theta}(t)$ with variations predicted by R_L model, and comparison of actual graphs $\sigma_{ef} - \varepsilon_{eq}$ with modeling graphs in the case of (a) Test at temperature 315°K, path No. 5. (b) Test at temperature 310.0°K, path No. 2.

The R_L model allows predicting the response of SMA materials at temperatures below A_s^{2-1} temperature. However, upon unloading to zero stress there remain residual inelastic strains. The R_L model does not allow resuming further modeling computations from such a state. In principle, the R_L model of pseudoelasticity requires the SMA material to be in a zero stress, zero strain state to start prediction of SMA material response submitted to thermomechanical loadings.

5.5 Summary

In this chapter, results of a well-planned experimental research program of biaxial, isothermal, and proportional tests on polycrystalline $Ni_{51at\%}$-Ti alloy have been presented. The experimental data processed allowed checking the admissibility of fundamental theoretical assumptions of phenomenological theory of pseudoelasticity—the R_L model. In particular, the admissibility of the assumption on elastic isotropy of polycrystalline NiTi alloy, inelastic isotropy has been validated; that is, the conjecture that macroscopic tensor of phase transition strains can be expressed with the aid of isotropic tensor function of stress tensor. The usefulness of the introduction of shape function $f(y)$ to describe pseudoelastic flow yield surface—isotropic J_3' theory—has been demonstrated. The admissibility of expressing macroscopic phase strain tensor $\boldsymbol{\varepsilon}^{pt}$ with an isotropic tensor function $\mathbf{G}^*(\boldsymbol{\sigma})$ of zero order with respect to stress deviator tensor has been validated. A comparison of modeling predictions with experimental data allows stating that with modeling conjectures made in the R_L models family and proper identification of material constants and functions, a discrepancy between actual and modeling predictions curves on the order of a dozen or so percent for a wide variety of thermomechanical loadings can be achieved. It seems that, taking into account the extreme complexity of shape memory alloys behavior, such accuracy of modeling predictions is quite satisfactory. A detailed procedure of identification of material constants and functions of the R_L model has been elucidated, as well as the complete set of the identified for $Ni_{51at\%}$-Ti alloy material values.

Kinetics models of thermoelastic martensitic phase transformation 6

6.1 Introduction

This chapter presents an analysis of experimental results obtained for a polycrystalline NiTi alloy, which in an indirect way allowed for gaining information on the kinetics of stress-induced thermoelastic martensitic transformation on the basis of the measured deformation effects. The analysis enabled experimental verification of models of thermoelastic martensitic phase transformation (p.t.) kinetics in the case of forward and reverse transformation proposed heuristically in the literature. The scrutiny brought two important observations, namely that specific rates of forward and reverse transformations do not grow monotonously with increase of thermodynamic driving force of p.t. (what indicates the depletion of the transforming phase is not the only factor for hampering stress induced thermoelastic martensitic p.t. at its more advanced stages), and that forward and reverse p.t. exhibit the property of kinetic reversibility. These observations allowed proposing a "triangle" kinetics model of p.t. an extension of the kinetics model proposed by Raniecki and Lexcellent (1994)—*R-L kinetics*.

The formal equations of thermoelastic martensitic phase transformation kinetics constitute a key element of any constitutive model of shape memory alloys. Metallurgical observations indicate that the martensitic phase transformation proceeds through nucleation and growth. The development works on a kinetics model of nucleation and growth phase transitions were initiated by Avrami, in 1939. Nowadays, the model is often referred as the *JMAK kinetics model*, to tribute its main contributors Johnson, Mehl, Avrami and Kolmogorov; see Porter and Easterling (1992). The JMAK kinetics model can be expressed with the following very general and simple formula:

$$\ln(1-z) \equiv f(x) = -k \cdot (x - x_s)^n, \tag{6.1}$$

where z is the volume fraction of the newly formed phase, x is variable in terms of which advancement of the phase transition is described (e.g., it can be time), x_s denotes critical value of initiation of active phase transformation, and k and n are parametric modeling constants characterizing phase transition kinetics and depending on actual physical processes of the nucleation and growth. When it can be assumed that phase transformation progress takes place at a constant nucleation rate of spherical nuclei with a constant rate of growth, then it can be shown that parameter n takes the value $n = 4$. In general, coefficients k and n depend on the actual state of the transforming material. Hence, usually they are not "constants" but rather functions of state (e.g., temperature). Formal derivation of formula (6.1) can be found, for example, in the

Pseudoelasticity of Shape Memory Alloys. http://dx.doi.org/10.1016/B978-0-12-801697-8.00006-5

book by Porter and Easterling (1992). The range of applications of the JMAK kinetics model goes far beyond the area of materials sciences, where it is used equally often for description of diffusional as well as diffusionless phase transformations. The use of the JMAK kinetics model can be spotted in quite distant domains such as biology, economics, chemistry, or social sciences.

An extensive collection of experimental data devoted to the kinetics of temperature-induced martensitic transformation taking place in carbon steels during hardening was published by Koistinen and Marburger (1959). They found that phase transition experimental data for steels in a wide variety of carbon content fraction and small amounts of alloying additives fit perfectly to the modeling curve predicted by formula (6.1) upon substitution of $k=0.011$ and $n=1$. The temperature was adopted as variable x $(x=T)$, and x_s denoted the temperature of the martensitic transformation start $x_s=M_s$. As perfect agreement of modeling predications has been obtained for a broad class of steel experimental data, Koistinen and Marburger posed a conjecture that the relation (6.1) with parameters $k=0.011$ and $n=1$ is universally valid for all types of martensitic steels. This hypothesis proved to be false in view of later experimental research.

The kinetics rule of Koistinen and Marburger, proposed on purely experimental grounds, was derived from theoretical grounds by Magee (1968). He assumed that the amount of newly forming martensitic plates per unit of volume of austenitic phase dN is directly proportional to the increase of the value of thermodynamic driving force of phase transformation $\mathrm{d}N=-\varphi \mathrm{d}\left(\Delta G_v^{\gamma\to\alpha'}\right)$, where φ denotes constant, proportionality factor, $\Delta G_v^{\gamma\to\alpha'}(T)$ is the difference between free energy of the martensitic and austenitic phase—by assumption dependent on temperature only $\mathrm{d}\left(\Delta G_v^{\gamma\to\alpha'}\right)=\left(\mathrm{d}\Delta G_v^{\gamma\to\alpha'}/\mathrm{d}T\right)\mathrm{d}T$. The expression $\mathrm{d}\left(\Delta G_v^{\gamma\to\alpha'}\right)/\mathrm{d}T=\Delta S_v^{\gamma\to\alpha'}=\mathrm{const}$ determines change of entropy resulting from the formation of the martensitic phase. The volume fraction increase of the martensitic phase can then be expressed as follows: $\mathrm{d}f=(1-f)\overline{V}\mathrm{d}N$, where $\overline{V}$ is an average volume of the newly formed plates. The above assumptions lead to the following relation specifying kinetics of martensitic phase transformation in steel:

$$\ln(1-z)=-\left(\overline{V}\varphi\Delta S_v^{\gamma\to\alpha'}\right)(T-M_s). \tag{6.2}$$

This relation reveals physical interpretation of constant k appearing in the Koistinen and Marburger kinetics model $\left(k=\overline{V}\varphi\Delta S_v^{\gamma\to\alpha'}\right)$.

In carbon steel, only forward martensitic transformation can be induced, by a quick enough lowering of temperature. The reverse transition is precluded due to plastic deformation of austenitic (parent) phase taking place during forward transformation. Austenitic phase recovery requires heating steel material to temperatures high enough to induce recrystallization of the material. In the case of shape memory alloys, thermoelastic martensitic transformation takes place. Forward and reverse thermoelastic martensitic transformation can be induced in these materials, reversibly, by the cooling-heating cycle and/or, alternatively, by application and release of stress (strain).

Tanaka et al. (1986) proposed adaptation of the Koistinen-Marburger (JMAK) kinetics rules suitable for description of thermoelastic martensitic transformation.

The kinetics rules for forward and reverse phase transformation were formulated, which can be induced by the combined action of temperature and/or stress (uniaxial). The authors have given explicit conditions that must be fulfilled by incremental thermomechanical loading to induce active phase transformation. However, they did not show conformity of the proposed rules of phase transformation kinetics with the second law of thermodynamics. The kinetics rules of Tanaka et al. (1986) are given below in slightly modified form.

$$\begin{aligned} \ln(1-z) &= a_M(T-M_s) - b_M\sigma_{11}, \quad \dot{z}/(1-z) = \left(b_M\dot{\sigma}_{11} - a_M\dot{T}\right), \\ \ln(z) &= -a_A(T-A_s) + b_A\sigma_{11}, \quad \dot{z}/z = \left(b_A\dot{\sigma}_{11} - a_A\dot{T}\right), \end{aligned} \tag{6.3}$$

where A_s, M_s denote austenitic and martensitic phase transformation start temperatures, and $a_M, b_M, a_A, b_A > 0$ are material constants of phase transformation kinetics. In accordance with the authors' requirement, active forward transformation $(\mathcal{A} \rightarrow \mathcal{M})$ can take place when $\left(b_M\dot{\sigma}_{11} - a_M\dot{T}\right) \geq 0$, and active reverse transformation $(\mathcal{M} \rightarrow \mathcal{A})$ can take place when $\left(b_A\dot{\sigma}_{11} - a_A\dot{T}\right) \leq 0$. When $\sigma_{11} = 0$ (stress-free state) is substituted in formula $(6.3)_1$, the Koistinen-Marburger kinetics rule is recovered. Tanaka et al. (1986) rules made substantial progress in the formulation of thermoelastic martensitic transformation kinetics relations, but they are not capable for modeling formation of so-called internal hysteresis loops. This last problem of p.t. kinetics was addressed by Liang and Rogers (1990), who proposed "cosine" phase transformation kinetics relations in the following form:

$$\begin{aligned} z &= \frac{1-z^*}{2}\cos\left[\alpha_M(T-M_f) + \beta_M\sigma_{11}\right] + \frac{1+z^*}{2} \quad \text{active forward p.f. if } M_f \leq T \leq M_s, \\ z &= \frac{z^*}{2}\left\{\cos\left[\alpha_A(T-A_s) + \beta_A\sigma_{11}\right] + \frac{1}{2}\right\} \quad \text{active reverse p.f. if } A_s \leq T \leq A_f, \end{aligned} \tag{6.4}$$

where z^* denotes starting value of volumetric phase fraction of martensitic phase in some monotonic process of forward or reverse partial phase transformation, M_f, A_f are martensite finish and austenite finish temperatures at zero stress, and $\alpha_M = \pi/(M_s - M_f)$, $\alpha_A = \pi/(A_f - A_s)$, and β_M, β_A are material constants of p.t. kinetics.

Active forward phase transformation at zero external stress $\sigma_{11} = 0$ can take place only in temperature range $M_f \leq T \leq M_s$, while active reverse transformation can take place only in temperature range $A_s \leq T \leq A_f$. Application of uniaxial external stress $\sigma_{11} = \sigma^*$ results in shifting of the characteristic temperatures of phase transformation toward higher temperatures. When in formula $(6.4)_1$ $z^* = 0$ is substituted, then the formula obtained is capable of describing the progress of monotonic, complete phase transformation $(\mathcal{A} \rightarrow \mathcal{M})$. When in formula $(6.4)_2$ $z^* = 1$ is substituted, then the formula obtained is enabling the description of monotonic, complete reverse phase transformation $(\mathcal{M} \rightarrow \mathcal{A})$.

The formation of *external (outermost)* and *internal hysteresis loops* during full and partial phase transformations in accordance with the kinetics rules of Liang and Rogers is schematically illustrated in Figure 6.1. The arrows to the right indicate

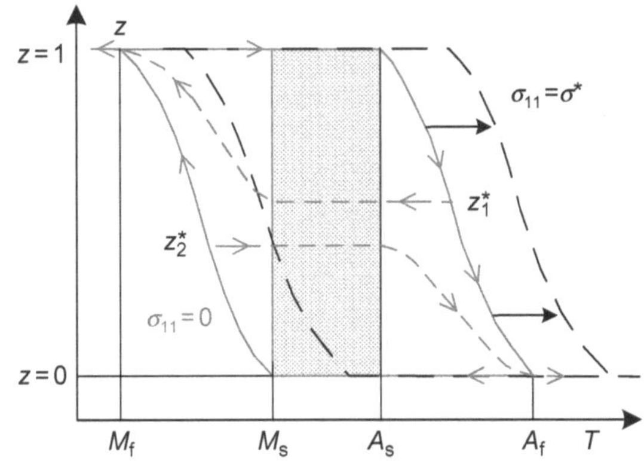

Figure 6.1 Schematic illustration of the formation of external and internal hysteresis loops at full and partial phase transformations in accordance with the kinetics equations of Liang and Rogers (1990).

direction of shifting of the domain of states in which active phase transformation can take place as a result of application of external stress. While the Liang and Rogers "cosine" kinetics model is mathematically and conceptually very simple, it constitutes a departure from metallurgical observations. These last indicate considerably better agreement of JMAK kinetics rules (logarithmic kinetics) modeling predictions with experimental data.

Further significant progress in modeling of thermoelastic martensitic phase transformation kinetics was achieved by Raniecki et al. (1992) and Raniecki and Lexcellent (1994). They formulated an extremely comprehensive model of rules of thermoelastic martensitic phase transformation kinetics basing on the formalism of nonequilibrium thermodynamics with internal state parameters. These rules can be conceptually divided into two parts. A general phase transformation framework defines domains of thermodynamic states from which active forward or reverse phase transformation can take place, and under what loading conditions of external state parameters this can happen. The general phase transformation kinetics rules assure conformity of phase transformation processes with the second law of thermodynamics. The second part of phase transformation kinetics rules, let us call them specific p.t. kinetics rules, deliver in parametric form mathematical relations between thermodynamic flux, that is, volume fraction of martensitic phase z and thermodynamic driving force of phase transformation π^{1-2} (see $(3.13)_3$). The applicability of the kinetics rules for modeling behavior of particular SMA material is assured by proper identification of specific kinetics rules material constants. The phase transition kinetics model of Raniecki and Lexcellent (R-L kinetics) enables modeling of SMA materials p.t. under 3D stress states loadings and includes rules for internal hysteresis loops formation. Raniecki and Lexcellent (1994) improved their kinetics model to enable better description of the initial stages of phase transformation and enriched it by introducing a new feature, so-called threshold functions. The threshold functions allow the model to observe experimentally the delayed start of phase transition in the case of some SMA materials. The authors delivered an incremental form of kinetics rules convenient for performing numerical computations as well as an integrated form valid for monotonic p.t. processes, convenient for material constants identification purposes. The basic weakness of all p.t. kinetics models proposed in recent years for SMA materials, especially those conjectured to be valid in multiaxial stress states and/or under combined thermomechanical loading-unloading paths, is that experimental validation supporting their assumptions is scarce or nonexistent. The efforts toward removing this weakness are presented in this chapter.

6.2 Transformation kinetics relations—thermodynamic foundations

Let us adopt the R-L kinetics model proposed in Raniecki and Lexcellent (1994) as the starting point for further development of a description of thermoelastic martensitic phase transformation kinetics. In accordance with the second law of thermodynamics, each thermodynamically irreversible process is accompanied by the dissipation of energy, which in the case of thermoelastic martensitic transformation can be assessed using formula $\dot{w}_{\mathrm{Disp}} = \pi^{1-2}\dot{z} \geq 0$ (see (3.20)). This expression indicates the admissible direction of active p.t. for SMA material remaining in a particular thermodynamic state. In order to satisfy inequality (3.20) during active p.t., the sign of the thermodynamic driving force must always be the same as the sign of the volume fraction of product phase increment; that is, $\mathrm{sign}(\pi^{1-2}) = \mathrm{sign}(\dot{z})$. On the other hand, it indicates that a certain functional relationship must exist between thermodynamic driving force of phase transformation and volumetric fraction of martensitic phase $\pi^{1-2} = f^{\mathrm{transf}}(z)$. The explicit form of such a relationship constitutes specific equations of phase transformation kinetics. Thermodynamic formalism indicates the existence of special states defined with the constraint relation $\pi^{1-2}(\boldsymbol{\sigma}, T, z^{\mathrm{start}}) = 0$ (see (3.14) and (3.35)). This relationship defines a set of "critical" states, at which forward $\mathcal{A} \to \mathcal{M}$ and reverse $\mathcal{M} \to \mathcal{A}$ phase transformations start ($z_{\mathrm{start}} = z^{*}_{\alpha}$; $0 \leq z^{*}_{\alpha} \leq 1$)—index $\alpha = 1$ denotes starting value for forward p.t. and $\alpha = 2$ denotes starting value for reverse p.t. The existence of critical states predicted on theoretical grounds has been elucidated experimentally for CuZnAl shape memory alloy by Müller and Xu (1991); see, for example, Figure 3.2. In the case of some SMA materials, phase transformation can start with some delay, after crossing the line $\pi^{1-2} = 0$; see, for example, Tanaka et al. (1995). Experimental results shown in Figure 6.4c show that in the case of NiTi alloy discussed here there exists rather advance then delay in the start of partial martensitic phase transitions with respect to line $\pi^{1-2} = 0$. Raniecki and Lexcellent (1994), knowing about the lag phenomenon in the start of phase transitions, introduced to their kinetics rule of thermoelastic martensitic transformation (R-L kinetics), so-called threshold values functions $Y_{(\alpha)}(\boldsymbol{\sigma}, T, z)$ $(\alpha = 1, 2)$. The operation of the R-L kinetics model in the case of monotonic, complete forward, complete reverse, and partial phase transformations is schematically illustrated in Figure 6.2. In Figure 6.2a, temperature-induced evolution of phase transitions are shown. On decreasing temperature, forward phase transition $(\mathcal{A} \to \mathcal{M})$ starts at the state marked 1 upon crossing the line $\pi^{1-2} = 0$. As soon as lowering of temperature is stopped at state 2, the forward phase transition also stops. Subsequent increasing of temperature results in initiation

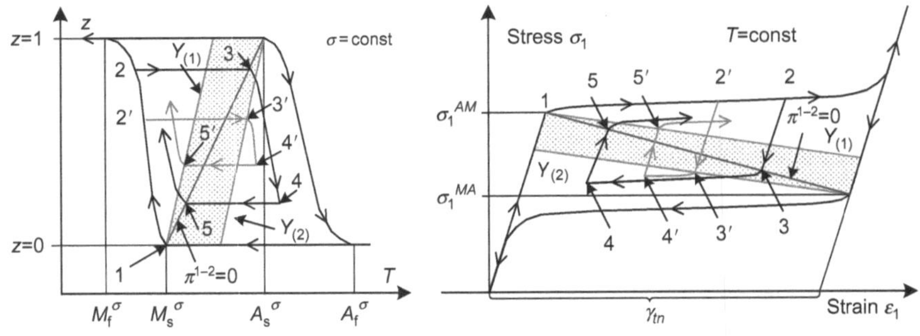

Figure 6.2 Schematic illustration of the formation of external and internal hysteresis loops at full and partial phase transformations in accordance with the kinetics equations of Raniecki et al. (1992) and Raniecki and Lexcellent (1998).

of reverse transition ($\mathcal{M}\rightarrow\mathcal{A}$) at state 3 upon crossing the line $\pi^{1-2}=0$. When increasing of temperature is stopped at state 4, reverse partial phase transition also stops. Consecutive decrease of temperature again induces partial forward phase transition at state 5 on crossing the line $\pi^{1-2}=0$. Such a scenario is realized for those shape memory alloys that do not exhibit lag threshold values for partial phase transitions; that is, those for which it can be with good approximation adopted that $Y_{(\alpha)}(\boldsymbol{\sigma},T,z)\equiv 0$ ($\alpha=1,2$) (see (3.26)). Instead of following path 1-2-3-4-5, some shape memory alloys instead follow behavior marked with path 1-2′-3′-4′-5′. For such alloys, the respective forward phase transformations start on crossing the lag threshold line $Y_{(1)}>0$, and respective reverse phase transformations start on crossing the lag threshold line $Y_{(2)}>0$.

The modeling predictions of isothermal loading and unloading paths involving stress-induced phase transformations are plotted schematically in stress-strain coordinates in Figure 6.2b. Generally, thermoelastic martensitic phase transformations can be induced by the combined action of stress and temperature.

It is convenient to rewrite an integral form of Raniecki and Lexcellent kinetics rules (3.25), describing monotonous, forward, and reverse phase transition processes such as those illustrated by paths $1\rightarrow 2$ or $3\rightarrow 4$ in Figure 6.2, in slightly modified form,

$$
\begin{aligned}
&(D(\sigma,T,z))^{m_1}=-r_1\ln\left((1-z)/\left(1-z_1^*\right)\right)+p_1\left(z-z_1^*\right),\\
&(-D(\sigma,T,z))^{m_2}=-r_2\ln\left(z/z_2^*\right)+p_2\left(z_2^*-z\right),\quad \left(D=\pi^{1-2}/\pi_N^{1-2}\right)\\
&f(y)\sigma_{\rm ef}^{\rm for}(z,T)=\frac{\phi_{12}\rho}{\gamma}\left\{\left[-r_1\ln\left(\frac{1-z}{1-z_1^*}\right)+p_1\left(z-z_1^*\right)\right]^{1/m_1}-\frac{\Delta s_0^2\left(M_s^0-T\right)}{\phi_{12}}-2z\right\},\\
&f(y)\sigma_{\rm ef}^{\rm rev}(z,T)=-\frac{\phi_{12}\rho}{\gamma}\left\{\left[-r_2\ln\left(\frac{z}{z_2^*}\right)-p_2\left(z-z_2^*\right)\right]^{1/m_2}+\frac{\Delta s_0^2\left(M_s^0-T\right)}{\phi_{12}}+2z\right\}.
\end{aligned}
\tag{6.5}
$$

m_1,r_1,p_1,m_2,r_2,p_2 denote phase transformation kinetics constants, which should be identified individually for each SMA material. The z_α^* ($\alpha=1,2$) denote initial volume fractions of martensitic phase at the start of the next segment of active, monotonous p.t. process. Formula $(6.5)_1$ is valid for forward phase transitions, and formula $(6.5)_2$ is valid for reverse phase transitions. After substitution into formulas $(6.5)_{1,2}$ the expression for thermodynamic driving force π^{1-2} (3.35), the expressions $(6.5)_{3,4}$ describing variation of effective stress variation during active phase transitions are obtained. These formulas are valid upon accepting zero lag threshold values $Y_{(\alpha)}(\boldsymbol{\sigma}, T, z) \equiv 0$, material constant $s_{12} = 0$, and normalizing factor $\pi_{\mathrm{N}}^{1-2} = \phi_{12}$. Please note that Raniecki and Lexcellent (1998) have adopted different scaling $\pi_{\mathrm{N}}^{1-2} = 2\bar{u}_0$. It is easy to notice that R-L kinetics model $(6.5)_{1,2}$ reduces to JMKA kinetics upon accepting constants $m_1 = m_2 = 1$ and $p_1 = p_2 = 0$.

The formulas (6.5) enable description of various possible subcategories of forward and reverse phase transformations such as "full," "partial," and/or "complete" (see Figure 6.3). By "full" forward phase transformation it is understood here a monotonic process of $(\mathcal{A} \rightarrow \mathcal{M})$ p.t., which starts from initial value of volume fraction of martensitic phase $z_1^* = 0$ (i.e., pure parent (austenitic) phase), regardless whether such transformation is "complete" (i.e., whether it runs monotonically to its very end reaching the value of $z = 1$ or conventionally $z = 0.99$). Similarly, by "full" reverse phase transformation it is understood that the monotonic p.t. process $(\mathcal{M} \rightarrow \mathcal{A})$, which starts from the initial value of volume fraction of martensitic phase $z_2^* = 1$ (pure product (martensitic) phase), regardless whether such reverse transformation is "complete" (i.e., whether it runs monotonically to reach pure parent (austenitic) state $z = 0$ or conventionally $z = 0.01$). The term "partial" transformation will be used in all other cases when forward or reverse transformation starts in any two-phase state $\left(z_{\text{start}} = z_\alpha^*;\ 0 < z_\alpha^* < 1\right)$. The closed path of complete, forward $\left(z = z_1^* = 0 \rightarrow z = 1;\ D_{\mathrm{s}}^{(1)} = 0\right)$, and consecutive complete reverse $\left(z = z_2^* = 1 \rightarrow z = 0;\ D_{\mathrm{s}}^{(2)} = 0\right)$ phase transitions encircle an envelope of thermodynamic states that are accessible for the particular SMA alloy during any isothermal phase transformation process.

Care must be exercised to correctly apply the R-L kinetics model for complex p.t. processes. For example, let us take a case when SMA material in pure parent phase state

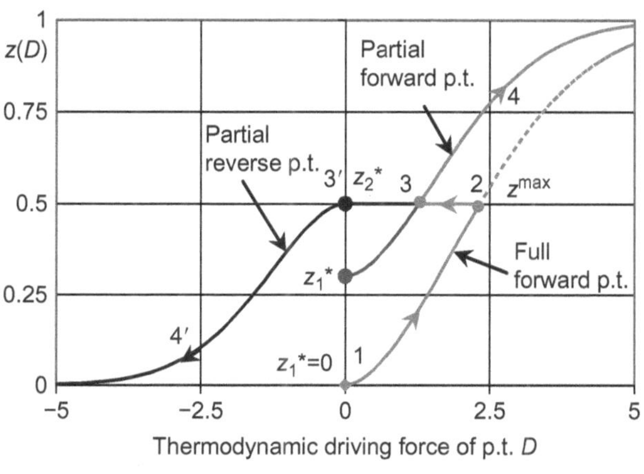

Figure 6.3 Schematic illustration of "full," "partial" forward and reverse phase transformations paths formation due to kinetics rule operation at multisegment loadings.

$(z=0)$ is submitted to loading, causing the start of active, full forward p.t. process $(z_1^*=0)$ at state 1, and its advances to some two-phase state 2 $(z=z^{\max})$ (see Figure 6.3). Next, the material is unloaded to state 3 where the value of t.d.f. is still greater than zero $\pi^{1-2}>0$. During process 2-3 volume fraction of martensite remains constant $z=z^{\max}$, because conditions for active p.t. are not fulfilled. In state 3, before initiation of reverse p.t.—possible at the earliest in state 3′—SMA material is again submitted to loading. Such loading immediately induces the start of partial forward p.t. and its progress to state 4 and beyond. The path of p.t. process 3-4 can be computed with formula $(6.5)_1$ when the correct initial value z_1^* is substituted. The proper value of z_1^* can be determined upon substitution of known state parameters in state 3 $(\sigma_3, T_3, z_3=z^{\max})$ to formula $(6.5)_1$. From such obtained constraint relation $D^{m_1}=-r_1 \ln\left((1-z^{\max})/(1-z_1^*)\right)+p_1\left(z^{\max}-z_1^*\right) \to z_1^*$ the value of z_1^* can be recovered. When z_1^* is known, the segment 3-4 of partial forward p.t. can be computed from $(6.5)_1$. In an alternative scenario when from state 2 only unloading of SMA material takes place in state 3′ partial reverse phase transition starts, and passes state 4′ to reach pure parent phase state again $(z=0)$. The path of p.t. process 3′-4′ can be computed with formula $(6.5)_2$ upon substitution of $z_2^*=z^{\max}$. All the difficulties indicated above, present when dealing with integral forms of p.t. kinetics model, are taken care of "automatically" when incremental R-L kinetics relations (3.26) are implemented into numerical code.

6.3 Parameters identification methodology of phase transition kinetics rules

In order to analyze the kinetics of thermoelastic martensitic phase transformation, the experimental data obtained in a series of isothermal tests performed on tubular specimen made of polycrystalline $Ni_{51at\%}$-Ti alloy are used. Results of fifteen isothermal tests are used, executed at three temperatures ($T=310$, 315, 322.5 K) with five proportional loadings paths at each temperature. Acquisition of the experimental data was described in detail in Chapter 5. The direct measurement of volume fraction of martensitic phase signal $z_{\exp}(t)$ has not been carried out; for example, by registering electrical resistivity variation of the sample. For that reason, in order to enable analyses of the kinetics of phase transformation a "semi-empirical" signal of the variation of volume fraction of martensitic phase in NiTi sample has been computed on the basis of measured deformation effects, taking advantage of the R_L model relation (3.40). The direct experimental confirmation on the existence of a proportionality relationship between inelastic strain and volume fraction of martensitic phase in the case of shape memory alloys was delivered by Vacher and Lexcellent (1991). They investigated the behavior of polycrystalline CuZnAl alloy during uniaxial loading; see Figure 3.1 and the accompanying text in Chapter 3. The plots of semi-empirical volume fraction of martensitic phase $z(D)$ in terms of normalized thermodynamic driving force of phase transformation are determined in accordance with the procedure of data elaboration presented schematically in Figure 6.4. This is possible only after identification of a complete set of material thermostatic constants and functions of the R_L model for

$$\left.\begin{array}{llll} \boldsymbol{\varepsilon} \overset{(5.4)}{\rightarrow} & \boldsymbol{\varepsilon}^{\mathrm{pt}} \overset{(5.5)}{\rightarrow} & \varepsilon_{\mathrm{eq}}^{\mathrm{pt}} \overset{(3.40)}{\rightarrow} & z = \varepsilon_{\mathrm{eq}}^{\mathrm{pt}} / \gamma F(y) \\ \boldsymbol{\sigma} \overset{(5.1)}{\rightarrow} & \sigma_{\mathrm{ef}} \overset{(5.12)}{\rightarrow} & f(y)\,\sigma_{\mathrm{ef}} \overset{(3.35)}{\rightarrow} & D = \pi^{1-2}(\sigma_{\mathrm{ef}}, T, z)/\phi_{12} \end{array}\right\} z \leftrightarrow D = \frac{\pi^{1-2}}{\phi_{12}}$$

Figure 6.4 Scheme of data elaboration leading to experimental relation between semi-empirical volume fraction of martensitic phase and normalized thermodynamic driving force of phase transformation *z(D)*.

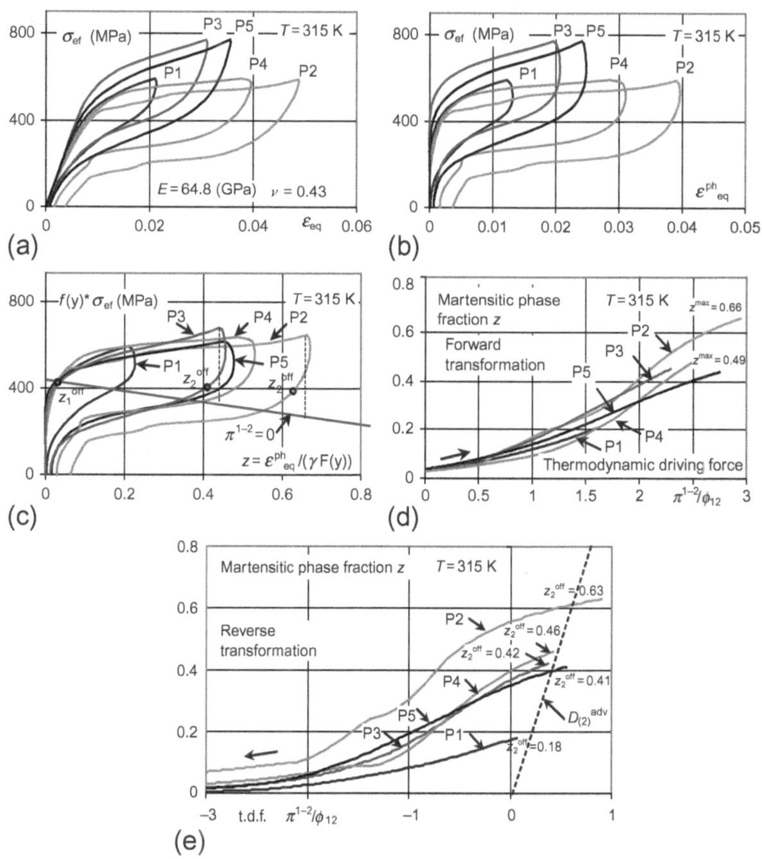

Figure 6.5 Illustration of the process of experimental data elaboration targeted at obtaining empirical graphs of phase transformation kinetics. Consecutively plotted graphs of (a) Effective stress ↔ equivalent strain $\sigma_{\mathrm{ef}} \leftrightarrow \varepsilon_{\mathrm{eq}}$; (b) Effective stress ↔ equivalent phase strain $\sigma_{\mathrm{ef}} \leftrightarrow \varepsilon_{\mathrm{eq}}^{\mathrm{pt}}$; (c) Reduced effective stress ↔ equivalent phase strain $f(y)\cdot\sigma_{\mathrm{ef}} \leftrightarrow \varepsilon_{\mathrm{eq}}^{\mathrm{pt}}$; (d) Graph of forward transformation kinetics $z_{\mathrm{f}} \leftrightarrow D$; (e) Curves of reverse transformation kinetics $z_{\mathrm{r}} \leftrightarrow D$.

investigated SMA material. Such material data for investigated NiTi alloy was identified in Chapter 5 (see Table 5.5).

The consecutive steps of experimental data processing, in accordance with the scheme shown in Figure 6.4, are shown graphically in Figure 6.5 on the example of data obtained in tests executed for five loading paths at temperature 315 K.

The obtained plots of semi-empirical, martensitic volume fraction versus thermodynamic driving force (t.d.f.) during monotonic, full forward p.t. $z_f(D)$ $(D \geq 0)$, are shown in Figure 6.5d, and the plots obtained for reverse p.t. $z_r(D)$ are shown in Figure 6.5e. Subscript "f" accompanying symbol z is added to distinguish forward p.t. path and subscript "r" is added to mark reverse p.t. path. In order to enable coherent comparison of experimental forward and reverse p.t. runs taking place during the same loading path, there has been adopted a 0.03 offset in the value of martensitic phase volume fraction z to mark the "practical" start of p.t. These offsets are preset in analogy to the classical plastic flow offset. They are marked by circle symbols and labels z_1^{off} and z_2^{off} in Figure 6.5c. The offset for reverse p.t. is taken with respect to maximum value of martensite volume fraction attained in particular test $z^{\max}$ $\left(z^{\max} - z_2^{\mathrm{off}} = 0.03\right)$. The values z_2^{off} shown in Figure 6.5e mark the "practical" values of volume fraction of martensitic phase when reverse phase transition starts during a specific test—in $\mathrm{R_L}$ model terms, it is $z^{\max}$.

The $\mathrm{R_L}$ model formulation predicts that active forward p.t. can take place only at states where $\pi^{1-2} \geq 0$ and reverse p.t. can take place only at states where $\pi^{1-2} \leq 0$. However, notice from Figure 6.5e that actual, partial reverse p.t. does not take place in the domain predicted by the R-L kinetics rule of $\mathrm{R_L}$ model $(\pi^{1-2} \leq 0)$; they take place already at states where $\pi^{1-2}\left(\sigma_{\mathrm{ef}}^{\exp}, T^{\exp}, z^{\exp}\right) > 0$. This experimental finding suggests the introduction of a new element into the thermoelastic martensite phase transition kinetics rules; namely, "advance" threshold value functions $\left(Y_{(\alpha)}^{\mathrm{adv}} \geq 0\right)$ in analogy to lag threshold functions already introduced in the R-L kinetics rule. Then, the condition $\pi^{1-2} \leq 0$ defining domain of states where active reverse martensitic phase transition is admissible, in modified criterion, is to be replaced with condition $D - D_{(2)}^{\mathrm{adv}} \leq 0$; $D_{(2)}^{\mathrm{adv}} \equiv Y_{(2)}^{\mathrm{adv}} / \phi_{12} \geq 0$. The new criterion will allow for partial reverse p.t. to take place at states where $D \geq 0$, as can be actually observed in Figure 6.5c. Data in Figure 6.5e prove that for investigated NiTi alloy it is reasonable to accept reverse "advance" threshold function to be a linear function of initial martensite volume fraction at the start of reverse p.t. and proportional to interaction energy; that is, $D_{(2)}^{\mathrm{adv}} = Y_{(2)}^{\mathrm{adv}} / \phi_{12} = 1 \cdot z_2^*$. Apparently, a still earlier start of p.t. visible for path No. 2, and also path No. 4, can be attributed to the disguising effect of R-phase transformation, pronounced in the case of these paths, taking place in NiTi alloys. No experimental data is available here for partial forward p.t., but it is reasonable to expect that corresponding forward advance thresholds $\left(Y_{(1)}^{\mathrm{adv}} \geq 0\right)$ also exist for them. Then, a modified criterion defining states where active forward partial martensitic phase transition is admissible could be defined with condition $D + D_{(1)}^{\mathrm{adv}} \geq 0$; $D_{(1)}^{\mathrm{adv}} \equiv Y_{(1)}^{\mathrm{adv}} / \phi_{12} \geq 0$. Such criterion will allow that partial forward p.t. can take place at states where $D \leq 0$.

Plots of evolution of martensitic phase volume fraction versus t.d.f. have been elaborated, also for data from tests executed at temperatures 310 and 322.5 K. They bear the same characteristics as plots shown in Figure 6.5.

The overall set of kinetics plots of forward and reverse phase transformations prepared on the basis of data originating from 15 isothermal tests executed at three temperatures for five loading paths is shown in Figure 6.6. In accordance with nomenclature introduced above, all graphs of phase transformation kinetics shown in

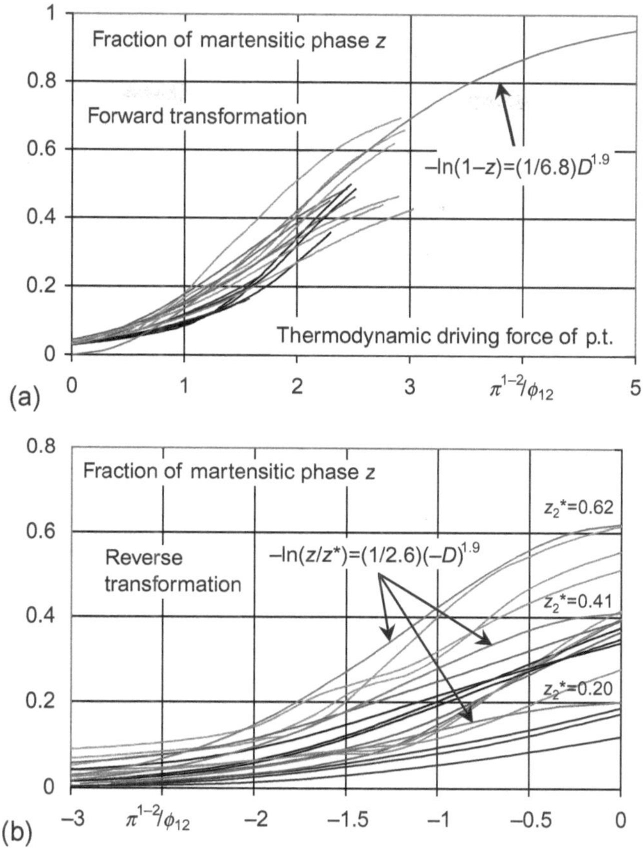

Figure 6.6 Overall collection of p.t. kinetics plots of stress-induced phase transformations for 15 tests embracing five loading paths at three temperatures: (a) Forward phase transitions; (b) Reverse phase transitions.

Figure 6.6a are full p.t. as they start from a single-phase austenitic state. The R-L kinetics model predicts that full forward phase transition path is described by a single modeling curve defined with formula $(6.5)_1$ $\left(z_1^* = 0\right)$, regardless of the direction of proportional loading path. Such an approach is motivated by the required simplicity of the kinetics model, ensuring numerical and cost effectiveness at the needed modeling predictions accuracy. In the evaluation procedure of a match between experimental and modeling stress-strain plots of forward p.t. it has been found that the best conformity is obtained when the following values of material parameters of R-L kinetics model are used: $m_1 = 1.9, r_1 = 6.8, p_1 = 0$. Experimental plots of volume fraction of martensite versus normalized t.d.f. $\left(z^{\text{for}} - D\right)$, together with the best match forward R-L kinetics modeling curve, are shown in Figure 6.6a. Notice that kinetics of thermoelastic martensitic transformation in NiTi alloy is quite a complex process. A scatter of the order of 0.1 can be observed between the actual values of volume fraction of martensitic phase for the same value of thermodynamic driving force of p.t. for the same loading path direction but different test temperatures. Still larger scatter of the order of 0.3 can be observed between actual values of volume fraction of martensitic phase at the same value of t.d.f. for the same test temperature but different loading paths directions. When the level of modeling accuracy achievable with a relatively uncomplicated kinetics rule (6.5) is unacceptable, then the kinetics model must be further complicated. The most obvious hint in this direction is the introduction into the kinetics model of a dependence on the direction of the loading path.

Plots of martensite volume fraction versus normalized t.d.f. during partial reverse phase transformations ($z^{\text{rev}} - D$) for 15 tests are shown in Figure 6.6b. Again, the best match procedure leads to adoption of the following R-L kinetics model material constants: $m_2 = 1.9, r_2 = 2.6, p_2 = 0$. Analysis of experimental data obtained for investigated NiTi alloy indicates that the terms proportional to constants p_1, p_2 in the form proposed in the R-L kinetics model seem not to play an important role in the p.t. kinetics of this material.

For illustrative purposes, in Figure 6.6b three partial reverse phase transition kinetics curves are plotted resulting from formula $(6.5)_2$ for three starting values of martensitic phase volume fractions $z_2^* = 0.2$, $z_2^* = 0.41$ and $z_2^* = 0.62$. Good agreement between experimental and modeling plots is obtained for the value $z_2^* = 0.62$. In the case of partial reverse p.t., starting from smaller initial fractions of martensitic phase considerable discrepancies appear between experimental and modeling curves. This gives an experimental hint that proposed in R-L kinetics model $(6.5)_2$ rule for internal hysteresis loops formation is imperfect and needs modification. This issue is addressed in the next section.

6.4 Experimental profiles of rates of thermoelastic martensitic transformation stress induced in polycrystalline NiTi alloy

6.4.1 *Property of kinetic reversibility of thermoelastic martensitic transformation*

More subtle information concerning thermoelastic martensitic phase transformation kinetics can be obtained upon analysis of phase transitions rate plots. In particular, an analysis of specific rate of forward p.t. is of interest referred to current value of volume fraction of parent phase, and specific rate of reverse p.t. referred to current value of volume fraction of product phase

$$\dot{z}_{\text{ext\_f}}(D) \equiv \frac{1}{1 - z(D)} \frac{\text{d}z}{\text{d}D}(D) \geq 0, \quad \dot{z}_{\text{ext\_r}}(D) \equiv \frac{1}{z(D)} \frac{\text{d}z}{\text{d}D}(D) \geq 0 \tag{6.6}$$

The specific rates defined above in essence allow separating out the factor of depletion of the transforming phase from the actual, overall rate of phase transformation. Speaking otherwise, these rates contain information on all other factors influencing p.t. rate except the factor of continuous depletion of the volumetric amount of transforming phase. The "rates" are understood here in a generalized sense, as thermoelastic martensitic p.t. is athermal; that is, it does not depend on the rate of time changes.

Laborious analytical work on experimental data of specific rates of forward and reverse p.t. computed with formulas (6.6) for individual loading paths allowed discovering an interesting pattern. It appears that with very good approximation the plot of specific rate of full forward p.t. very nicely matches the corresponding plot of specific rate of partial reverse p.t., both taking place in the same test run. This fit is obtained after executing two operations:

(i) First, t.d.f. of partial reverse p.t. is shifted by experimental threshold value $D_{(2)}^{\mathrm{adv}}$—depending on the advancement of previous full forward p.t. (i.e., the value $z^{\max}$). This operation has to be done in order that actual t.d.f. at the start of full forward p.t. and at the start of partial reverse p.t. have the same value. Speaking more simply, actual start of forward p.t. must correspond to actual start of reverse p.t in order to make a reasonable comparison.

(ii) Second, partial reverse p.t. rate is scaled by multiplying it with the initial value of the volume fraction of martensitic phase at the start of such partial reverse phase transition z_2^*.

The observed regularity is illustrated graphically in the left-hand side charts in Figure 6.7, showing experimental specific rate plots elaborated for data originating from tests executed at temperature 322.5 K for five different loading paths, accepted here in plots preparation $z_2^* = z^{\max} - 0.03$. The needed derivatives were obtained numerically with commercial software DADISP, which also proved very convenient for analysis purposes. Respective charts showing experimental phase transition kinetics curves in integral form are plotted in the right-hand side charts of Figure 6.7. The R-L kinetics modeling curves identified for investigated NiTi alloy are shown in charts prepared for path No. 2. Similar regularity to the one shown in Figure 6.7 is also observed for data from tests executed at temperatures 310 and 315 K for five different loading paths.

The pattern elucidated in the charts of Figure 6.7 can be expressed in mathematical terms with the relation

$$\dot{z}_{\mathrm{ext\_f}}(D) = \frac{1}{(1-z_{\mathrm{f}})} \cdot \frac{\mathrm{d}z_{\mathrm{f}}}{\mathrm{d}D}(D) = z_2^* \cdot \frac{1}{z_{\mathrm{r}}} \cdot \frac{\mathrm{d}z_{\mathrm{r}}}{\mathrm{d}D}\left(-D + D_{(2)}^{\mathrm{adv}}\right) = z_2^* \cdot \dot{z}_{\mathrm{ext\_r}}\left(-D + D_{(2)}^{\mathrm{adv}}\right), \tag{6.7}$$

where z_2^* denotes the value of volume fraction of martensitic phase at the start of the monotonic, partial, reverse phase transition process and $-D_{(2)}^{\mathrm{adv}}$ denotes value of normalized thermodynamic driving force corresponding to it. The subscripts "f" and "r" have been added to the martensite volume fraction symbol in order to distinguish between forward and reverse transformation segments.

The observation (6.7) together with the former observation for early initiation of partial reverse phase transitions delivers experimental premises for the formulation of modified version of kinetics rule (6.5) to better reflect actually observed physical

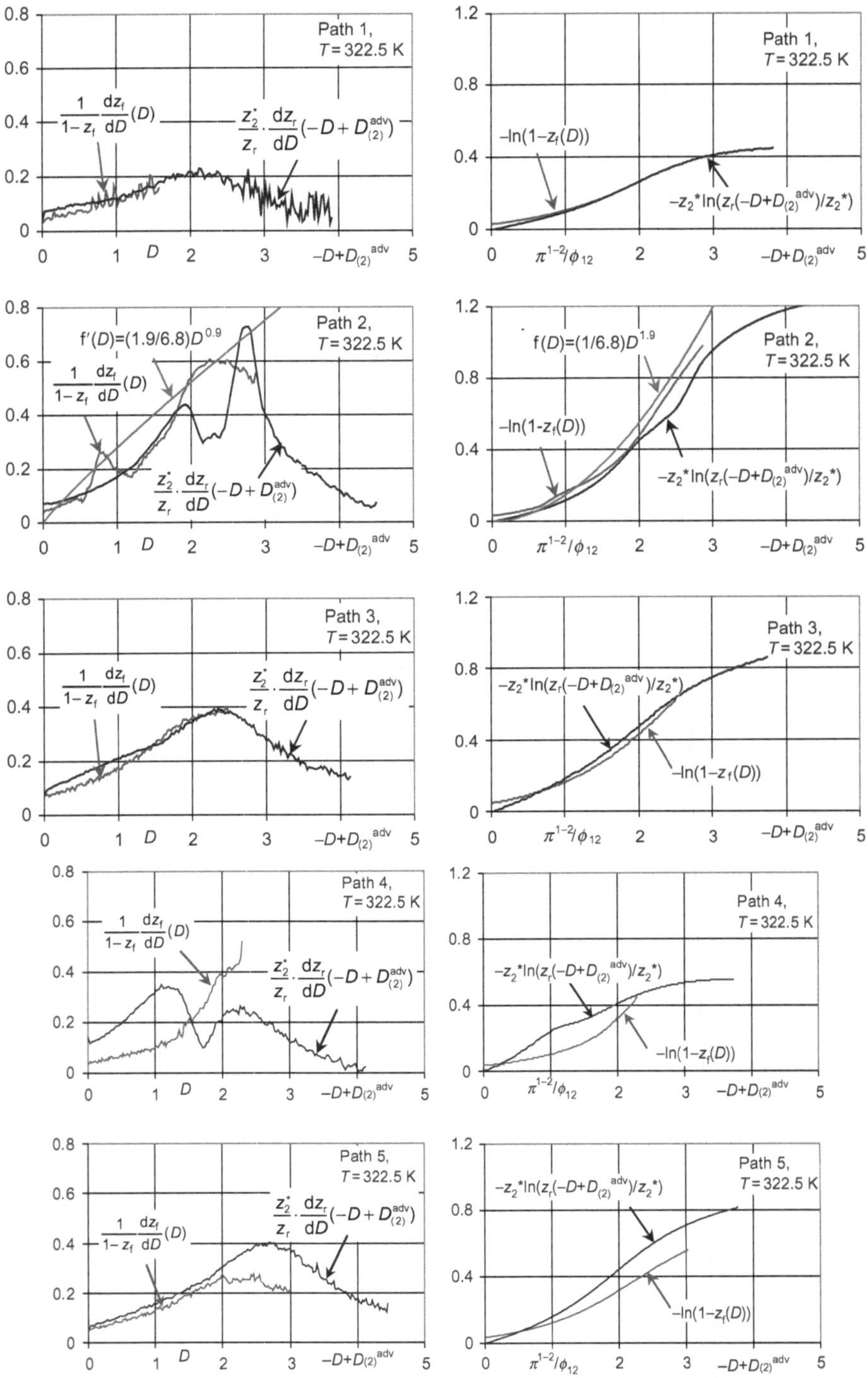

Figure 6.7 Left-hand side charts show specific p.t. rate of forward p.t. with overlaid scaled by the value z_2^* specific rate of respective reverse p.t. elaborated for data originating from tests executed at temperature $T=322.5$ K for five loading paths. Right-hand side charts show respective plots of p.t. kinetics in integral form.

phenomena. Generalization of experimental rate relation $(6.7)_2$ to the form postulated to be valid not only for reverse but also for forward phase transitions takes the form

$$\frac{(1-z_1^*)}{(1-z)}\cdot\frac{dz}{dD}=\frac{df_f\left(D+D_{(1)}^{adv}(z_1^*)\right)}{dD},\quad D+D_{(1)}^{adv}\geq 0,\ 0\leq z_1^*\leq z\leq 1,$$
$$\frac{z_2^*}{z}\cdot\frac{dz}{dD}=\frac{df_r\left(-D+D_{(2)}^{adv}(z_2^*)\right)}{d(-D)},\quad D-D_{(2)}^{adv}\leq 0,\ 0\leq z\leq z_2^*\leq 1,$$
$$D_{(1)}^{adv}(z_1^*)=\frac{Y_{(1)}^{adv}}{\phi_{12}},\ Y_{(1)}^{adv}\geq 0,\ D_{(2)}^{adv}(z_2^*)=\frac{Y_{(2)}^{adv}}{\phi_{12}},\ Y_{(2)}^{adv}=d_2\phi_{12}z_2^*\geq 0. \tag{6.8}$$

z_1^*, z_2^* denote starting values of martensitic phase volume fractions for partial, monotonous, forward, and reverse partial p.t., respectively. They are treated as constants during any active, monotonous p.t. segment. For full forward and full reverse p.t. they take the values $z_1^*=0, z_2^*=1$. Functions $Y_{(\alpha)}^{adv}, D_{(\alpha)}^{adv}$ $(\alpha=1,2)$ denote "advance" threshold functions for forward and reverse partial phase transitions and their normalized counterparts, d_2, is a material constant. Experimental data shown in Figure 6.5e deliver a hint to postulate advance threshold function $Y_{(2)}^{adv}$ to be linear functions of "initial" volume fraction of martensite z_2^*, and proportional to interaction energy $\phi_{12}\left(Y_{(2)}^{adv}(z_2^*=0)=0\right)$. No experimental data is available here to postulate an explicit form of $Y_{(1)}^{adv}$. Integral counterparts of relations (6.8) take the form

$$-(1-z_1^*)\cdot\ln\left(\frac{1-z\left(D+D_{(1)}^{adv}\right)}{1-z_1^*}\right)=f_f\left(D+D_{(1)}^{adv}(z_1^*)\right)\geq 0,\ D+D_{(1)}^{adv}\geq 0,$$
$$-z_2^*\cdot\ln\left(\frac{z\left(-D+D_{(2)}^{adv}\right)}{z_2^*}\right)=f_r\left(-D+D_{(2)}^{adv}(z_2^*)\right)\geq 0,\ D-D_{(2)}^{adv}\leq 0$$
$$df_f(D)/dD=(1-z_1^*)\cdot\dot{z}_{ext\_f},\quad df_r(D)/d(-D)=z_2^*\cdot z_{ext\_r} \tag{6.9}$$

In formulas (6.8) and (6.9) two generic functions f_f and f_r are present, which define actual profiles of forward and reverse phase transitions, full and/or partial. They are functions of thermodynamic driving force of phase transition D and possible threshold values $D_{(\alpha)}^{adv}$.

Discovered upon analysis of experimental data regularity (6.7) concerning full forward phase transitions $(z_1^*=0)$ and partial reverse phase transitions (z_2^*) can be recovered in modeling terms, upon validity of kinetics model (6.8), (6.9), only then when generic functions f_r and f_f fulfill the relation

$$\frac{df_f\left(D+D_{(1)}^{adv}\right)}{dD}\overset{(7.8)_1}{\overbrace{=}}\frac{1}{1-z}\frac{dz}{dD}\overset{(7.7)}{\overbrace{=}}z_2^*\cdot\frac{1}{z}\frac{dz}{dD}\overset{(7.8)_2}{\overbrace{=}}\frac{df_r\left(-D+D_{(2)}^{adv}\right)}{d(-D)}. \tag{6.10}$$

Integrating outermost left and right hand sides of (6.10) with initial conditions $f_f\left(D+D^{adv}_{(1)}=0\right)=f_r\left(-D+D^{adv}_{(2)}=0\right)=0$ leads to the relation

$$f_f\left(D+D^{adv}_{(1)}\right)=f_r\left(-D+D^{adv}_{(2)}\right). \tag{6.11}$$

The property (6.11) means that functions f_r and f_f must be mutually mirror reflections with respect to z-axis. In the case of full forward p.t. and, following it, subsequent partial reverse p.t., an experimental situation for data shown in Figure 6.7, relation (6.11), leads to the modeling constrain relation

$$-\ln(1-z_f)=f_f\left(D+D^{adv}_{(1)}\right)=f_r\left(-D+D^{adv}_{(2)}\right)=-z_2^*\cdot\ln\left(z_r/z_2^*\right). \tag{6.12}$$

It conforms to experimental plots shown in the right-hand side charts of Figure 6.7 and relation (6.7) upon acceptance of advance threshold value for full forward phase transition $D^{adv}_{(1)}\left(z_1^*\right)=0$.

Let us call the symmetry constraint relation (6.11) the property of *kinetic reversibility* of thermoelastic martensitic transformation. This macroscopic property might have its roots in the property of *crystallographic reversibility* of thermoelastic martensitic transformation observed on a microscopic scale. This last property manifests itself in the observation that the sequence of disappearance of martensitic plates during reverse transformation proceeds in exactly the reverse order to their prior appearance during forward p.t.— like in a movie set in reverse order. The feature described with formula (6.11) does not bear the character of irrevocable principle. Rather, it is a manifestation of a certain pattern resulting from averaging to macroscopic scale of observation of some mesoscopic processes of statistical character, as is indicated by plots in Figure 6.6. Identification of the features of a physical nature responsible for the appearance of special symmetry of forward and reverse p.t. described in macroscopic scale with relation (6.11) constitutes an open scientific problem requiring further investigation.

The explicit mathematical form of skeletal functions f_f, f_r has not been as yet discussed within the context of experimental results presented here for $Ni_{51at\%}$-Ti alloy. In the case of the R-L kinetics model, the functions $f_r(D)$ and $f_f(D)$ defining profiles of full forward and full reverse phase transition kinetics are postulated in the following explicit forms: $f_f(D)=D^{m_1}/r_1$ and $f_r(D)=(-D)^{m_2}/r_2$ $(p_1=p_2=0)$ (see (6.5)). No advance threshold functions $Y^{adv}_{(\alpha)}$ are present in the R-L kinetics model. Hence, the feature (6.11) will take place when mutually corresponding kinetic model parameters m_α, r_α defining profiles of forward and reverse kinetics of p.t. have the same values. This is straightforward to demonstrate as follows:

$$df_f/dD=\frac{m_1}{r_1}D^{m_1-1}=\frac{m_2}{r_2}(-D)^{m_2-1}=df_r/d(-D);\ \ m_1=m_2,\ \ r_1=r_2$$

$$f_f(D)=D^{m_1}/r_1,\ \ f_r(D)=(-D)^{m_2}/r_2,\ \ \left(f_{sc}(x)=\frac{1}{r}x^m\right)$$

$$\dot{z}_{ext\_f}=df_f/dD=\frac{m_1}{r_1}D^{m_1-1},\ \ \dot{z}_{ext\_r}=df_r/d(-D)=\frac{m_2}{r_2}(-D)^{m_2-1}. \tag{6.13}$$

The above discussion of experimental results for polycrystalline NiTi alloy suggests taking the same values of kinetics model parameters in relations describing forward and reverse p.t. under the condition that instead of relations (6.5), modified kinetics relations (6.9) are used. Hence, kinetics constants $m_1 = m_2 = 1.9$ $r_1 = 6.8$, $r_2 = 2.6$ identified for model (6.5) (see also Table 5.5) are to be replaced with kinetics constants $m_1 = m_2 = 1.9$, $r_1 = r_2 = 6.8$ valid for better reflecting actual physical phenomena model (6.9). The plots of modeling curves for monotonic, partial reverse phase transitions predicted by the R-L kinetics model (see (6.5)), and predicted by the new kinetics model (6.9), for three starting values of volume fraction of martensitic phase at the beginning of monotonic reverse transformation $z = 0.62$, $z = 0.41$, $z = 0.2$ are plotted in Figure 6.8. In both cases, the same skeletal power function of phase transformations is used: $f_{sc}(x) = x^{1.9}/6.8$ (see also $(6.13)_{2,3}$). Figure 6.8 also contains plots of p.t. kinetics obtained upon experimental data elaboration. A comparison of plots obtained with modified kinetics model (6.9) of thermoelastic martensitic transformation with an experimental one allows stating that with the aid of this kinetics model considerable improvement can be obtained regarding compatibility between modeling predictions and experimental data with respect to the original R-L kinetics model.

6.4.2 Improved model of thermoelastic phase transformation kinetics

The skeletal functions f_f, f_r, describing rate profiles of thermoelastic martensitic phase transformation proposed in the literature, predict constant value of such rates; for example, this is the case in the Tanaka et al. (1986) model or they predict monotonic increase of the rate value with increase of thermodynamic driving force of p.t like in the case of the R-L kinetics model (see $(6.13)_{5,6}$). Accepting this kind of modeling assumption causes the complete depletion of transforming phase to be the only factor

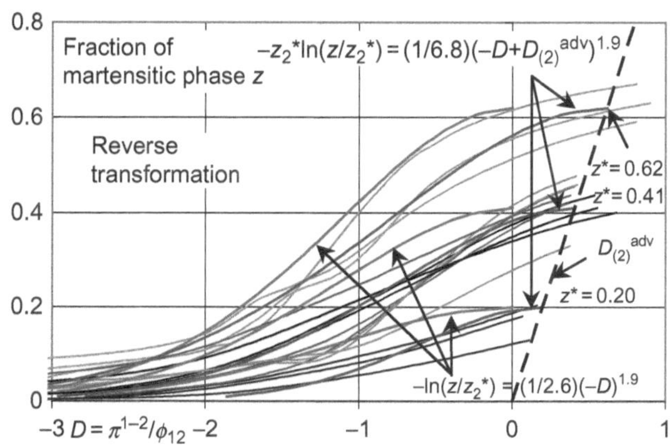

Figure 6.8 Predictions of R-L kinetics model and new kinetics model (6.9) overlaid on experimental curves of p.t. kinetics.

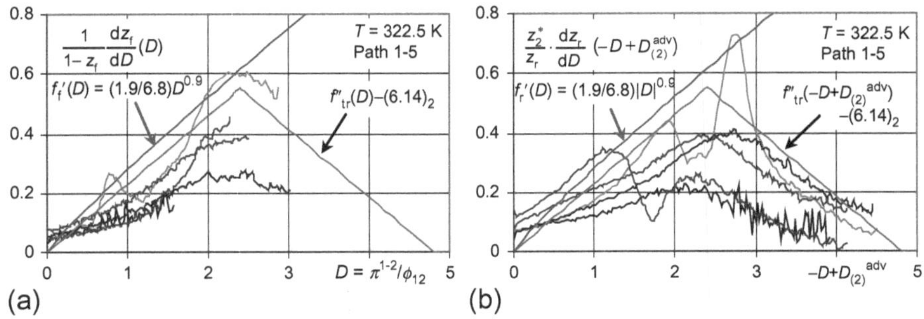

Figure 6.9 (a) Plots of experimental specific rates of full forward p.t. for tests executed at 322.5 K. (b) Plots of specific rates of respective partial reverse p.t. for tests executed at 322.5 K. The plots of specific rate predicted by R-L kinetics model (see $(6.13)_{1,2}$) are labeled (1.9/6.8) $D^{0.9}$ and $(1.9/6.8)|D|^{0.9}$, respectively. The plots of specific rate predicted by new kinetics model (see (6.14)) are labeled $f_f'(D)$ and $=f_r'(-D+D_{(2)}^{\text{adv}})$.

which can effect the termination of phase transformation, upon a monotonic increase of thermodynamic driving force of p.t.. However, experimental p.t. rates grow with increasing thermodynamic driving force D only to a certain value, as can be observed in Figure 6.9a, where the plots of specific rates for full forward transformations taking place during tests performed at temperature 322.5 K are plotted. The rates reach a maximum for the value of t.d.f. of about $D \simeq 2.4$, and next they start to decrease despite a further increase of thermodynamic driving force D.

The plot of specific rate predicted by the R-L kinetics model with the use of kinetics material parameters identified for investigated NiTi alloy is also plotted in Figure 6.9a, marked with label $f_f' = \frac{1.9}{6.8}D^{0.9}$. The experimental data described above indicate that in the case of stress-induced thermoelastic martensitic transformation in NiTi alloy, it is more and more difficult to bring this phase transformation to its completion $(z = 0.99)$ despite reaching very high loading stresses, and existence of pool of untransformed austenitic phase. Analysis of profiles of specific rates shown in Figure 6.7 and, corresponding to them, values of phase transformation volume fractions of martensitic phase plotted in Figure 6.6a, allows judging that even at very high value of stress inducing martensitic p.t., the value of martensitic phase volume fraction during forward transformation would hardly exceed about $z = 0.75$ for investigated NiTi alloy. A similar observation on incomplete progress of thermoelastic martensitic phase transformation in NiTi alloy has been reported by Brinson et al. (2004). These authors using an optical method (hence, different from the one based exclusively on deformation effects used here) evaluated that maximum volume fraction of martensitic phase after passing pseudoelastic plateau reached the value of around 0.7 for the NiTi alloy investigated by them. They also found that further increasing of force load of the sample does not lead to a substantial increase of the amount of martensitic phase. Hence, two independent methods suggest that it is rather difficult to bring about the complete thermoelastic martensitic transformation in NiTi alloy when it is induced by stress at higher temperatures.

The plots of rescaled by z_2^* rates of partial reverse transformations elaborated for data obtained in tests executed at temperature 322.5 K are collected together in Figure 6.9b. In accordance with the hypothesis conjectured in the previous section (see formula (6.11)), all these plots can be generated with the aid of one skeletal function, which defines the profiles of kinetics of forward and reverse phase transformations. It is clear that the power law skeletal function proposed in the R-L kinetics model ($f_{sc}(x) = \frac{1}{r}x^m$) does not allow correctly reproducing experimental data shown in Figure 6.9, as it does not predict the existence of a maximum of specific rate of p.t.. The experimental rates of full forward p.t. actually exhibit maximum at $D_{max} \approx 2.4$. The experimental rates of partial reverse p.t. for five loading paths also exhibit maximum. However, the location of this maximum depends on the starting value from which reverse transformation starts z_2^*. It is also reached at approximately the value of 2.4 of the thermodynamic driving force but shifted with advance threshold value in accordance with the realtion $D_{max} = -D + D_{(2)}^{adv} = -D + 1 \cdot \phi_{12} \cdot z_2^* \approx 2.4$. This is in accordance with the hypothesis on kinetic reversibility of phase transition.

Reasonable and effective modification of the R-L kinetics model to take into account the experimental facts indicated above is not an easy task, and requires more experimental evidence to confirm noticed regularities. The most straightforward modification is to use skeletal functions defining profiles of thermoelastic martensitic transformation to predict appearance of a maximum of specific rates of phase transition (6.6). In order to improve the R-L kinetics model, the acceptance of a *triangle skeletal functions* is proposed, defining profiles of forward and reverse phase transitions, in the parametric form

$$f_f(D) = \begin{cases} \frac{1}{2}a_1 \cdot D^2 & 0 \le D \le b_1 \\ \frac{1}{2}a_1 \cdot \left[D^2 - 2(b_1 - D)^2\right] & b_1 < D \le 2b_1 \\ 0 & 2b_1 < D \end{cases},$$

$$f_f'(D) = \begin{cases} a_1 \cdot D & 0 \le D \le b_1 \\ a_1 \cdot (2b_1 - D) & b_1 < D \le 2b_1 \\ 0 & 2b_1 < D \end{cases},$$

$$f_r(D) = \begin{cases} \frac{1}{2}a_2 \cdot (-D)^2 & 0 \le -D \le b_2 \\ \frac{1}{2}a_2 \cdot \left[(-D)^2 - 2(b_2 + D)^2\right] & b_2 < -D \le 2b_2 \\ 0 & 2b_2 < -D \end{cases},$$

$$f_r'(D) = \begin{cases} -a_2 \cdot (-D) & 0 \le -D \le b_2 \\ -a_2 \cdot (2b_2 - (-D)) & b_2 < -D \le 2b_2 \\ 0 & 2b_2 < -D \end{cases}. \qquad (6.14)$$

where a_α, b_α; $\alpha = 1,2$ are material parameters of the *triangle kinetics model*. They must be identified for each SMA material from experimental data.

Functions f_f, f_r, defined with formula (6.14), lead to a triangle shape of plots of specific rates, which are determined by functions f_f', f_r'. Parameters a_α define pace of increase (decrease) of specific rate and parameters b_α determine the coordinate of location of specific rate maximum. The modeling profiles of specific rates of forward and reverse transformations produced by triangle generic function defined with formulas (6.14) are plotted in Figure 6.9, on the background of profiles obtained on the basis of experimental data from the tests performed at a temperature of 322.5 K. The following values $a_1 = a_2 = 0.23$, $b_1 = b_2 = 2.4$ of material parameters of the modified kinetics model were accepted as the most appropriate for the investigated NiTi alloy. Slightly too high a value of the maximum of proposed modeling profiles in comparison to experimental data for the temperature 322.5 K results from the requirement that the adopted material parameters values of the new kinetics model provide possibly the best conformance of modeling profiles with experimental data for all studied temperatures (i.e., 310, 315, and 322.5 K). Specific rates of stress-induced p.t. for a temperature of 310 K indicate that p.t. proceeds noticeably more quickly at that temperature than at a temperature of 322.5 K. This temperature dependence is not introduced into the kinetic model in order to make it as simple as possible. In spite of experimental premises indicating that thermoelastic martensitic transformations forward and reverse can be modeled with good approximation by using only one set of material kinetic parameters consisting of only two values, due to pragmatic reasons it was decided here to retain two such sets (i.e., four values of kinetics model constants).

It is convenient to express the improved kinetics model of thermoelastic martensitic transformation (6.8), (6.14), with zero advance threshold functions $D^{\mathrm{adv}}_{(\alpha)} = 0$, $\alpha = 1,2$, in the form of state functions $\lambda^{\sigma}_{(\alpha)}(\boldsymbol{\sigma}, T, z)$ appearing in the general rate kinetics rule (3.26), so as it is possible to replace them with the state functions defined by formulas (3.29). Multiplying both sides of equation (6.8) by $\dot{D}$ and taking advantage of the expression for $\dot{D}$ obtained from formula $(3.35)_2$ after simple manipulations, one obtains

$$\begin{aligned}\lambda^{\sigma}_{(1)} &= \frac{(1-z)\cdot f_{\mathrm{f}}'(D)}{2\phi_{12}\cdot z_1^* - (1-z)\cdot f_{\mathrm{f}}'(D)\cdot 2\phi_{\mathrm{it}}(T)},\\ \lambda^{\sigma}_{(2)} &= \frac{z\cdot f_{\mathrm{r}}'(D)}{2\phi_{12}\cdot z_2^* - z\cdot\left[-f_{\mathrm{r}}'(D)\right]\cdot 2\phi_{\mathrm{it}}(T)}.\end{aligned} \tag{6.15}$$

Appearing in the above formulas, functions $f_{\mathrm{f}}', f_{\mathrm{r}}'$ and parameters z_1^*, z_2^* are defined by formulas (6.14) and (6.8). In a similar manner, it is possible to obtain formulas for functions $\lambda^{\varepsilon}_{(\alpha)}(\boldsymbol{\varepsilon}, T, z)$ (see $(3.26)_{2,3}$), useful when studying strain-controlled phase transition processes

$$\begin{aligned}\lambda^{\varepsilon}_{(1)} &= \frac{(1-z)\cdot f_{\mathrm{f}}'(D)}{2\cdot\phi_{12}\cdot z_1^* - (1-z)\cdot f_{\mathrm{f}}'(D)\cdot[2\phi_{\mathrm{it}}(T) - \boldsymbol{\kappa}\cdot\mathbf{L}\boldsymbol{\kappa}/\rho]},\\ \lambda^{\varepsilon}_{(2)} &= \frac{z\cdot f_{\mathrm{r}}'(D)}{2\cdot\phi_{12}\cdot z_2^* - z\cdot\left[-f_{\mathrm{r}}'(D)\right]\cdot[2\phi_{\mathrm{it}}(T) - \boldsymbol{\kappa}\cdot\mathbf{L}\boldsymbol{\kappa}/\rho]}.\end{aligned} \tag{6.16}$$

Operation of the triangle kinetics model is illustrated in Figure 6.10 where plots of full, monotonic phase transformations predicted by R-L model kinetics and by a new triangle kinetics model (6.15), (6.14), (6.8) are plotted together on the background of

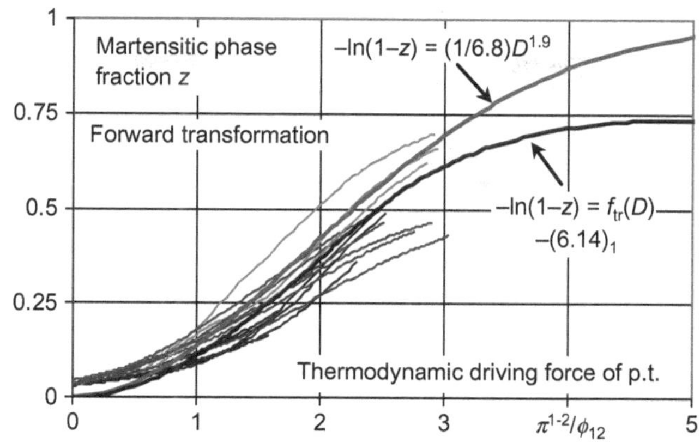

Figure 6.10 Experimental plots of full forward martensitic phase transformation kinetics in integral form with overlaid plots of rates predicted by the R-L kinetics model and new "triangle" specific kinetics model $(6.14)_1$.

graphs obtained from experimental data. Notice that in the initial stage of phase transformation, modeling plots of both kinetics models do not differ substantially. The essential difference in operation of both kinetics models lies in the fact that the R-L kinetics model predicts the possibility of bringing the phase transformation to the completion ($z = 1$) as a result of applying sufficiently large stress in isothermal test conditions. The triangle kinetics model, on the other hand, predicts that phase transformation remains incomplete even for very large stresses reaching some saturation value of martensitic phase fraction ($z_{sat} < 1$). This is illustrated in the case of investigated NiTi alloy in Figure 6.10.

6.5 Summary

An analysis of experimental data obtained for $Ni_{51at\%}$-Ti alloy shows that during thermoelastic martensitic transformation, specific rates of forward and reverse transformations initially grow with an increase of thermodynamic driving force of phase transformation to reach a maximum and next decrease despite further continuous increase of the value of thermodynamic driving force of phase transformation. This observation allows concluding that depletion of the transforming phase is not the only factor for hampering stress-induced thermoelastic martensitic p.t. at its more advanced stages. The other observation reveals that specific rate of forward transformation and rate of subsequent partial reverse transformation taking place in the same stress loading-unloading test match to each other under the condition that this second rate is scaled by the initial value of volume fraction of martensitic phase at the start of partial reverse phase transition. This macroscopic property, expressed mathematically with formulas (6.11) and (6.7), and describing certain symmetry in runs of forward and reverse transformations, can be called kinetic reversibility of thermoelastic martensitic transformation. The appearance of symmetry of this kind can be expected, remembering about mezoscopic scale property of crystallographic reversibility of thermoelastic martensitic transformation. Taking advantage of the above observations, modification of R-L model kinetics formulation is proposed to better reflect actual physical phenomena taking place during thermoelastic martensitic transformation. The modified "triangle" kinetics model, expressed in rate form with the formulas (6.14), (6.15), and in integral form by formulas (6.9), allows for a more accurate description of specific rates profiles than the R-L kinetics model can do. The model allows for prediction of full and partial phase transformations; that is, those responsible for internal hysteresis loops formation. It predicts the possibility of incomplete phase transformation; that is, the existence of untransformed parent phase even at very high stresses. Experimental premises indicating that stress-induced martensitic phase transformation in NiTi alloy cannot be brought to a single-phase martensitic state (i.e., in conventional terms $z = 0.99$) has been already reported by Brinson et al. (2004), who made such observation using microscopic optical investigations. Combined mechanical and metallurgical research work is needed to establish the physical mechanisms responsible for the appearance of maximum in profiles of specific rates of forward phase transitions, and thus to understand the reasons for incomplete progress of thermoelastic martensitic transformation induced by stress in NiTi alloys.

Thermodynamic model of SMA pseudoelasticity based on multiplicative decomposition of deformation gradient tensor

7

7.1 Introduction

A member of the R_L family of macroscopic, phenomenological, thermodynamic models of pseudoelasticity is discussed in this chapter, taking into account geometrical effects; see also Ziółkowski (2007). The concept of multiplicative decomposition of the total deformation gradient is used together with the concept of a family of virtual, mobile, isoclinic, unloaded elastically configurations serving as reference configurations for a description of elastic deformation—Mandel formalism (Mandel, 1974)—to extend modeling predictions into the finite deformations regime. The term *Mandel formalism description* is used interchangeably with the term *mobile Lagrangean description;* they are synonymous in this chapter below. The Lagrangean logarithmic (Hencky) elastic strain tensor, temperature, and volume fraction of oriented martensite phase $\{\boldsymbol{E}^{e}(0), T, z\}$ make a set of state parameters entering the macroscopic free energy function describing two-phase states of an SMA macroelement in mobile Lagrangean description. In Eulerian description, the $\{\boldsymbol{e}^{e}(0), T, z\}$ is the set of state parameters corresponding to $\{\boldsymbol{E}^{e}(0), T, z\}$, where $\boldsymbol{e}^{e}(0)$ denotes spatial *logarithmic (Hencky) elastic strain.* In order to describe deformation effects connected with martensitic phase transition or/and martensitic phase reorientation, a formal pseudoelastic flow rule for the Eulerian inelastic stretching rate tensor $\boldsymbol{d}^{in}$ is proposed. The formal phase transformation kinetics equations and criteria describing conditions for the active phase transformation progress—forward and reverse—are specified in the objective form; that is, invariant with respect to rigid body rotations. Rate constitutive relations of elasticity are formulated in Eulerian description using the Jaumann elastic corotational rate of stress in view of the favorable properties of this particular rate as elucidated in the text below. Next, they are transformed to the form adequate for application of the updated Lagrangean computational technique. Closing the thermomechanical problem of SMA finite deformation thermomechanics in mathematical terms requires specification of the evolution of inelastic Eulerian spin $\boldsymbol{\omega}^{in}$. Very little if any experimental information on inelastic spin evolution in deforming metallic materials exists. For that reason it is adopted in computational mechanics of inelastic processes conjecture most often that $\boldsymbol{\omega}^{in}$ is identically equal to zero. This conjecture is also made here. A specific

Pseudoelasticity of Shape Memory Alloys. http://dx.doi.org/10.1016/B978-0-12-801697-8.00007-7

hypothesis on $\boldsymbol{\omega}^{\text{in}}$ defines a family of unloaded elastically reference configurations. The selection of the isoclinic family of unloaded elastically reference configurations by accepting the assumption $\boldsymbol{\omega}^{\text{in}} = \mathbf{0}$ leads to considerable simplification of the mathematical formulation of the proposed SMA materials model. Information on the material behavior and motion inseparably enters a global weak form of rate mechanical balance equation through a familiar bridging equation involving the Eulerian rate of the nominal stress tensor. Again, an especially simple form of the bridging equation is obtained by such selection of a family of unloaded elastically reference configurations that at each time instant inelastic Eulerian spin $\boldsymbol{\omega}^{\text{in}}$ is identically equal to zero.

Apparently, the first macroscopic constitutive model of SMA written in finite deformations formalism is that proposed by Auricchio and Taylor (1997). Unfortunately, the theoretical development of the model itself is borrowed directly from generalized plasticity theory, with the smallest possible alterations. This means that the model is not well matched for the description of shape memory alloys behavior, as the physical mechanisms underlying plasticity—plastic slip—and shape memory effects—martensitic phase transformation and motion of highly mobile twin boundaries—are different. Additionally, this model is not very well suited for description of phase transitions based on nonequilibrium thermodynamics formalism. An interesting proposition for description of shape memory alloys behavior written in large deformations formalism is the comprehensive work by Annand and Gurtin (2003). This work remains within the stream of micromechanical constitutive models originating in a series of papers by Patoor et al. (1988). The prevailing majority of micromechanical models of SMA materials are based on the well-known methodology of crystal plasticity theories. The key point is in the replacement of the plastic flow law $\boldsymbol{d}^{\text{p}} = \Sigma\dot{\gamma}^{\alpha}\boldsymbol{s}^{\alpha}\otimes\boldsymbol{m}^{\alpha}; \alpha = 1,\ldots,N$ with the phase transformation/twinning of martensitic phase flow law $\boldsymbol{d}^{\text{in}} = \Sigma\dot{z}^{\alpha}\boldsymbol{t}^{\alpha}\otimes\boldsymbol{h}^{\alpha}; \alpha = 1,\ldots,M$ supplemented with some micromechanical criteria for active forward, reverse phase transition/reorientation of martensitic variants processes. The dyadics $\boldsymbol{s}^{\alpha}\otimes\boldsymbol{m}^{\alpha}$ denote well-known slip systems, N is the number of available slip systems, and the dyadics $\boldsymbol{t}^{\alpha}\otimes\boldsymbol{h}^{\alpha}$ denote "phase transformation" systems or martensitic variants eigenstrains "generators." Sometimes, self-accommodating groups of martensitic variants eigenstrain generators are used in place of a habit plane variant eigenstrain generator. In principle, micromechanical models of shape memory alloys can bring a lot of useful information, for example, on evolution of texture, material properties of the phase mixture, and so forth. However, it seems that they rely on more constitutive assumptions, in some cases very difficult for experimental verification, than the macroscopic models. Micromechanical models are extremely needy regarding computational resources, as their prediction of macroscopic behavior is effectively obtained through numerical averaging. For that reason they are not adequate for the purposes of design of SMA engineering structures or devices. Phenomenological, macroscopic models of SMA materials' behavior are most appropriate for engineering applications purposes. The model discussed in this chapter, upon its implementation into finite element code, is very well suited for performing engineering design computations of structures containing elements made of SMA, in analyzing situations when large rotations of structural elements appear during nominal operating conditions, or in analyzing emergency situations when large strains may also appear.

7.2 Kinematic relations at finite deformations

Let us consider a polycrystalline body $\mathcal{B}$ made of SMA material occupying at time instant $t = t_0$ configuration ${}^0\mathcal{B}$ (see Figure 7.1). The location of material points in the configuration ${}^0\mathcal{B}$ at time t_0 is characterized by their position $\boldsymbol{X}$. Due to applied thermomechanical loads the body $\mathcal{B}$ undergoes motion, $\boldsymbol{x}(\boldsymbol{X},t)$, and at generic time instant $t = t$ it takes actual, deformed configuration ${}^t\mathcal{B}$. For the purposes of building a macroscopic constitutive model of SMA materials, it is assumed that in the neighborhood of arbitrary material point of body $\mathcal{B}$ there can be distinguished the so-called representative volume element (RVE). The RVE is treated as a macroelement or a material point in the macroscopic constitutive model. The RVE must be large enough to grasp essential mesostructural features of the material and small enough, with respect to characteristic dimensions of the structure for description of which constitutive material model is used, that the fields in the RVE volume could be treated as homogeneous. The material macroelement remaining at location $\boldsymbol{X}$ in initially undeformed configuration is symbolically represented with a hexahedron in Figure 7.1.

The macroelement undergoes deformation as a result of applied external loadings, and at time instant t takes the location $\boldsymbol{x}(\boldsymbol{X},t)$. In the sequence, the deformation gradient $\boldsymbol{F} = \partial \boldsymbol{x}/\partial \boldsymbol{X}$ is taken as homogeneous within the volume of RVE at all time instants of the deformation process. Definition and instructive discussion on the RVE concept can be found in Nemat and Hori (1993). It is known that shape memory alloys when loaded thermo-mechanically can undergo forward or reverse martensitic phase transformation, which results in a two-phase SMA macroelement mesostructure.

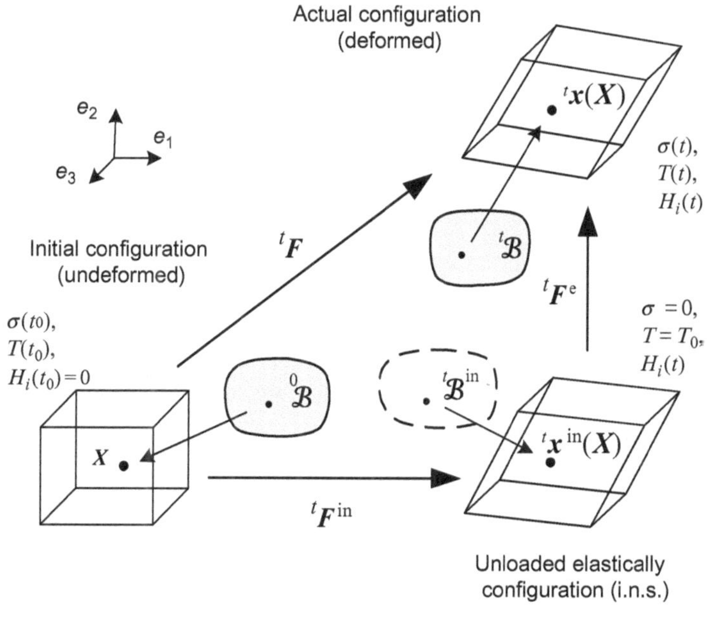

Figure 7.1 Kinematics of finite deformations.

The stress-induced martensitic phase transition is accompanied by inelastic deformation effects resulting from material macroelement mesostructure evolution, for example, generation or reorientation of martensitic variants. In the macroscopic model of SMA materials discussed here, inelastic deformation effects—called "phase" deformations—are described with the aid of two macroscopic internal state parameters, that is, volume fraction of oriented martensite, z, and isotropic ultimate eigenstrain tensor, $\boldsymbol{\kappa}$, treated as a function of a certain number of scalar internal (hidden) parameters, h_K, characterizing the mesostructure of the macroelement. Similar to the case of all members of the R_L models family, it is adopted here as conjecture that an SMA macroelement always tends to reach thermodynamic mesostructural equilibrium; that is, its mesostructure "exhibits tendency to optimally adapt to changing external loading." In the case of small deformations, Raniecki and Lexcellent (1994) formally showed that in states of thermodynamic mesostructural equilibrium $(h_K = h_K^{\mathrm{eq}})$ the free energy reaches minimum with respect to internal state parameters h_K. As a consequence, the requirement of "optimal adaptation of mesostructure" leads those authors to the conclusion that tensor $\boldsymbol{\kappa}(h_K)$ in states of mesostructural equilibrium $(h_K = h_K^{\mathrm{eq}})$ must be a certain function of external macroscopic state parameters—$\boldsymbol{\kappa}(h_K^{\mathrm{eq}}) = \boldsymbol{\kappa}^{\mathrm{eq}}(h_K^{\mathrm{eq}}(\boldsymbol{\sigma}, T))$ $= \boldsymbol{\kappa}^{\mathrm{eq}}(\boldsymbol{\sigma}, T)$—see also formula (3.6) in Chapter 3. The same argument delineated above and used in small deformations theory is adopted here to postulate that tensor $\boldsymbol{\kappa}$ is a function of macroscopic stress $\boldsymbol{\sigma}$ through equilibrium values of internal state parameters $h_K = h_K^{\mathrm{eq}}$. The possible dependence of $\boldsymbol{\kappa}$ on temperature is omitted here to keep the constitutive model as simple as practical. Thus, it is assumed in the discussed model that the mesostructure of an SMA macroelement optimally adapts to external thermomechanical loadings also when finite deformations are involved. Similar to the case of other members of the R_L models family, it is conjectured that there exists inelastic potential g^{ult} generating ultimate phase eigenstrain tensor $\boldsymbol{\kappa}$. Incremental numerical methods prove to be extremely efficient in solving nonlinear problems of mechanics, in view of the computer power available today. Thus, the rate form of constitutive model specification is the most useful. In the rate formulation of the finite deformations model of pseudoelasticity discussed here, evolution of the ultimate phase eigenstrain tensor due to the applied loading program is described with the aid of the inelastic stretching rate tensor $\boldsymbol{d}^{\mathrm{in}}$.

In order to properly describe inelastic deformation effects connected with phase transition, it is convenient to recall the known concept of virtual, unloaded elastically, instantaneous, "natural" reference configurations. The instantaneous, natural reference configuration is a configuration, which is attained by an SMA macroelement at particular time instant t, when the macroscopic surface tractions $\boldsymbol{t}^n$ are momentarily reduced to zero $(\boldsymbol{\sigma} = \mathbf{0})$ and temperature T is brought to certain fixed reference temperature T_0 while keeping the internal mesostructure of the macroelement fixed, "frozen"; that is, values of the set of internal parameters characterizing mesostructure are kept constant during the elastic unloading process $(z = \mathrm{const},$ $h_K = \mathrm{const}) \Leftrightarrow (\boldsymbol{\kappa} = \mathrm{const})$. The natural configuration at time t is symbolically marked with symbol ${}^t\mathcal{B}^{\mathrm{in}}$ in Figure 7.1. The state of the macroelement (RVE) remaining in the virtual, unloaded elastically configuration described above would be called *instantaneous, natural state* $(\boldsymbol{\sigma} = \mathbf{0}, T = T_0, z = z_t, h_K = h_K(t))$. Instantaneous, natural states should not be mistaken for the *thermodynamic reference state* $(\boldsymbol{\sigma}=\mathbf{0}, T=T_0, z=0, h_K=0)$. In the case of the classical theory of plasticity, natural states can be, at least theoretically, attained physically, namely, through the process of thermomechanical unloading of a deformed elastoplastically macroscopic body and next by dividing it into small pieces in such a way as to release all possible residual stresses. In the presently discussed case, instantaneous, natural configurations are entirely virtual and cannot be attained physically in any way. The reason for that is that in the pseudoelastic range of SMA alloys behavior $T > A_f^{2-1}$ it is impossible to constrain the progress of the reverse martensitic phase transition during thermomechanical unloading of an SMA macroelement to the natural state, that is, to maintain fixed values of internal state parameters $z, h_K = \text{const}$. Nevertheless, the concept of natural states is extremely useful when building a macroscopic model of shape memory alloys valid at finite strains. It conveniently allows uncoupling effects of elastic deformation from an inelastic one, here connected with martensitic phase transformation or possible evolution of a two-phase/martensitic mesostructure. The modeling approach is that instead of investigating the actual deformation process of an SMA macroelement when elastic and inelastic deformations take place simultaneously, it is investigated as two virtual, sequential processes. First, a pure inelastic deformation process from initial configuration to instantaneous natural configuration is accepted to occur ${}^0\mathcal{B} \rightarrow {}^t\mathcal{B}^{in}$, described by deformation gradient $\boldsymbol{F}^{\text{in}}$, and the next subsequent virtual pure elastic deformation process is adopted to occur from natural configuration to actual configuration ${}^t\mathcal{B}^{\text{in}} \rightarrow {}^t\mathcal{B}$, described with deformation gradient $\boldsymbol{F}^{\text{e}}$. Effectively, in this way deformation is obtained from initial configuration to actual one ${}^0\mathcal{B} \rightarrow {}^t\mathcal{B}$, described by actual deformation gradient $\boldsymbol{F}$. This construction leads to the known concept of multiplicative decomposition of the total deformation gradient

$$\boldsymbol{F} = \boldsymbol{F}^{\text{e}}\boldsymbol{F}^{\text{in}}. \tag{7.1}$$

Variable in time configurations ${}^t\mathcal{B}^{\text{in}}$ constitute convenient, reference configurations for an operational definition of elastic deformation. Multiplicative decomposition of the deformation gradient into elastic and inelastic parts still leaves several degrees of freedom in a further approach toward the description of material behavior. For example, a number of elastic strain measures can be defined on the finite elastic deformation gradient $\boldsymbol{F}^{\text{e}}$. Reasonable selection of such a measure, which is a state parameter appearing in free energy thermodynamic potential, can facilitate convenient description of the elastic branch of material behavior. This problem is addressed in more detail in the next section. The other issue is that, in accordance with its definition, any natural configuration ${}^t\mathcal{B}^{\text{in}}$ defining tensor $\boldsymbol{F}^{\text{e}}$ must be a stress-free configuration that imposes constraints on its selection (fixes its shape). But still, orientation of the configuration ${}^t\mathcal{B}^{\text{in}}$ with respect to fixed laboratory coordinates frame can be selected at free will. Freedom in selection of the orientation of natural configurations ${}^t\mathcal{B}^{\text{in}}$ finds its reflection in nonuniqueness of multiplicative decomposition of the deformation gradient into elastic and inelastic parts. For any rotation tensor $\boldsymbol{Q}$ $(\boldsymbol{Q}^{\text{T}}\boldsymbol{Q}=\mathbf{I})$, it is $\boldsymbol{F}=\boldsymbol{F}^{\text{e}}\mathbf{I}\boldsymbol{F}^{\text{in}}=(\boldsymbol{F}^{\text{e}}\boldsymbol{Q})(\boldsymbol{Q}^{\text{T}}\boldsymbol{F}^{\text{in}})=\boldsymbol{F}_1^{\text{e}}\boldsymbol{F}_1^{\text{in}}$, where $\boldsymbol{F}_1^{\text{e}}$, $\boldsymbol{F}_1^{\text{in}}$ are alternative gradients of deformation defined through new natural configuration ${}^t\mathcal{B}_1^{\text{in}}$ differing in rotation defined by tensor Q in comparison to configuration ${}^t\mathcal{B}^{\text{in}}$. Naturally, nonuniqueness in the definition of unloaded elastically configuration appears only when there appears whatever inelastic deformation $F^{\text{in}} \neq \mathbf{I}$. Before such an instance, $F = F^{\text{e}}$, and hence natural configuration is defined uniquely in view of initial conditions imposed for the field problem of mechanics. Premises of a physical nature may predetermine the selection of specific orientation of natural configuration with respect to laboratory coordinates frame, like it is in the case of crystallographic theory of plasticity. Then such orientation is an element of constitutive model. Alternatively, if premises of a physical nature for selection of orientation of natural configuration are nonexistent, as it is in the case of the macroscopic isotropic model of material behavior, then selection of specific orientation is dictated by the effort to obtain mathematical simplicity of the material behavior description. The constraint relationship determining the unique family of natural configurations for the model of SMA materials discussed here constitutes the latter case and has the form $\boldsymbol{\omega}^{\text{in}} \equiv \mathbf{0}$. This issue is discussed in more detail in Section 7.8.2 and Annex 7.4. The multiplicative decomposition of total deformation gradient (7.1) leads to the following known kinematical relations:

$$L = \partial \mathrm{v}/\partial x = \dot{F}F^{-1} = L^{\text{e}} + L^{\text{in}}, \quad L^{\text{e}} = \dot{F}^{\text{e}} \overset{-1}{F^{\text{e}}}, \quad L^{\text{in}} = F^{\text{e}} L^{*} \overset{-1}{F^{\text{e}}}, \quad L^{*} = \dot{F}^{\text{in}} \overset{-1}{F^{\text{in}}}, \tag{7.2}$$

$$d = \frac{1}{2}\left[L + \overset{T}{L}\right] = d^{\text{e}} + d^{\text{in}}, \quad d^{\text{e}} = \frac{1}{2}\left[L^{\text{e}} + \overset{T}{L^{\text{e}}}\right], \quad d^{\text{in}} = \frac{1}{2}\left[L^{\text{in}} + \overset{T}{L^{\text{in}}}\right],$$

$$D^{*} = \frac{1}{2}\left[L^{*} + \overset{T}{L^{*}}\right], \tag{7.3}$$

$$\boldsymbol{\omega} = \frac{1}{2}\left[L - L^{T}\right] = \boldsymbol{\omega}^{\text{e}} + \boldsymbol{\omega}^{\text{in}}, \quad \boldsymbol{\omega}^{\text{e}} = \frac{1}{2}\left[L^{\text{e}} - \overset{T}{L^{\text{e}}}\right], \quad \boldsymbol{\omega}^{\text{in}} = \frac{1}{2}\left[L^{\text{in}} - \overset{T}{L^{\text{in}}}\right],$$

$$\boldsymbol{\Omega}^{*} = \frac{1}{2}\left[L^{*} - \overset{T}{L^{*}}\right], \tag{7.4}$$

where L denotes velocity gradient tensor, d is stretching rate tensor, and $\boldsymbol{\omega}$ is material spin (vorticity) tensor. Upper indexes denote objects connected with the respective type of deformation: "e" purely elastic, "in" elastic-inelastic, "*" purely inelastic—here connected with martensitic phase transformation and/or martensitic phase reorientation.

7.3 Strain measures at finite deformations

The following relations result from polar and spectral decomposition of tensor of elastic deformation gradient

$$F^{\text{e}} = R^{\text{e}}U^{\text{e}} = V^{\text{e}}R^{\text{e}}, \quad U^{\text{e}} = \sum \lambda_i^{\text{e}} N_i^{\text{e}} \otimes N_i^{\text{e}}, \quad V^{\text{e}} = \sum \lambda_i^{\text{e}} n_i^{\text{e}} \otimes n_i^{\text{e}}, \tag{7.5}$$

where R^{e} is proper orthogonal rotation tensor. The positive definite, symmetric tensors U^{e}, V^{e} are right and left elastic stretch tensors, respectively. The scalars λ_i^{e} are the principal elastic stretches common for tensors $\boldsymbol{U}^{\mathrm{e}}$ and $\boldsymbol{V}^{\mathrm{e}}$. They are eigenvalues of the characteristic equation $\det\left(\boldsymbol{F}^{\mathrm{e}T}\boldsymbol{F}^{e} - \lambda^2\mathbf{I}\right) = 0$. Corresponding to them, eigenvectors $\boldsymbol{N}_i^{\mathrm{e}}$ and $\boldsymbol{n}_i^{\mathrm{e}}$ are Lagrangean (material) and Eulerian (spatial) principal directions of elastic stretch tensors $\boldsymbol{U}^{\mathrm{e}}$ and $\boldsymbol{V}^{\mathrm{e}}$. Principal directions vectors $\boldsymbol{N}_i^{\mathrm{e}}$ can be rotated to principal directions $\boldsymbol{n}_i^{\mathrm{e}}$ of tensor $\boldsymbol{V}^{\mathrm{e}}$ with the aid of rotation tensor $\boldsymbol{R}^{\mathrm{e}}$

$$\boldsymbol{n}_i^{\mathrm{e}} = \boldsymbol{R}^{\mathrm{e}}\boldsymbol{N}_i^{\mathrm{e}},\;\; \boldsymbol{R}^{\mathrm{e}} = \boldsymbol{n}_i^{\mathrm{e}}\otimes\boldsymbol{N}_i^{\mathrm{e}},\;\; (\boldsymbol{R}^{\mathrm{e}})^{\mathrm{T}}\boldsymbol{R}^{\mathrm{e}} = \mathbf{I},\;\; \det(\boldsymbol{R}^{\mathrm{e}}) = \mathbf{I}. \tag{7.6}$$

The logarithmic (Hencky) strain measure is used in the present model as a measure of elastic deformation. In mobile Lagrangean description, this measure is based on tensor $\boldsymbol{U}^{\mathrm{e}}$, while in Eulerian description on tensor $\boldsymbol{V}^{\mathrm{e}}$

$$\begin{aligned}&\boldsymbol{E}^{\mathrm{e}}(0) = \ln(\boldsymbol{U}^{\mathrm{e}}) = \sum \ln\left(\lambda_i^{\mathrm{e}}\right)\boldsymbol{N}_i^{\mathrm{e}}\otimes\boldsymbol{N}_i^{\mathrm{e}},\;\; \boldsymbol{e}^{\mathrm{e}}(0) = \ln(\boldsymbol{V}^{\mathrm{e}}) = \sum \ln\left(\lambda_i^{\mathrm{e}}\right)\boldsymbol{n}_i^{\mathrm{e}}\otimes\boldsymbol{n}_i^{\mathrm{e}},\\ &\boldsymbol{e}^{\mathrm{e}}(0) = \boldsymbol{R}^{\mathrm{e}}\boldsymbol{E}^{\mathrm{e}}(0)(\boldsymbol{R}^{\mathrm{e}})^{\mathrm{T}}\end{aligned} \tag{7.7}$$

The *Lagrangean (material) elastic strain* measure $\boldsymbol{E}^{\mathrm{e}}(0)$ is connected with its counterpart *Eulerian (spatial) elastic strain* $\boldsymbol{e}^{\mathrm{e}}(0)$ by formula $(7.7)_3$. Strain measure $\boldsymbol{E}^{\mathrm{e}}(0)$ is an element of parametric family of Lagrangean elastic strain measures—for parameter $n = 0$, analogical to the family of total strain measures introduced originally by Hill (1978),

$$\begin{aligned}&\boldsymbol{E}^{\mathrm{e}}(n) = \left((\boldsymbol{U}^{\mathrm{e}})^{2n} - \mathbf{I}\right)/(2n);\; n\neq 0,\;\; \boldsymbol{E}^{\mathrm{e}}(0) = \ln(\boldsymbol{U}^{\mathrm{e}});\; n = 0,\\ &\boldsymbol{E}^{\mathrm{e}}(n) = \sum f\left(\lambda_i^{\mathrm{e}}\right)\boldsymbol{N}_i^{\mathrm{e}}\otimes\boldsymbol{N}_i^{\mathrm{e}},\\ &f\left(\lambda_i^{\mathrm{e}}\right) = \left(\left(\lambda_i^{\mathrm{e}}\right)^{2n} - 1\right)/2n,\; n\neq 0,\;\; f\left(\lambda_i^{\mathrm{e}}\right) = \ln\left(\lambda_i^{\mathrm{e}}\right),\; n = 0.\end{aligned} \tag{7.8}$$

The scale function $f(\lambda_i^{\mathrm{e}})$ of principal elastic stretches has the properties $f\left(\lambda_i^{\mathrm{e}} = 1\right) = 0$ and $\mathrm{d}f\left(\lambda_i^{\mathrm{e}} = 1\right)/\mathrm{d}\lambda_i^{\mathrm{e}} = 1$. These properties assure that all strain measures from the family (7.8) converge to *small strains tensor* $\boldsymbol{\varepsilon}^{\mathrm{e}} \equiv 0.5\left(\boldsymbol{F}^{\mathrm{e}} + {}^{T}\boldsymbol{F}^{e} - 2\mathbf{I}\right)$ when elastic deformation gradient tends to unity $(\boldsymbol{F}^{\mathrm{e}} - \mathbf{I}) \to \mathbf{0}$.

The logarithmic elastic strain measure $\boldsymbol{E}^{\mathrm{e}}(0)$ possesses the known, valuable property that its spherical part describes purely dilatational deformation of the material element, while its deviatoric part describes purely distortional deformation at finite deformations

$$\begin{aligned}&\boldsymbol{E}^{\mathrm{e}}(0) = \ln(\boldsymbol{U}^{\mathrm{e}}) = \widetilde{\boldsymbol{E}}^{\mathrm{e}}(0) + \tfrac{1}{3}\mathrm{tr}(\boldsymbol{E}^{\mathrm{e}}(0))\mathbf{I},\;\; E_{ij}^{\mathrm{e}} = \ln\left(\lambda_i^{\mathrm{e}}\right) = \widetilde{E}_{ij}^{\mathrm{e}} + \tfrac{1}{3}\ln\left(\lambda_1^{\mathrm{e}}\lambda_2^{\mathrm{e}}\lambda_3^{\mathrm{e}}\right)\delta_{ij},\\ &\mathrm{tr}(\boldsymbol{E}^{\mathrm{e}}(0)) = \ln\left(\lambda_1^{\mathrm{e}}\right) + \ln\left(\lambda_2^{\mathrm{e}}\right) + \ln\left(\lambda_3^{\mathrm{e}}\right) = \ln(J^{\mathrm{e}}),\\ &J^{\mathrm{e}} \equiv \det(\boldsymbol{F}^{\mathrm{e}}) = \rho^{\mathrm{in}}/\rho = \det(\boldsymbol{U}^{\mathrm{e}}) = \lambda_1^{\mathrm{e}}\lambda_2^{\mathrm{e}}\lambda_3^{\mathrm{e}},\end{aligned} \tag{7.9}$$

where ρ^{in} and ρ denote density in stress free ("natural") and actual configuration, respectively. The $\widetilde{\boldsymbol{E}}^{\mathrm{e}}(0)$ is the deviator of logarithmic elastic strain $\left(\widetilde{E}_1^{\mathrm{e}} + \widetilde{E}_2^{\mathrm{e}} + \widetilde{E}_3^{\mathrm{e}} = 0\right)$.

It is this property, which is not possessed by other strain measures from the family (7.8), for example, the Green-Lagrange strain $\boldsymbol{E}^{\mathrm{e}}(1)$, that prevailed in selection of $\boldsymbol{E}^{\mathrm{e}}(0)$ as a state parameter in the SMA materials model discussed here. The principal values of elastic stretch tensor $\boldsymbol{U}^{\mathrm{e}}$ can be determined if there are known components of elastic logarithmic strain with the aid of the formula $\lambda_i^{\mathrm{e}} = (J^{\mathrm{e}})^{1/3}\exp\left(\widetilde{E}_i^{\mathrm{e}}\right)$.

The experiments show that under arbitrary loadings metallic materials exhibit only small distortional elastic strains, as they start to flow inelastically at strains of the order of 0.002. Only dilatational elastic strains can possibly be large, for example, at dynamic loadings due to high pressures. Hence, the special property of logarithmic strain measure $\boldsymbol{E}^{e}(0)$ of uncoupling dilatational and distortional deformation effects makes this measure particularly useful in building constitutive models of metallic materials. It straightforwardly allows postulating independently constitutive relations for distortional and dilatational deformations of a material.

7.4 Work conjugate stress measures

In the course of elastic deformation of macroelement, the power of elastic work per unit of mass of the deforming material $\dot{w}_e$ can be expressed as

$$\rho^0 \dot{w}_e = \mathrm{tr}(\boldsymbol{\tau}\boldsymbol{d}^e), \quad \boldsymbol{\tau} \equiv J\boldsymbol{\sigma}, \quad \boldsymbol{\tau} = \sum \tau_{ij}\boldsymbol{n}_i^e \otimes \boldsymbol{n}_j^e, \quad \boldsymbol{d}^e = \Sigma d_{ij}^e \boldsymbol{n}_i^e \otimes \boldsymbol{n}_j^e, \tag{7.10}$$

where $\boldsymbol{\tau}$ denotes classical *Kirchoff stress*, $\boldsymbol{\sigma}$ is *Cauchy stress*, $J \equiv \det(\boldsymbol{F}) = \rho^0/\rho$ denotes determinant of total deformation gradient, $\boldsymbol{d}^e$ is rate of elastic stretching tensor (see $(7.3)_1$). It is worth noting here that experimental evidence shows that the density of thermoelastic martensite and parent phase is practically the same. Hence, the density of a two-phase mixture in any unloaded elastically configuration is the same as density of the parent phase. Thus, $\rho^0 = \rho^{in}$, and $J = J^e J^{in} = \frac{\rho^{in}}{\rho}\frac{\rho^0}{\rho^{in}} = \rho^0/\rho = J^e$, where ρ^0 denotes density in initial undeformed configuration.

Using total Lagrangean description, Hill (1978) introduced the concept of stress measure $\boldsymbol{T}(n)$ *work conjugate* to any Lagrangean total strain measure $\boldsymbol{E}(n)$ from the family analogical to (7.8), but defined on total stretch tensor $\boldsymbol{U}$ $(\boldsymbol{F} = \boldsymbol{R}\boldsymbol{U})$. Hill relied on the presumption that the power of total work in the deformation process cannot depend on the way it is described; it remains invariant under change of strain measure. Adopting the presumption that the power of elastic work in the deformation process also cannot depend on the way of its description, and following Hill, the concept of stress measure $\boldsymbol{T}^e(n)$ elastic work conjugate to any Lagrangean elastic strain measure $\boldsymbol{E}^e(n)$ it is introduced here according to the definition

$$\rho^0 \dot{w}_e = \mathrm{tr}(\boldsymbol{\tau}\boldsymbol{d}^e) = \mathrm{tr}\left(\boldsymbol{T}^e(n)\dot{\boldsymbol{E}}^e(n)\right), \quad \boldsymbol{T}^e(n) = T_{ij}^e(n)\boldsymbol{N}_i^e \otimes \boldsymbol{N}_j^e, \tag{7.11}$$

where $\dot{\boldsymbol{E}}^e(n)$ denotes the usual material derivative of Lagrangean elastic strain measure from the family (7.8), and $\boldsymbol{T}^e(n)$ is the stress measure elastic work conjugate to $\boldsymbol{E}^e(n)$. Base free (index free) formulas for stress measures $\boldsymbol{T}^e(n)$ work conjugate to strains $\boldsymbol{E}^e(n)$ are rather complex except for a few cases. For example, simple base free formula connects *Green-Lagrange strain* $\boldsymbol{E}^e(1)$ with its work conjugate stress measure, that is, elastic *II Piola-Kirchoff stress* $\boldsymbol{T}^e(1)$,

$$\boldsymbol{T}^e(1) \equiv {}^{-1}\boldsymbol{F}^e \boldsymbol{\tau} {}^{-\mathrm{T}}\boldsymbol{F}^e, \tag{7.12}$$

The proof is simple: $\left(\operatorname{tr}(\boldsymbol{\tau d}^{e}) = \operatorname{tr}\left({}^{-T}\boldsymbol{F}^{e}\boldsymbol{\tau}^{T}\boldsymbol{F}^{e-T}\boldsymbol{F}^{e}\boldsymbol{d}^{eT}\boldsymbol{F}^{e}\right) = \operatorname{tr}\left(\boldsymbol{T}^{e}(1)\dot{\boldsymbol{E}}^{e}(1)\right), \dot{\boldsymbol{E}}(1) = {}^{T}\boldsymbol{F}^{e}\boldsymbol{d}^{e}\boldsymbol{F}^{e}\right)$. The disadvantage in using Green-Lagrange strain measure is that the spherical and deviatoric part of Green-Lagrange strain describes dilatational and distortional effects in a coupled way.

Elastic work conjugate stress tensors $\boldsymbol{T}^{e}(n)$ appearing in mobile Lagrangean description should be carefully distinguished from stress tensors $\boldsymbol{T}(n)$ total work conjugate to total strain tensors appearing in classical total Lagrangean description used by Hill and Rice (1973). The former stress measures depend on $\boldsymbol{F}^{e}$ $(\boldsymbol{T}^{e}(n) = \mathbf{f}(\boldsymbol{\tau}, \boldsymbol{F}^{e}))$, while the later on $\boldsymbol{F}$ $(\boldsymbol{T}(n) = \mathbf{f}(\boldsymbol{\tau}, \boldsymbol{F}))$; for example, classical II Piola Kirchoff stress is expressed with the formula $\boldsymbol{T}(1) \equiv {}^{-1}\boldsymbol{F}\boldsymbol{\tau}^{-T}\boldsymbol{F}$, while the elastic one with formula (7.12). Respective work conjugate stress measures are equal to each other $\boldsymbol{T}^{e}(n) = \boldsymbol{T}(n)$ only when $\boldsymbol{F}^{in} = \mathbf{I}$.

In general base free expressions for stresses $\boldsymbol{T}^{e}(n)$, elastic work conjugate with strains $\boldsymbol{E}^{e}(n)$ are rather complicated. Hill (1978), using the "principal axes method" elaborated by himself, derived simple formulas for components of stress tensors total work conjugate with respective total strain measures, expressed on principal axes. In order to reach this goal, he derived relations between components of $\dot{\boldsymbol{E}}(n)$ and $\boldsymbol{d}$ expressed on Lagrangean and Eulerian principal axes, respectively. It is enough to substitute in Hill's formulas objects originating from $\boldsymbol{F}$ with objects originating from $\boldsymbol{F}^{e}$ to obtain relations valid for $\dot{\boldsymbol{E}}^{e}(n)$ and $\boldsymbol{d}^{e}$. Here, relations linking material derivate of logarithmic elastic strain $\dot{\boldsymbol{E}}^{e}(0)$ with elastic stretching rate $\boldsymbol{d}^{e}$ are only specified (see $(7.7)_1$)

$$
\begin{aligned}
&\overline{\dot{E}}(0)^{e}_{ij} = d^{e}_{ij} = \dot{\lambda}^{e}_{i}/\lambda^{e}_{j} && \text{for } i=j \\
&\overline{\dot{E}}(0)^{e}_{ij} = \frac{2\left(\lambda^{e}_{i}/\lambda^{e}_{j}\right)\ln\left(\lambda^{e}_{i}/\lambda^{e}_{j}\right)}{\left(\lambda^{e}_{i}/\lambda^{e}_{j}\right)^{2}-1} d^{e}_{ij} && \text{for } i\neq j \text{ and } \lambda^{e}_{i}\neq\lambda^{e}_{j}, \\
&\dot{\boldsymbol{E}}^{e}(0) = \dot{\overline{\ln(\boldsymbol{U}^{e})}} = \sum \overline{\dot{E}}(0)^{e}_{ij}\boldsymbol{N}^{e}_{i}\otimes\boldsymbol{N}^{e}_{j}
\end{aligned}
\tag{7.13}
$$

The explicit derivation of the above formulas is delivered in Annex 7.1 in order to make this work as self-contained as possible (see (A7.24)).

Formulas (7.11) specified for logarithmic strain $(n=0)$ to take the form

$$
\tau_{ij}d^{e}_{ij} = T^{e}_{ij}(0)\overline{\dot{E}}^{e}_{ij}, \quad \operatorname{tr}(\boldsymbol{\tau d}^{e}) = \operatorname{tr}\left(\boldsymbol{T}^{e}(0)\dot{\boldsymbol{E}}^{e}(0)\right). \tag{7.14}
$$

The above relation is valid for whatever value of elastic stretching rate $\boldsymbol{d}^{e}$. Hence, components of stress tensor $\boldsymbol{T}^{e}(0)$ elastic work conjugate with elastic logarithmic strain $\boldsymbol{E}^{e}(0)$ written on Lagrangean elastic principal axes must fulfill the relations (see (7.13))

$$
T^{e}_{ij}(0) = \tau_{ij} \; i=j, \quad T^{e}_{ij}(0)\frac{2\lambda^{e}_{i}\lambda^{e}_{j}\ln\left(\lambda^{e}_{i}/\lambda^{e}_{j}\right)}{\left(\lambda^{e}_{i}\right)^{2}-\left(\lambda^{e}_{j}\right)^{2}} = \tau_{ij} \; i\neq j. \tag{7.15}
$$

The above formulas are valid irrespective of the symmetries that the constitutive law posses; that is, it is valid for isotropic and anisotropic materials.

When investigating materials that can be treated as isotropic elastically, it is convenient to introduce the concept of "rotated" elastically Kirchoff stress tensor $\boldsymbol{T}^{\mathrm{Re}}$ and rotated elastically elastic stretching rate tensor $\boldsymbol{D}^{\mathrm{e}}$ defined as

$$\boldsymbol{T}^{\mathrm{Re}} \equiv {}^{T}\boldsymbol{R}^{e}\boldsymbol{\tau}\boldsymbol{R}^{e},\ \ \boldsymbol{T}^{\mathrm{Re}} = T_{ij}^{\mathrm{Re}}\boldsymbol{N}_{i}^{\mathrm{e}}\otimes\boldsymbol{N}_{j}^{\mathrm{e}},\ \ \boldsymbol{D}^{\mathrm{e}} = {}^{T}\boldsymbol{R}^{e}\boldsymbol{d}^{\mathrm{e}}\boldsymbol{R}^{e},\ \ \boldsymbol{D}^{\mathrm{e}} = \Sigma D_{ij}^{\mathrm{e}}\boldsymbol{N}_{i}^{\mathrm{e}}\otimes\boldsymbol{N}_{j}^{\mathrm{e}}. \tag{7.16}$$

As can be easily noticed, components of $\boldsymbol{T}^{\mathrm{Re}}$ and $\boldsymbol{D}^{\mathrm{e}}$ on Lagrangean elastic principal axes are equal to the components of Kirchoff stress $\boldsymbol{\tau}$ and components of rotated elastically elastic stretching rate tensor $\boldsymbol{d}^{\mathrm{e}}$ written on Eulerian elastic principal axes (see (7.10))

$$T_{ij}^{\mathrm{Re}} = \tau_{ij},\ \ D_{ij}^{\mathrm{e}} = d_{ij}^{\mathrm{e}}. \tag{7.17}$$

Substituting expressions (7.16) for $\boldsymbol{T}^{\mathrm{Re}}$ and $\boldsymbol{D}^{\mathrm{e}}$ to (7.11) it is obtained that

$$\mathrm{tr}\left(\boldsymbol{T}^{\mathrm{Re}}\boldsymbol{D}^{\mathrm{e}}\right) = \mathrm{tr}\left({}^{T}\boldsymbol{R}^{e}\boldsymbol{\tau}\boldsymbol{R}^{\mathrm{e}}\,{}^{T}\boldsymbol{R}^{e}\boldsymbol{d}^{\mathrm{e}}\boldsymbol{R}^{\mathrm{e}}\right) = \mathrm{tr}\left(\boldsymbol{\tau}\boldsymbol{d}^{\mathrm{e}}\right) = \rho^{0}\dot{w}_{e} = \mathrm{tr}\left(\boldsymbol{T}^{\mathrm{e}}(0)\dot{\boldsymbol{E}}^{\mathrm{e}}(0)\right). \tag{7.18}$$

When the material is isotropic elastically, that is, its elastic properties do not depend on orientation in space, then constitutive relation linking tensors $\boldsymbol{T}^{\mathrm{e}}(n)$ and $\boldsymbol{E}^{\mathrm{e}}(n)$ must take the form of isotropic tensor function. Isotropic tensor function $\boldsymbol{\chi}(\boldsymbol{T})$ of tensorial argument $\boldsymbol{T}$ by definition fulfills the relation $\boldsymbol{\chi}\left(\boldsymbol{Q}\boldsymbol{T}^{T}\boldsymbol{Q}\right) = \boldsymbol{Q}\boldsymbol{\chi}(\boldsymbol{T})^{T}\boldsymbol{Q}$ for all proper orthogonal tensors $\left(\boldsymbol{Q}^{T}\boldsymbol{Q} = \boldsymbol{I}, \det\boldsymbol{Q} = 1\right)$. Hence, all stress tensors $\boldsymbol{T}^{\mathrm{e}}(n)$ must have principal directions common with $\boldsymbol{E}^{\mathrm{e}}(n)$; that is, those of tensor $\boldsymbol{U}^{\mathrm{e}}$ $\left(\boldsymbol{N}_{i}^{\mathrm{e}}\right)$. Formal proof of this statement can be found in many manuals; see, for example, the monograph by Ogden (1984). The consequence is that their only nonzero components are diagonal ones, when they are expressed on Lagrangean elastic principal axes. Any two pair of mentioned above tensors commutes,

$$\boldsymbol{T}^{\mathrm{e}}(n) = \Sigma T_{i}^{\mathrm{e}}(n)\boldsymbol{N}_{i}^{\mathrm{e}}\otimes\boldsymbol{N}_{i}^{\mathrm{e}},\ \ \boldsymbol{T}^{\mathrm{e}}(n)\boldsymbol{E}^{\mathrm{e}}(n) = \boldsymbol{E}^{\mathrm{e}}(n)\boldsymbol{T}^{\mathrm{e}}(n). \tag{7.19}$$

Similarly, in the case of elastic isotropy when Eulerian description is used, Cauchy stress $\boldsymbol{\sigma}$, Kirchoff stress $\boldsymbol{\tau}$, and elastic strain tensors $\boldsymbol{e}^{\mathrm{e}}(n)$ all must have common principal directions; that is, those of tensor $\boldsymbol{V}^{\mathrm{e}}$ $\left(\boldsymbol{n}_{i}^{\mathrm{e}}\right)$ (see $(7.7)_3$). As a consequence, any tensor from the group $\boldsymbol{\sigma}$, $\boldsymbol{\tau}$, $\boldsymbol{e}^{\mathrm{e}}(n)$ has only diagonal components that are nonzero when it is expressed on Eulerian elastic principal axes, and any two pair of them commutes, for example,

$$\boldsymbol{\tau} = \Sigma\tau_{i}\boldsymbol{n}_{i}^{\mathrm{e}}\otimes\boldsymbol{n}_{i}^{\mathrm{e}},\ \ \boldsymbol{\tau}\boldsymbol{e}^{\mathrm{e}}(n) = \boldsymbol{e}^{\mathrm{e}}(n)\boldsymbol{\tau}. \tag{7.20}$$

In the case of elastic isotropy, rotated Kirchoff stress tensor $\boldsymbol{T}^{\mathrm{Re}}$ also has common principal directions with tensor $\boldsymbol{U}^{\mathrm{e}}$ and only its diagonal components are nonzero in representation on elastic Lagrangean principal axes (see $(7.17)_1$ and (7.20)),

$$\boldsymbol{T}^{\mathrm{Re}} = T_{i}^{\mathrm{Re}}\boldsymbol{N}_{i}^{\mathrm{e}}\otimes\boldsymbol{N}_{i}^{\mathrm{e}},\ \ T_{i}^{\mathrm{Re}} = \tau_{i}\ i = j,\ \ T_{ij}^{\mathrm{Re}} = 0\ i \neq j,\ \ \boldsymbol{T}^{\mathrm{Re}}\boldsymbol{U}^{\mathrm{e}} = \boldsymbol{U}^{\mathrm{e}}\boldsymbol{T}^{\mathrm{Re}}. \tag{7.21}$$

From (7.15) and (7.20) it can be deduced that only diagonal components of tensor $\boldsymbol{T}^{e}(0)$ are nonzero in representation on Lagrangean elastic principal axes $\left(T_i^{e}(0)=\tau_i\ i=j\ \ T_{ij}^{e}(0)=0\ i\neq j\right)$.

The final conclusion is that in the case of material elastic isotropy stress tensor $\boldsymbol{T}^{e}(0)$ elastic work conjugate to elastic logarithmic strain $\boldsymbol{E}^{e}(0)$ is identical with Kirchoff stress tensor $\boldsymbol{T}^{Re}$ rotated elastically and the following relations are valid:

$$\boldsymbol{T}^{e}(0)=\boldsymbol{T}^{Re},\quad \boldsymbol{T}^{e}(0)\boldsymbol{U}^{e}=\boldsymbol{U}^{e}\boldsymbol{T}^{e}(0)=\boldsymbol{T}^{Re}\boldsymbol{U}^{e}=\boldsymbol{U}^{e}\boldsymbol{T}^{Re}. \tag{7.22}$$

In the literature devoted to computational mechanics, tensor $\boldsymbol{T}^{Re}$ is very often presented as stress tensor elastic work conjugate with elastic logarithmic strain $\boldsymbol{E}^{e}(0)$ without adequately underlying that it is true only in the case of isotropic elasticity. This can mislead the less-experienced reader that this is true in the case of any material symmetry, which is not true (see (7.15)).

7.5 Equations of state—elastically isotropic SMA materials

The SMA macroelement (RVE) of shape memory alloy material is treated as a two-phase compound thermodynamic system, which in general remains in constrained thermodynamic equilibrium at any stage of the deformation process. The logarithmic elastic strain $\boldsymbol{E}^{e}(0)$ as a function of elastic stretch tensor $\boldsymbol{U}^{e}$, temperature T, and volume fraction of martensitic phase z are adopted as state variables characterizing the thermodynamic state of SMA macroelement in mobile Lagrangean description—$\{\boldsymbol{E}^{e}(0),T,z\}$ or equivalently $\{\boldsymbol{e}^{e}(0),T,z\}$ in Eulerian description. The logarithmic strain measure is selected due to its property of separation of dilatational and distortional deformations and favorable property resulting from estimate (7.34), which is very convenient in rate formulation of constitutive relations. In the reminder of this chapter the discussion is limited to such SMA materials, which with good approximation can be treated as elastically isotropic; that is, their elastic properties are the same with respect to any orientation in space. Experimental evidence shows that elastic isotropy conjecture applies for many polycrystalline SMA materials. As a consequence, it is conjectured that free energy potential ϕ characterizing SMA macroelement is isotropic scalar function with respect to each of its canonic arguments (i.e., SMA macroelement state variables). Please note that isotropic scalar function $\phi(\boldsymbol{T})$ of tensorial argument $\boldsymbol{T}$ by definition fulfills the relation $\phi\left(\boldsymbol{Q}\boldsymbol{T}\boldsymbol{Q}^{T}\right)=\phi(\boldsymbol{T})$ for all proper orthogonal tensors $\boldsymbol{Q}$ $\left(\boldsymbol{Q}^{T}\boldsymbol{Q}=\boldsymbol{I}\right)$. Upon adopting the above assumptions, Gibbs fundamental equation of state determining total differential of specific Helmholtz free energy function $\phi(\boldsymbol{E}^{e}(0),T,z)$ takes the form

$$\begin{aligned}&\dot{\phi}=-s\dot{T}+\dot{w}_r\\&\dot{w}_r=\mathrm{tr}\left(\boldsymbol{T}^{Re}\dot{\boldsymbol{E}}^{e}(0)\right)/\rho^0-\pi_u^{1-2}\dot{z},\quad \dot{w}_r=\mathrm{tr}(\boldsymbol{\tau}\dot{\boldsymbol{e}}^{e}(0))/\rho^0-\pi_u^{1-2}\dot{z}\ (\mathrm{J/kg}),\end{aligned} \tag{7.23}$$

where s denotes specific entropy, π_u^{1-2} is the thermodynamic driving force of martensitic phase transformation in macroscopic stress-free state (see $(7.24)_3$ and (7.40)), $\dot{w}_r$ is the specific power of work per unit of mass, which would have to be performed over the macroelement in a thermodynamically reversible process run through the same path as the actual one. Formula $(7.23)_2$ defines power of reversible work $\dot{w}_r$ in mobile Lagrangean description, and formula $(7.23)_3$ defines $\dot{w}_r$ in Eulerian description.

The following equations of state for macroelement made of isotropic elastically SMA material are obtained in states of constrained thermodynamic equilibrium in mobile Lagrangean description:

$$\begin{aligned} \boldsymbol{T}^{\mathrm{Re}} &= \rho^0 \frac{\partial\phi(\boldsymbol{E}^{\mathrm{e}}(0),T,z)}{\partial \boldsymbol{E}^{\mathrm{e}}(0)}, \quad s = -\frac{\partial\phi(\boldsymbol{E}^{\mathrm{e}}(0),T,z)}{\partial T}, \quad \pi_{\mathrm{u}}^{1-2} = -\frac{\partial\phi(\boldsymbol{E}^{\mathrm{e}}(0),T,z)}{\partial z}, \\ \dot{\phi} &= \frac{\partial\phi}{\partial \boldsymbol{E}^{\mathrm{e}}(0)}\dot{\boldsymbol{E}}^{\mathrm{e}}(0) + \frac{\partial\phi}{\partial T}\dot{T} + \frac{\partial\phi}{\partial z}\dot{z}. \end{aligned} \tag{7.24}$$

The function ϕ is conjectured isotropic, hence its mathematical form does not change after expressing it in variables $\{\boldsymbol{e}^{\mathrm{e}}(0),T,z\}$, as tensors $\boldsymbol{E}^{\mathrm{e}}(0)$ and $\boldsymbol{e}^{\mathrm{e}}(0)$ differ only by rotation (see $(7.7)_3$). Due to this property, state equations of SMA macroelement in Eulerian description can be immediately obtained

$$\boldsymbol{\tau} \equiv \rho^0 \frac{\partial\phi(\boldsymbol{e}^{\mathrm{e}}(0),T,z)}{\partial \boldsymbol{e}^{\mathrm{e}}(0)}, \quad s \equiv -\frac{\partial\phi(\boldsymbol{e}^{\mathrm{e}}(0),T,z)}{\partial T}, \quad \pi_{\mathrm{u}}^{1-2} \equiv -\frac{\partial\phi(\boldsymbol{e}^{\mathrm{e}}(0),T,z)}{\partial z}. \tag{7.25}$$

The property of elastic isotropy results in colinearity of respective pairs of tensors $\boldsymbol{\tau},\boldsymbol{e}^{\mathrm{e}}(n)$ and $\boldsymbol{T}^{\mathrm{Re}},\boldsymbol{E}^{\mathrm{e}}(0)$ (see (7.20), (7.22)). Hence, functional relations $\boldsymbol{\tau} = \boldsymbol{\tau}(\boldsymbol{e}^{\mathrm{e}}(0))$, $\boldsymbol{T}^{\mathrm{Re}} = \boldsymbol{T}^{\mathrm{Re}}(\boldsymbol{E}^{\mathrm{e}}(0))$ must be isotropic tensor valued functions of tensorial argument. The known property of such functions (see note below), leads to the following formulas if conditions of their differentiability are fulfilled:

$$\begin{aligned} (\partial\boldsymbol{\tau}/\partial\boldsymbol{e}^{\mathrm{e}}(0))\left[\boldsymbol{e}^{\mathrm{e}}(0)\boldsymbol{\omega}^{Q} - \boldsymbol{\omega}^{Q}\boldsymbol{e}^{\mathrm{e}}(0)\right] &= \left[\boldsymbol{\tau}\boldsymbol{\omega}^{Q} - \boldsymbol{\omega}^{Q}\boldsymbol{\tau}\right], \\ (\partial\boldsymbol{T}^{\mathrm{Re}}/\partial\boldsymbol{E}^{\mathrm{e}}(0))\left[\boldsymbol{E}^{\mathrm{e}}(0)\boldsymbol{\omega}^{Q} - \boldsymbol{\omega}^{Q}\boldsymbol{E}^{\mathrm{e}}(0)\right] &= \left[\boldsymbol{T}^{\mathrm{Re}}\boldsymbol{\omega}^{Q} - \boldsymbol{\omega}^{Q}\boldsymbol{T}^{\mathrm{Re}}\right], \end{aligned} \tag{7.26}$$

Note: For any isotropic tensor function of tensorial argument it is $\mathrm{d}(\boldsymbol{Q}\boldsymbol{\chi}(\boldsymbol{T})\boldsymbol{Q}^{\mathrm{T}})/\mathrm{d}t = \mathrm{d}(\boldsymbol{\chi}(\boldsymbol{Q}\boldsymbol{T}\boldsymbol{Q}^{\mathrm{T}}))/\mathrm{d}t$. Hence, $\boldsymbol{Q}\left[\dot{\boldsymbol{\chi}} + \boldsymbol{\chi}\boldsymbol{\omega}^{Q} - \boldsymbol{\omega}^{Q}\boldsymbol{\chi}\right]\boldsymbol{Q}^{\mathrm{T}} = \boldsymbol{Q}(\partial\boldsymbol{\chi}/\partial\boldsymbol{T})$ $\boldsymbol{Q}^{\mathrm{T}}\boldsymbol{Q}\left[\dot{\boldsymbol{T}} + \boldsymbol{T}\boldsymbol{\omega}^{Q} - \boldsymbol{\omega}^{Q}\boldsymbol{T}\right]\boldsymbol{Q}^{\mathrm{T}}$. This relation is valid for any spin $\boldsymbol{\omega}^{Q} = \dot{\boldsymbol{Q}}\boldsymbol{Q}^{\mathrm{T}}$, where $\boldsymbol{Q}$ is proper orthogonal tensor ($\boldsymbol{Q}\boldsymbol{Q}^{\mathrm{T}} = \mathbf{I}$, $\det(\boldsymbol{Q}) = 1$) and for any rate $\dot{\boldsymbol{T}}$. Substituting $\dot{\boldsymbol{T}} = \mathbf{0}$, the property used in (7.26) is obtained.

7.6 Rate equations of state

Rate equations of state of SMA macroelement in Eulerian description can be obtained by calculating any corotational derivative of state variables (see (7.25)). Corotational derivative of any second order tensor $\boldsymbol{T}$ is defined as follows: $\overset{\circ}{\boldsymbol{T}}{}^{Q} \equiv \boldsymbol{Q}\frac{\mathrm{D}}{\mathrm{D}t}({}^{T}\boldsymbol{Q}\boldsymbol{T}\boldsymbol{Q})^{\mathrm{T}}\boldsymbol{Q} = \dot{\boldsymbol{T}} + \boldsymbol{T}\boldsymbol{\omega}^{Q} - \boldsymbol{\omega}^{Q}\boldsymbol{T}$. After substituting respective spin there can be obtained various corotational derivatives; for example, Zaremba-Jaumann—$\overset{\circ}{\boldsymbol{\tau}}^{(J)} = \dot{\boldsymbol{\tau}} + \boldsymbol{\tau}\boldsymbol{\omega} - \boldsymbol{\omega}\boldsymbol{\tau} = (\rho^0/\rho)[\dot{\boldsymbol{\sigma}} + \boldsymbol{\sigma}\boldsymbol{\omega} - \boldsymbol{\omega}\boldsymbol{\sigma} + \boldsymbol{\sigma}\mathrm{tr}(\boldsymbol{d})]$ $\quad (\boldsymbol{\tau} = (\rho^0/\rho)\boldsymbol{\sigma})$, elastic Zaremba-Jaumann—$\overset{\circ}{\boldsymbol{\tau}}^{(Je)} = (\rho^0/\rho)[\dot{\boldsymbol{\sigma}} + \boldsymbol{\sigma}\boldsymbol{\omega}^e - \boldsymbol{\omega}^e\boldsymbol{\sigma} + \boldsymbol{\sigma}\mathrm{tr}(\boldsymbol{d}^e)]$, Green-Naghdi, and so forth. If in place of a tensor $\boldsymbol{T}$ some scalar quantity is substituted then corotational derivative of a scalar is obtained. It is easy to verify through direct calculation that such derivative is simply equal to usual material derivative; for example, in the case of temperature it is $\overset{\circ}{T}^{Q} = \dot{T}$. For reasons elucidated in the sequence below, SMA material constitutive relations will be expressed with the use of elastic Zaremba-Jaumann corotational derivative. In such a case, rate constitutive relations in Eulerian description take the following form:

$$\begin{aligned}
\overset{\circ}{\boldsymbol{\tau}}^{(Je)} &= J\mathbf{L}\,\overline{\overset{\circ}{\boldsymbol{e}^e(0)}}^{(Je)} - J\mathbf{L}\boldsymbol{\alpha}_0\dot{T} - \rho^0\left(\partial\pi_u^{1-2}/\partial\boldsymbol{e}^e(0)\right)\dot{z},\\
T\dot{s} &= (1/\rho)T\mathbf{L}\boldsymbol{\alpha}_0\overline{\overset{\circ}{\boldsymbol{e}^e(0)}}^{(Je)} + c_\varepsilon\dot{T} + T\left(\partial\pi_u^{1-2}/\partial T\right)\dot{z},\\
\dot{\pi}_u^{1-2} &= \frac{\partial\pi_u^{1-2}}{\partial\boldsymbol{e}^e(0)}\overline{\overset{\circ}{\boldsymbol{e}^e(0)}}^{(Je)} + \frac{\partial s}{\partial z}\dot{T} + \frac{\partial\pi_u^{1-2}}{\partial z}\dot{z}, \quad (J = \rho^0/\rho),
\end{aligned} \tag{7.27}$$

where $\mathbf{L}$ is isotropic tensor of elastic stiffness modules, $\mathbf{L}\boldsymbol{\alpha}_0$ is tensor of elastic thermal stresses ($\boldsymbol{\alpha}_0 = \alpha_0\mathbf{I}$—isotropic tensor of thermal expansion), and $c_\varepsilon \equiv T(\partial s/\partial T)$ denotes specific heat capacity per unit of mass at constant strain. In order to obtain relations (7.27) the property that corotational derivative of a scalar function is just material derivative, for example, $\overset{\circ}{T}^{(Je)} = \dot{T}$, and relation $J\mathbf{L}[\boldsymbol{e}^e(0)\boldsymbol{\omega}^e - \boldsymbol{\omega}^e\boldsymbol{e}^e(0)] = [\boldsymbol{\tau}\boldsymbol{\omega}^e - \boldsymbol{\omega}^e\boldsymbol{\tau}]$ resulting from property $(7.26)_1$ upon substitution of elastic spin $\boldsymbol{\omega}^e$ (see $(7.4)_2$) and using definition of $\mathbf{L}$ (see $(7.28)_1$) is used. In $(7.27)_3$, reciprocal Maxwell relation has been used (see $(7.28)_4$).

Knowledge of explicit form of free energy function ϕ allows for determination of instantaneous values of $\mathbf{L}$, $\mathbf{L}\boldsymbol{\alpha}_0$ and $\partial\pi^{1-2}/\partial\boldsymbol{e}^e(0)$ with the aid of the formula

$$\begin{aligned}
&J\mathbf{L} \equiv \rho^0\frac{\partial^2\phi}{\partial\boldsymbol{e}^e(0)\partial\boldsymbol{e}^e(0)} = \frac{\partial\boldsymbol{\tau}}{\partial\boldsymbol{e}^e(0)}, \quad J\mathbf{L}\boldsymbol{\alpha}_0 \equiv -\rho^0\frac{\partial^2\phi}{\partial T\partial\boldsymbol{e}^e(0)} = \frac{\partial\boldsymbol{\tau}}{\partial T}\\
&\rho^0\frac{\partial\pi_u^{1-2}}{\partial\boldsymbol{e}^e(0)} = -\rho^0\frac{\partial^2\phi}{\partial z\partial\boldsymbol{e}^e(0)} = -\frac{\partial\boldsymbol{\tau}}{\partial z}, \quad \frac{\partial\pi_u^{1-2}}{\partial T} = -\frac{\partial^2\phi}{\partial T\partial z} = \frac{\partial s}{\partial z}.
\end{aligned} \tag{7.28}$$

Relations $(7.28)_{3,4}$ result from reciprocal Maxwell relations. Sometimes in the literature so-called energetic tensors of elastic properties $\mathbf{L}^E$, $\mathbf{L}^E\boldsymbol{\alpha}_0$ are used related with classical tensors of elastic properties by the formulas

$$\mathbf{L}^E = \mathbf{L}/\rho, \quad \mathbf{L}^E\boldsymbol{\alpha}_0 = \mathbf{L}\boldsymbol{\alpha}_0/\rho, \tag{7.29}$$

where ρ denotes density in actual configuration.

The multiplicative decomposition of deformation gradient on which the presently discussed model of pseudoelasticity is based relies on the condition that inelastic spin is identically equal to zero $\boldsymbol{\omega}^{in} = \mathbf{0}$. This condition unambiguously defines a family of so-called isoclinic, mobile, natural configurations ${}^t\mathcal{B}^{in}$ and evolution of their orientations in fixed global laboratory coordinates frame. In fact, this condition means that director triad versors $\{\boldsymbol{m}_i\}$, $i = 1,2,3$ determining orientation of natural configurations ${}^t\mathcal{B}^{\text{in}}$ remains invariant in fixed coordinates frame; that is, material derivatives of triad versors are zero ($\dot{\boldsymbol{m}}_i = \mathbf{0}$) at each time instant of the deformation process. Thus, configurations ${}^t\mathcal{B}^{\text{in}}$ have the same orientation with respect to laboratory axes at any time instant of inelastic deformation process; see also Annex 7.4 for a broader discussion. Because configurations ${}^t\mathcal{B}^{\text{in}}$, used as reference configurations for specification of equations of state (7.24) do not rotate, hence, corresponding rate equations of state defined with respect to these configurations can be calculated as usual material derivatives. These equations take the form

$$\begin{aligned}
\dot{\boldsymbol{T}}^{\text{Re}} &= J\mathbf{L}\dot{\boldsymbol{E}}^{\text{e}}(0) - J\mathbf{L}\boldsymbol{\alpha}_0\dot{T} - \rho^0\left(\partial\pi_{\text{u}}^{1-2}/\partial\boldsymbol{E}^{\text{e}}(0)\right)\dot{z}, \\
T\dot{s} &= (1/\rho)T(\mathbf{L}\boldsymbol{\alpha}_0)\dot{\boldsymbol{E}}^{\text{e}}(0) + c_\varepsilon\dot{T} + T\left(\partial\pi_{\text{u}}^{1-2}/\partial T\right)\dot{z}, \\
\dot{\pi}_{\text{u}}^{1-2} &= \frac{\partial\pi_{\text{u}}^{1-2}}{\partial\boldsymbol{E}^{\text{e}}(0)}\dot{\boldsymbol{E}}^{\text{e}}(0) + \frac{\partial s}{\partial z}\dot{T} + \frac{\partial\pi_{\text{u}}^{1-2}}{\partial z}\dot{z}.
\end{aligned} \tag{7.30}$$

Where,

$$\begin{aligned}
J\mathbf{L} = J\mathbf{L}(\boldsymbol{E}^{\text{e}}(0),T,z) &\equiv \rho^0\frac{\partial^2\phi}{\partial\boldsymbol{E}^{\text{e}}(0)\partial\boldsymbol{E}^{\text{e}}(0)} = \frac{\partial\boldsymbol{T}^{\text{Re}}}{\partial\boldsymbol{E}^{\text{e}}(0)}, \\
J\mathbf{L}\boldsymbol{\alpha}_0 = J\mathbf{L}\boldsymbol{\alpha}_0(\boldsymbol{E}^{\text{e}}(0),T,z) &\equiv -\rho^0\frac{\partial^2\phi}{\partial T\partial\boldsymbol{E}^{\text{e}}(0)} = \frac{\partial\boldsymbol{T}^{\text{Re}}}{\partial T}, \\
\rho^0\frac{\partial\pi_{\text{u}}^{1-2}}{\partial\boldsymbol{E}^{\text{e}}(0)} &= -\rho^0\frac{\partial^2\phi}{\partial z\partial\boldsymbol{E}^{\text{e}}(0)} = -\frac{\partial\boldsymbol{T}^{\text{Re}}}{\partial z}.
\end{aligned} \tag{7.31}$$

Tensors $\mathbf{L}$, $\mathbf{L}\boldsymbol{\alpha}_0$ appearing in formulas (7.28), (7.31) are the same tensors but treated as functional relations they are expressed in terms of different variables.

The following kinematical relation exists between elastic Zaremba-Jaumann derivative $\overset{\circ}{\overline{\boldsymbol{e}^{\text{e}}(0)}}{}^{(\text{Je})} \equiv \dot{\boldsymbol{e}}^{\text{e}}(0) + \boldsymbol{e}^{\text{e}}(0)\boldsymbol{\omega}^{\text{e}} - \boldsymbol{\omega}^{\text{e}}\boldsymbol{e}^{\text{e}}(0)$ and elastic stretching rate tensor $\boldsymbol{d}^{\text{e}}$ (see $(7.4)_2$):

$$\begin{aligned}
&\overset{\circ}{\overline{\boldsymbol{e}^{\text{e}}(0)}}{}^{(\text{Je})} = \left[\mathbf{I}^{(4)} + \boldsymbol{\mathcal{E}}^{\text{de}}\right]\boldsymbol{d}^{\text{e}}, \quad \overset{\circ}{\overline{\boldsymbol{e}^{\text{e}}(0)}}{}^{(\text{Je})} = \sum \overset{\text{Je}}{\overset{\circ}{e}}{}^{\text{e}}_{ij}\boldsymbol{n}^{\text{e}}_i \otimes \boldsymbol{n}^{\text{e}}_i \\
&\mathcal{E}^{\text{de}}_{ijkl}d^{\text{e}}_{kl} = \overset{\text{Je}}{\overset{\circ}{e}}{}^{\text{e}}_{ij} - d^{\text{e}}_{ij} = \begin{cases} 0, \; i=j \\ \left[\dfrac{\left(\lambda^{\text{e}}_i/\lambda^{\text{e}}_j\right)^2 + 1}{\left(\lambda^{\text{e}}_i/\lambda^{\text{e}}_j\right)^2 - 1}\ln\left(\lambda^{\text{e}}_i/\lambda^{\text{e}}_j\right) - 1\right]d^{\text{e}}_{ij}; \; i \neq j, \end{cases}
\end{aligned} \tag{7.32}$$

where $\mathbf{I}^{(4)}$ denotes unit tensor of the fourth order. The components of the fourth order *connection tensor of elastic Zaremba-Jaumann and elastic stretching rate* $\boldsymbol{\mathcal{E}}^{\text{de}}$ depend only on the state of elastic deformation of the material (λ^{e}_i) but not on the components of elastic stretching tensor $\boldsymbol{d}^{\text{e}}$. The derivation of relation (7.32) is delivered in Annex 7.3 (see (A7.31)–(A7.36)). The tensor $\boldsymbol{\mathcal{E}}^{\text{de}}$ has the useful property that it commutes with isotropic tensor of elasticity $\mathbf{L}\boldsymbol{\mathcal{E}}^{\text{de}} = \boldsymbol{\mathcal{E}}^{\text{de}}\mathbf{L}$. The other useful property is that $\boldsymbol{\mathcal{E}}^{\text{de}}\mathbf{s} = \mathbf{0}$ for any tensor that is collinear with tensor $\boldsymbol{V}^{\text{e}}$. Hence, in the case of elastic isotropy $\mathcal{E}^{de}\boldsymbol{\tau} = \mathcal{E}^{de}\boldsymbol{\sigma} = \mathcal{E}^{de}\boldsymbol{e}^{e}(0) = \mathbf{0}$; see Annex 7.3 for more discussion. It seems that Raniecki and Nguyen (1984) were the first to discuss in an explicit manner the mathematical relationship between elastic Zaremba-Jaumann derivative and elastic stretching rate $\boldsymbol{d}^{e}$ (7.32) and the constitutive modeling and computational advantages resulting from it. Substituting relation (7.32) for (7.27), the following rate constitutive relations in Eulerian description are obtained:

$$\overset{\circ}{\boldsymbol{\tau}}{}^{(\mathrm{Je})} = J\mathbf{L}^{(\mathrm{Je})}\boldsymbol{d}^{e} - J\mathbf{L}\boldsymbol{\alpha}_0\dot{T} - \rho^0\frac{\partial \pi_{\mathrm{u}}^{1-2}}{\partial \boldsymbol{e}^{e}}\dot{z}, \quad \mathbf{L}^{(\mathrm{Je})} \equiv \mathbf{L}\left[\mathbf{I}^{(4)} + \mathcal{E}^{de}\right],$$

$$\dot{s} = (1/\rho)\mathbf{L}^{(\mathrm{Je})}\boldsymbol{\alpha}_0\boldsymbol{d}^{e} + \frac{c_{\varepsilon}}{T}\dot{T} + \frac{\partial \pi_{\mathrm{u}}^{1-2}}{\partial T}\dot{z}, \quad \dot{s} = \boldsymbol{\alpha}_0\cdot\overset{\circ}{\boldsymbol{\tau}}{}^{(\mathrm{Je})} + \frac{c_p}{T}\dot{T} + \frac{\partial \pi_{\mathrm{u}}^{1-2}}{\partial z}\dot{z},$$

$$\dot{\pi}_{\mathrm{u}}^{1-2} = \frac{\partial \pi_{\mathrm{u}}^{1-2}}{\partial \boldsymbol{e}^{e}}\left[\mathbf{I}^{(4)} + \boldsymbol{\varepsilon}^{de}\right]\boldsymbol{d}^{e} + \frac{\partial s}{\partial z}\dot{T} + \frac{\partial \pi_{\mathrm{u}}^{1-2}}{\partial z}\dot{z}, \tag{7.33}$$

where $J\mathbf{L}^{(\mathrm{Je})}$ is the current, *instantaneous, elastic Zaremba-Jaumann rate tangent*. It must be strongly underlined that tensor $\mathbf{L}^{(\mathrm{Je})}$ does not depend on elastic stretching rate $\boldsymbol{d}^{e}$ but it is only a function of the elastic deformation state of SMA macroelement (λ_i^{e}). Relation (7.32) is of the utmost importance for modeling constitutive behavior of metallic materials at finite deformations. Experimental investigations show that all metallic materials undergo only small elastic distortional deformations at arbitrary multiaxial loadings, while at the same time they can undergo large dilatational elastic deformations (e.g., at impact loadings). This special physical property of metallic materials allows for employment of known estimation of elastic Zaremba-Jaumann derivative of logarithmic strain with the aid of elastic stretching rate tensor (Raniecki and Nguyen, 1984). This in turn allows for considerable simplification of rate constitutive relations at the cost of small error acceptable in the majority of cases of engineering computations. If ratios of elastic principal stretches fulfill condition $5/6 \leq \lambda_i^{e}/\lambda_j^{e} \leq 7/6$, which in the case of simple tension corresponds to logarithmic strain 16% $\left(\boldsymbol{e}^{e}(0)\left(\lambda_i^{e}/\lambda_j^{e} = 0.85\right) = -0.16\right)$ then a simple substitution to formula (7.32) allows finding out that component-wise differences between components of elastic Zaremba-Jaumann derivative of logarithmic strain and components of elastic stretching tensor on Eulerian principal axes fulfill the condition $|e_{ij}^{e\,\mathrm{Je}} - d_{ij}^{e}| \leq 1.1\%$. The stress corresponding to 16% of elastic strain upon the assumption of linear elastic behavior is $\sigma = 2\mu\cdot(-0.16) = -0.32\mu$, where μ denotes shear modulus. In a majority of cases elastic strains of shape memory alloy materials do not exceed 1%. At larger strains, the material starts to flow pseudoelastically due to the initiation and progress of martensitic phase transformation. At this level of elastic strain, the following inequality is valid:

$$\begin{aligned}&\left(e_{\mathrm{eq}}^{e} = 0.01\right) \Rightarrow \left(|\overset{\mathrm{Je}}{e_{ij}^{e}} - d_{ij}^{e}| \leq 3\times10^{-5} = 0.003\ (\%)\right),\\ &\left(5/6 \leq \lambda_i^{e}/\lambda_j^{e} \leq 7/6,\ e_{\mathrm{eq}}^{e} = -0.16 \Rightarrow |\overset{\mathrm{Je}}{e_{ij}^{e}} - d_{ij}^{e}| \leq 0.011 = 1.1\ (\%)\right)\end{aligned} \tag{7.34}$$

Estimate $(7.34)_2$ means that in the practice of engineering design computations of shape memory alloy materials, as well as in the case of other metallic alloys for which strain elasticity limit is adopted at $\varepsilon_{\text{offset}} = 0.2\ (\%)$, it can be accepted with perfect accuracy that

$$\overline{\overset{\text{o}}{\boldsymbol{e}^{\text{e}}(0)}}^{\text{Je}} = \boldsymbol{d}^{\text{e}}, \quad \mathbf{L}^{(\text{Je})} = \mathbf{L}. \tag{7.35}$$

The above relations indicate that elastic Zaremba-Jauman derivative is especially useful in the task of formulating rate constitutive relations of elasticity. As in the majority of practical cases, the same instantaneous tensor of modules of elasticity present in nonrate form of equations of state (i.e., $\mathbf{L} = \text{const}$) can be immediately used in corresponding rate equations of state expressed with the use of elastic stretching rate $\boldsymbol{d}^{\text{e}}$ with very good accuracy. This allows for large computational savings.

It is worth noting that equation $(7.33)_1$ constitutes an exact mathematical transition from mechanical equations of *hyperelasticity* $(7.25)_1$ to fully equivalent equations of hypoelasticity $(7.33)_1$. In particular taking the special case of isothermal elasticity ($\dot{T} = 0,\ \dot{z} = 0,\ \boldsymbol{d}^{\text{in}} = 0$), the exact mathematical counterpart of possibly nonlinear *hyperelasticity* law (7.25) is the rate hypoelastic law (7.33)

$$\begin{aligned} &\overset{\text{o}}{\boldsymbol{\tau}}(\text{Je}) = J\mathbf{L}^{(\text{Je})}\boldsymbol{d}^{\text{e}} \Leftrightarrow \boldsymbol{\tau} = \rho^0 \partial\phi/\partial\boldsymbol{e}^{\text{e}}(0), \\ &\mathbf{L}^{(\text{Je})} = \mathbf{L}\left[\mathbf{I}^{(4)} + \boldsymbol{\mathcal{E}}^{\text{de}}\right], \mathbf{L} = \rho\partial^2\phi/\partial\boldsymbol{e}^{\text{e}}(0)\partial\boldsymbol{e}^{\text{e}}(0). \end{aligned} \tag{7.36}$$

Approximation (7.35) has been used with great success in computational mechanics for many years now based on the estimations delivered by Hill and Rice (1973). However, it was Raniecki and Nguyen (1984) who delivered the exact mathematical relations between the Zaremba-Jauman derivative of logarithmic strain and stretching rate; thus enabling among the other elucidation of exact conditions that must be fulfilled to obtain equivalence in description of material behavior with hyperelastic and hypoelastic constitutive laws. If approximation (7.35) leads to unacceptable errors in numerical computations, then exact tangent should be used $\mathbf{L}^{(\text{Je})}$ defined with the formula $(7.33)_2$. Such necessity arises, for example, when numerical computations are performed for a simple shear problem in the finite deformations regime. In many publications devoted to modeling of this specific deformation regime, oscillatory stress predictions of material response were reported when hypoelastic constitutive law is used together with Zaremba-Jaumann rate and constant tangent $\mathbf{L}$ in modeling computations. Such oscillatory stresses are not observed experimentally.

7.7 Work dissipation

Specific power of total work $\dot{w}$ and reversible work $\dot{w}_{\text{r}}$ spent on SMA macroelement can be expressed with the following formulas in Eulerian description:

$$\dot{w} = \text{tr}(\boldsymbol{\tau}\boldsymbol{d})/\rho^0\ (\text{J/kg}), \quad \dot{w}_{\text{r}} = \text{tr}(\boldsymbol{\tau}\boldsymbol{d}^{\text{e}})/\rho^0 - \pi_{\text{u}}^{1-2}\dot{z}. \tag{7.37}$$

Formula $(7.37)_2$ results from formulas $(7.23)_3$ and (7.11). According to the concepts of nonequilibrium thermodynamics, specific power of *work dissipation* per unit of mass $\dot{w}_{\text{Disp}}$ is a difference between specific power of total and reversible work (see also $(7.3)_1$)

$$\dot{w}_{\text{Disp}} \equiv \dot{w} - \dot{w}_{\text{r}} = \text{tr}\left(\boldsymbol{\tau}\boldsymbol{d}^{\text{in}}\right)/\rho^0 + \pi_{\text{u}}^{1-2}\dot{z} \geq 0. \tag{7.38}$$

In accordance with the second law of thermodynamics, mechanical work dissipation in any thermodynamic process must be greater or equal to zero. Its magnitude is determined by the relations between respective thermodynamic forces and fluxes $\boldsymbol{\tau} \leftrightarrow \boldsymbol{d}^{\text{in}}$, $\pi^{1-2} \leftrightarrow \dot{z}$—at the same time the second law of thermodynamics indicates that such relations must exist. It should be pointed out that no limitations are imposed by the second law of thermodynamics on inelastic spin $\boldsymbol{\omega}^{\text{in}}$ (see $(7.4)_3$).

7.8 Special linear elastic model of pseudoelasticity of shape memory alloys

A very comprehensive, macroscopic thermodynamic framework has been discussed in the previous section enabling the formulation of a phenomenological model of SMA materials pseudoelastic behavior; for example, taking into account nonlinear elasticity, different elastic properties of austenitic and martensitic phase, or material damaging effects. In the present section a complete 3D, SMA materials model of pseudoelasticity is specified liner elastic in logarithmic strain measure. The following elements of the model are discussed:

- free energy potential—$\phi(\boldsymbol{e}^{\text{e}},T,z)$,
- Eulerian representation of inelastic spin—$\boldsymbol{\omega}^{\text{in}}$,
- pseudoelastic flow rule—$\boldsymbol{d}^{\text{in}}$ $\left(\text{tr}\left(\boldsymbol{d}^{\text{in}}\right)=0\right)$,
- phase transformation kinetics relations written in objective form—$\dot{z}$.

The model is a member of the R_{L} models family, constituting an extension of the Raniecki and Lexcellent (1998) model (see Section 3.3 in Chapter 3) to the finite deformations regime.

7.8.1 Free energy function

The specific free energy function of a two-phase shape memory alloy macroelement (RVE) remaining in state of "constrained phase equilibrium" and undergoing finite deformations is conjectured in mobile Lagrangean description in the form

$$\begin{aligned}
\phi(\boldsymbol{E}^{\text{e}}(0),T,z) &= \phi^{\text{th}}(T,z) + \phi^{\text{tm}}(\boldsymbol{E}^{\text{e}}(0),T) + \phi^{\text{coh}}(T,z),\\
\phi^{\text{th}}(T,z) &= \left(u_1^0 - Ts_1^0\right) - \left[\Delta u^0 - T\Delta s^0\right]z + c_v[(T-T_0) - T\ln(T/T_0)],\\
\rho\phi^{\text{tm}}(\boldsymbol{E}^{\text{e}}(0),T) &= \boldsymbol{E}^{\text{e}}(0)\cdot\mathbf{L}\boldsymbol{E}^{\text{e}}(0)/2 - (T-T_0)\boldsymbol{\alpha}_0\cdot\mathbf{L}\boldsymbol{E}^{\text{e}}(0),\\
\phi^{\text{coh}}(T,z) &= z\phi_2^{\text{st}} + \phi_{\text{it}}(T)(1-z)z,\quad \phi_2^{\text{st}} = \Delta u_2^{\text{st}} - T\Delta s_2^{\text{st}},\quad \phi_{\text{it}}(T) = \phi_{12} - Ts_{12},\\
\Delta u^0 &= u_1^0 - u_2^0,\ \Delta s^0 = s_1^0 - s_2^0,\ \Delta u_2^0 = \Delta u^0 - \phi_2^0,\ \Delta s_2^0 = \Delta s^0 - \Delta s_2^{\text{st}},\\
\pi_0^{1-2}(T) &= \Delta u_2^0 - T\Delta s_2^0.
\end{aligned} \tag{7.39}$$

the terms $\phi^{\mathrm{th}}(T,z)$ and $\phi^{\mathrm{tm}}(\boldsymbol{E}^{\mathrm{e}}(0),T)$ denote thermal and thermoelastic part of SMA macroelement free energy, while $\phi^{\mathrm{coh}}(T,z)$ denotes so-called coherency energy, u_{α}^{0}, s_{α}^{0}, $\alpha=1,2$ are specific internal energies and entropies of austenitic $(\alpha=1)$ and martensitic $(\alpha=2)$ phase formation at mesoscopic thermodynamic reference state (m.r.s), that is, in mesostress stress-free state $\breve{\boldsymbol{\sigma}}(\boldsymbol{x})=\mathbf{0}$, and at some thermodynamic reference temperature $T=T_0$, z is mass fraction of martensitic phase, the term $(\Delta u^0-T\Delta s^0)$ is so-called chemical thermodynamic driving force of phase transformation, $c_{\varepsilon}=\text{const}$ is heat capacity at fixed strain, $\mathbf{L}$ is isotropic tensor of linear elastic modules assumed to be constant $(L_{ijkl}=L_{ijkl}=\left[\mu\left(\delta_{ik}\delta_{jl}+\delta_{il}\delta_{jk}\right)+\lambda\delta_{ij}\delta_{kl}\right]=\text{const}$, $\lambda=(K-2\mu/3))$, $\boldsymbol{\alpha}_0=\alpha_0\mathbf{I}$ is constant isotropic thermal expansion coefficient $\left((\alpha_0)_{ij}=\alpha_0\delta_{ij}\right)$, and ϕ_{12}, s_{12} are material constants.

The above explicit form of free energy potential corresponds to the one postulated by Raniecki and Lexcellent for small deformations in a 3D case; see also Raniecki et al. (1992), Raniecki and Lexcellent (1994, 1998), and Müller and Xu (1991) in a 1D case. The difference between specification of free energy potential given here and that given by Raniecki and Lexcellent is that here elastic strain is used directly as a state variable, while the previous authors expressed this potential with the aid of total and phase transition strains tensors. After using additive decomposition of total strain tensor into elastic and inelastic part, valid in small deformations theory, Raniecki and Lexcellent potential takes formal mathematical form identical with that specified here by formulas (7.39). In order to obtain the free energy potential in Eulerian description it is sufficient to replace Lagrangean logarithmic strain $\boldsymbol{E}^{\mathrm{e}}(0)$ with Eulerian logarithmic strain $\boldsymbol{e}^{\mathrm{e}}(0)$, in view of elastic isotropy.

The inverted strain-stress constitutive relations corresponding to stress-strain relations originating from free energy potential (7.39) can be easily obtained by formal differentiation of Gibbs potential. The Gibbs potentials on natural and actual configurations, respectively, can be constructed in a classical way with the aid of Legendre transformations $g\left(\boldsymbol{T}^{\mathrm{Re}},T,z\right)\equiv\phi-(1/\rho^0)\boldsymbol{T}^{\mathrm{Re}}\boldsymbol{E}^{\mathrm{e}}(0)$, $g(\boldsymbol{\tau},T,z)\equiv\phi-(1/\rho^0)\boldsymbol{\tau}\cdot\boldsymbol{e}^{\mathrm{e}}(0)$ (J/kg).

In accordance with formalism of thermodynamics, the equations of state in Eulerian description resulting from potential ϕ defined with (7.39) take the following form—they constitute explicit specification of relations (7.25):

$$\begin{aligned}
\boldsymbol{\sigma}&=J^{-1}\boldsymbol{\tau}=\mathbf{L}[\boldsymbol{e}^{\mathrm{e}}(0)-\boldsymbol{\alpha}_0(T-T_0)],\quad J=\rho^0/\rho,\\
s&=c_v\ln(T/T_0)+s_1^0-z\Delta s_2^0+s_{12}(1-z)z+\boldsymbol{\alpha}_0\cdot\mathbf{L}\boldsymbol{e}^{\mathrm{e}}(0)/\rho,\\
\pi_{\mathrm{u}}^{1-2}&=\pi_0^{1-2}(T)-(1-2z)\phi_{\mathrm{it}}(T).
\end{aligned}\tag{7.40}$$

7.8.2 Tensor of inelastic spin ω^{in}

The SMA model discussed here is assumed to be valid for isotropic elastically SMA materials. For that reason there are no physical premises indicating that selection of some specific orientation of the family of isoclinic, natural, reference configurations would be more advantageous from a physical point of view. In fact from the physical constitutive model point of view it is quite free. However, the selection of a specific orientation of isoclinic configurations is important from a pragmatic point of view. Namely, "proper" selection of such orientation leads to simplicity of mathematical formulas and consequently to better numerical effectiveness of the model with the same quality of modeling predictions. Such mathematical simplicity can be obtained upon accepting the assumption that Eulerian inelastic spin $\boldsymbol{\omega}^{\text{in}}$ is identically equal to zero during any deformation process

$$
\begin{aligned}
&\boldsymbol{\omega}^{\text{in}} = \mathbf{0}, \\
&\left(\boldsymbol{\omega}^{\text{in}} = \mathbf{0}\right) \Leftrightarrow \left(\boldsymbol{\omega}^{\text{e}} = \boldsymbol{\omega}, \overset{\circ}{\boldsymbol{\tau}}^{(\text{Je})} = \overset{\circ}{\boldsymbol{\tau}}^{(\text{J})}\right).
\end{aligned}
\tag{7.41}
$$

Then, elastic Zaremba-Jaumann derivative defined with spin $\boldsymbol{\omega}^{\text{e}}$ becomes identically equal to Zaremba-Jaumann derivative defined with spin $\boldsymbol{\omega}$ (see $(7.4)_1$). This leads to considerable mathematical simplifications of the model without losing its generality. Selection of other than (7.42) family of isoclinic configurations results in the necessity of continuous tracking of spin $\boldsymbol{\omega}^{\text{in}}$ evolution, in order to be able to determine elastic spin $\boldsymbol{\omega}^{\text{e}} = \boldsymbol{\omega} - \boldsymbol{\omega}^{\text{in}}$. Naturally, condition (7.41) removes ambiguity in multiplicative decomposition of total deformation gradient into elastic and inelastic parts (see (7.1)). This condition effectively leads to the selection of the specific family of isoclinic, natural configurations discussed below formula (7.1). The proof of this last statement is in Annex 7.4. The concept of isoclinic configurations allows for harmonious and efficient introduction into macroscopic model elements of description of mesoscopic behavior of the material such as evolution of texture, or effect of large inelastic deformations induced anisotropy. Such elements can also be introduced to the model of shape memory alloys presented here, when such a need arises, but first of all when credible experimental data regarding, for example, the influence of evolution of microstructure on material properties are available. Such data must enable among the other formulation of evolution rule for inelastic spin analogous to (7.41); then it would be true physical constitutive information.

7.8.3 Tensor of velocity of phase strains in Eulerian description $\mathbf{d}^{in}$

Results of isothermal tests of polycrystalline NiTi alloy submitted to stress-controlled multiaxial proportional loading paths presented and discussed in Chapter 5 show that with good approximation ratios of velocities of components of macroscopic phase transformation strain tensor remained constant, in response to keeping constant ratios of velocities of components of inducing stress tensor (see Figures 5.5 and 5.12). Hence, it can be indirectly presumed that principal axes of macroscopic phase transformation strain tensor remained constant when principal axes of inducing phase transition stress tensor remained constant. Experimental work done on CuZnAl alloy by Rogueda et al. (1996), on CuAlBe alloy by Bouvet et al. (2002) (see Figures 17, 18, and 19; see also Bouvet et al. (2004), on NiTi alloy by Helm and Haupt (2003) (see Figures 7 and 8) for nonproportional loading paths indicate that ratios of increments of components of macroscopic phase transformation strain tensor followed after ratios of increments of components of inducing macroscopic stress tensor. The above listed experimental evidence constitutes, partly quantitative, partly qualitative support for the theoretical concept put forward by Raniecki and Lexcellent of "optimal adaptation of the SMA alloy microstructure" to changing external loadings. They also deliver an important premise indicating that it is well accepted to express macroscopic phase transformation strain tensor, describing inelastic deformation effects connected with phase transition with the aid of tensor function isotropic with respect to macroscopic stress tensor. The concept of "optimal adaptation of the SMA alloy microstructure" together with investigations regarding conditions for microstructural thermodynamic equilibrium led Raniecki et al. to the supposition that macroscopic phase eigenstrains tensor $\boldsymbol{\kappa}$ admits existence of a generating it potential. The potential g^{ult} generating phase eigenstrains tensor $\boldsymbol{\kappa}$ is called *ultimate phase eigenstrain potential*. Thermodynamic considerations show that it can be conveniently postulated as homogeneous function of the first order with respect to stress tensor (see formula $(3.9)_1$). Further, yet unpublished theoretical analyses indicated that phase eigenstrain potential g^{ult} plays an essential role not only in the pseudoelastic behavior of shape memory alloys modeling discussed here but also in the case of one-way memory effect, as it constitutes a macrscopic energy measure of phases mesostructural interactions.

Due to the premises of an experimental and theoretical nature recalled above. the concept of "optimal adaptation of the SMA alloy microstructure" together with all the consequences resulting from it is accepted as valid when building a model of SMA behavior at finite deformations. In particular, hereby, the conjecture is adopted that there exists ultimate phase eigenstrain potential g^{ult}, which is isotropic function, homogeneous of the first order with respect to Cauchy (Kirchoff) stress, in Eulerian description. Its explicit form proposed here is identical with that proposed in the R_L model (see (3.32))

$$\rho^0 g^{\mathrm{ult}}(\boldsymbol{\tau}) = \boldsymbol{\kappa}\cdot\boldsymbol{\tau} = \eta f(y)\tau', \quad \rho g^{\mathrm{ult}}(\boldsymbol{\sigma}) = \rho^0 g^{\mathrm{ult}}(\boldsymbol{\tau})|_{\tau'=\sigma'} = \boldsymbol{\kappa}\cdot\boldsymbol{\sigma} = \eta f(y)\sigma' \tag{7.42}$$

The ultimate phase eigenstrain potential (7.42) generates macroscopic ultimate phase eigenstrain tensor $\boldsymbol{\kappa}$

$$\begin{aligned} &\boldsymbol{\kappa} \equiv \rho\partial g^{\mathrm{ult}}(\boldsymbol{\sigma})/\partial\boldsymbol{\sigma} = \rho^0\partial g^{\mathrm{ult}}(\boldsymbol{\tau})/\partial\boldsymbol{\tau} = \overline{\boldsymbol{\kappa}} + \overline{\overline{\boldsymbol{\kappa}}}, \\ &\overline{\boldsymbol{\kappa}} = \eta f(y)\boldsymbol{n}^\sigma, \quad \overline{\overline{\boldsymbol{\kappa}}} = \eta\tau'\frac{\mathrm{d}f(y)}{\mathrm{d}y}\frac{\mathrm{d}y}{\mathrm{d}\boldsymbol{\tau}} = 3\eta\frac{\mathrm{d}f(y)}{\mathrm{d}y})\left[\sqrt{6}\left((\boldsymbol{n}^\sigma)^2 - \tfrac{1}{3}\mathbf{1}\right) - y\boldsymbol{n}^\sigma\right], \end{aligned} \tag{7.43}$$

where η denotes pseudoelastic flow amplitude treated as material constant, and $\boldsymbol{n}^\sigma$ denotes versor of "direction" of Cauchy (Kirchoff) stress tensor

$$\begin{aligned} &\boldsymbol{n}^\sigma \equiv \overline{\boldsymbol{\sigma}}/\sigma' = \overline{\boldsymbol{\tau}}/\tau' \equiv \boldsymbol{n}^\tau, \quad \boldsymbol{n}^\sigma\cdot\boldsymbol{n}^\sigma = \mathbf{I}, \quad y \equiv \sqrt{6}\mathrm{tr}\left((\boldsymbol{n}^\sigma)^3\right), \\ &\sigma' = (\overline{\boldsymbol{\sigma}}\cdot\overline{\boldsymbol{\sigma}})^{1/2}, \quad \tau' = (\overline{\boldsymbol{\tau}}\cdot\overline{\boldsymbol{\tau}})^{1/2}, \quad \overline{\boldsymbol{\sigma}} = \mathrm{dev}(\boldsymbol{\sigma}), \quad \overline{\boldsymbol{\tau}} = \mathrm{dev}(\boldsymbol{\tau}). \end{aligned} \tag{7.44}$$

In view of isotropy of potential g^{ult} with respect to stress tensor, this potential and Eulerian ultimate phase eigenstrain tensor $\boldsymbol{\kappa}$ can be immediately expressed in mobile Lagrangean description as

$$
\begin{aligned}
\rho^0 g^{\text{ult}}\left(\overline{\boldsymbol{T}}^{\text{Re}}\right) &= \boldsymbol{T}^{\text{Re}}\cdot\boldsymbol{K}^* = \eta f(y)\underset{T}{(T^{\text{Re}})'},\ \ \boldsymbol{K}^* \equiv \rho^0 \partial g^{\text{ult}}\left(\boldsymbol{T}^{\text{Re}}\right)/\partial \boldsymbol{T}^{\text{Re}},\\
\boldsymbol{K}^*\cdot\boldsymbol{T}^{\text{Re}} &= \boldsymbol{T}^{\text{Re}}\cdot\boldsymbol{K}^*,\ \ \boldsymbol{K}^* = \mathbf{R}^e\boldsymbol{\kappa}\mathbf{R}^e,\ \ \overline{\boldsymbol{T}}^{\text{Re}} = \operatorname{dev}\left(\boldsymbol{T}^{\text{Re}}\right)\\
\overline{\boldsymbol{K}}^* &= \eta f(y)\mathbf{N}_T,\ \ \overline{\overline{\boldsymbol{K}}}^* = 3\eta(\mathrm{d}f(y)/\mathrm{d}y)\left[\sqrt{6}\left(\mathbf{N}_T^2 - \tfrac{1}{3}\mathbf{1}\right) - y\mathbf{N}_T\right]\\
(T^{\text{Re}})' &= \left(\overline{\boldsymbol{T}}^{\text{Re}}\cdot\overline{\boldsymbol{T}}^{\text{Re}}\right)^{1/2},\ \ \mathbf{N}_T \equiv \overline{\boldsymbol{T}}^{\text{Re}}/(T^{\text{Re}})' = {}^T\mathbf{R}^e\boldsymbol{n}^\sigma\mathbf{R}^e,\ \ y = \sqrt{6}\operatorname{tr}\left(\mathbf{N}_T^3\right).
\end{aligned}
\tag{7.45}
$$

The property of isotropy of potential $g^{\text{ult}}(\boldsymbol{\sigma})$ with respect to stress tensor leads to colinearity of ultimate phase eigenstrain tensor $\boldsymbol{\kappa}$ and Kirchoff stress $\boldsymbol{\tau}$ (see (7.43)). In view of that, the following equations are satisfied identically for any spin tensor $\boldsymbol{\omega}^Q$, which can be shown by direct calculation

$$
\boldsymbol{\kappa}\cdot\left[\boldsymbol{\tau}\boldsymbol{\omega}^Q - \boldsymbol{\omega}^Q\boldsymbol{\tau}\right] = 0,\ \ \boldsymbol{\tau}\cdot\left[\boldsymbol{\kappa}\boldsymbol{\omega}^Q - \boldsymbol{\omega}^Q\boldsymbol{\kappa}\right] = 0,\ \ \boldsymbol{\kappa}\boldsymbol{\tau} = \boldsymbol{\tau}\boldsymbol{\kappa}. \tag{7.46}
$$

The relation $\boldsymbol{\kappa} = \boldsymbol{\kappa}(\boldsymbol{\tau})$ is tensor function isotropic with respect to its tensorial argument. Hence, for any spin $\boldsymbol{\omega}^Q = \dot{\boldsymbol{Q}}\boldsymbol{Q}^{\mathrm{T}}$ $\left({}^T\boldsymbol{\omega}^Q = -\boldsymbol{\omega}^Q\right)$ the following relation is valid:

$$
(\partial\boldsymbol{\kappa}/\partial\boldsymbol{\tau})\left[\boldsymbol{\tau}\boldsymbol{\omega}^Q - \boldsymbol{\omega}^Q\boldsymbol{\tau}\right] = \left[\boldsymbol{\kappa}\boldsymbol{\omega}^Q - \boldsymbol{\omega}^Q\boldsymbol{\kappa}\right]. \tag{7.47}
$$

Indeed, for any isotropic tensor function it is

$$
\overset{\bullet}{\overline{\boldsymbol{Q}\boldsymbol{\kappa}(\boldsymbol{\tau})\boldsymbol{Q}^{\mathrm{T}}}} = \overset{\bullet}{\overline{\boldsymbol{\kappa}\left(\boldsymbol{Q}\boldsymbol{\tau}\boldsymbol{Q}^{\mathrm{T}}\right)}}, \tag{7.48}
$$

hence,

$$
\boldsymbol{Q}\left[\dot{\boldsymbol{\kappa}} + \boldsymbol{\kappa}\boldsymbol{\omega}^Q - \boldsymbol{\omega}^Q\boldsymbol{\kappa}\right]\boldsymbol{Q}^{\mathrm{T}} = \boldsymbol{Q}(\partial\boldsymbol{\kappa}/\partial\boldsymbol{\tau})\boldsymbol{Q}^{\mathrm{T}}\boldsymbol{Q}\left[\dot{\boldsymbol{\tau}} + \boldsymbol{\tau}\boldsymbol{\omega}^Q - \boldsymbol{\omega}^Q\boldsymbol{\tau}\right]\boldsymbol{Q}^{\mathrm{T}}. \tag{7.49}
$$

The above relation is valid for any spin $\boldsymbol{\omega}^Q$ and for any $\dot{\boldsymbol{\tau}}$. Substituting $\dot{\boldsymbol{\tau}} = \mathbf{0}$, relation (7.47) is recovered.

It is convenient to introduce the following denotation of the derivative $\partial\boldsymbol{\kappa}/\partial\boldsymbol{\tau}$:

$$
\mathcal{P}(\boldsymbol{\tau}) \equiv \rho^0\frac{\partial^2 g^{\text{ult}}(\boldsymbol{\tau})}{\partial\boldsymbol{\tau}\partial\boldsymbol{\tau}} = \frac{\partial\boldsymbol{\kappa}}{\partial\boldsymbol{\tau}},\ \ \frac{\partial\boldsymbol{\kappa}}{\partial\boldsymbol{\sigma}} = \rho\frac{\partial^2 g^{\text{ult}}(\boldsymbol{\sigma})}{\partial\boldsymbol{\sigma}\partial\boldsymbol{\sigma}} = J\mathcal{P}. \tag{7.50}
$$

The explicit form of the fourth order tensor $\mathcal{P}$ in the case of ultimate phase eigenstrain potential $g^{\text{ult}}(\boldsymbol{\sigma})$ is given below:

$$
\begin{aligned}
\mathcal{P}(\boldsymbol{\tau}) &= -2\eta\frac{\mathrm{d}f(y)}{\partial y}\frac{\partial y}{\partial\boldsymbol{\tau}}\otimes\boldsymbol{n}^\tau + \left[\eta f(y) - 3\eta y\frac{\mathrm{d}f(y)}{\partial y}\right]\frac{\partial\boldsymbol{n}^\tau}{\partial\boldsymbol{\tau}}\\
&\quad + \eta\tau'\frac{\mathrm{d}^2 f(y)}{\partial y^2}\frac{\partial y}{\partial\boldsymbol{\tau}}\otimes\frac{\partial y}{\partial\boldsymbol{\tau}} + 6\sqrt{6}\eta\frac{\mathrm{d}f(y)}{\partial y}\boldsymbol{n}^\tau\frac{\partial\boldsymbol{n}^\tau}{\partial\boldsymbol{\tau}},\\
\frac{\partial y}{\partial\tau} &= \frac{3}{\tau'}\left[\sqrt{6}\left((\boldsymbol{n}^\tau)^2 - \tfrac{1}{3}\mathbf{1}\right) - y\boldsymbol{n}^\tau\right] = J^{-1}\frac{\partial y}{\partial\sigma},\\
\frac{\partial\boldsymbol{n}^\tau}{\partial\boldsymbol{\tau}} &= \frac{1}{\tau'}\left[\mathbf{1}^{(4)} - \frac{1}{3}\mathbf{1}\otimes\mathbf{1} - \frac{\overline{\boldsymbol{\tau}}\otimes\overline{\boldsymbol{\tau}}}{(\tau')^2}\right] = J^{-1}\frac{\partial\boldsymbol{n}^\sigma}{\partial\boldsymbol{\sigma}}.
\end{aligned}
\tag{7.51}
$$

The postulated form of potential $g^{\mathrm{ult}}(\boldsymbol{\tau})$ is a function homogeneous of order one with respect to $\boldsymbol{\tau}$ ($\boldsymbol{\sigma}$). Hence, the functional relation $\boldsymbol{\kappa}(\boldsymbol{\tau})$ (see (7.43)) is isotropic function, homogeneous of order zero with respect to $\boldsymbol{\tau}$. As a consequence, tensor $\boldsymbol{\mathcal{P}}$ has the property

$$\boldsymbol{\mathcal{P}}\boldsymbol{\tau}=\frac{\partial\boldsymbol{\kappa}}{\partial\boldsymbol{\tau}}\boldsymbol{\tau}=\mathbf{0},\quad (J\boldsymbol{\mathcal{P}})\boldsymbol{\sigma}=\frac{\partial\boldsymbol{\kappa}}{\partial\boldsymbol{\sigma}}\boldsymbol{\sigma}=\mathbf{0}, \tag{7.52}$$

which can be verified by direct calculation (see also (3.9)).

As results from formula (7.49), any corotational derivative of tensor $\boldsymbol{\kappa}$ can be expressed with the aid of the corotational derivative of Kirchoff stress tensor corresponding to it as follows (using denotation (7.50))

$$\overset{\circ}{\boldsymbol{\kappa}}{}^{(\mathrm{Q})}=\dot{\boldsymbol{\kappa}}+\boldsymbol{\kappa}\boldsymbol{\omega}^{Q}-\boldsymbol{\omega}^{Q}\boldsymbol{\kappa}=\boldsymbol{\mathcal{P}}\overset{\circ}{\boldsymbol{\tau}}{}^{(\mathrm{Q})},\quad \overset{\circ}{\boldsymbol{\kappa}}{}^{(\mathrm{Q})}=(J\boldsymbol{\mathcal{P}})\overset{\circ}{\boldsymbol{\sigma}}{}^{(\mathrm{Q})}, \tag{7.53}$$

In the case of the linear elastically model of SMA materials thermodynamic force of p.t. at zero macroscopic stress proposed here, π_{u}^{1-2} does not depend on elastic strain $\boldsymbol{e}^{\mathrm{e}}$. Hence, it is

$$\frac{\partial\pi_{\mathrm{u}}^{1-2}}{\partial\boldsymbol{e}^{\mathrm{e}}}=-\frac{\partial\boldsymbol{\sigma}}{\partial z}=0. \tag{7.54}$$

The rate form of mechanical equation of state $(7.40)_1$ takes the following form upon taking advantage of (7.54) (see also $(7.33)_1$)

$$\overset{\circ}{\boldsymbol{\tau}}{}^{(\mathrm{Je})}=J\mathbf{L}^{(\mathrm{Je})}\left(\boldsymbol{d}-\boldsymbol{d}^{\mathrm{in}}\right)-J\mathbf{L}\boldsymbol{\alpha}_0\dot{T},\quad \overset{\circ}{\boldsymbol{\sigma}}{}^{(0)}=\mathbf{L}^{(\mathrm{Je})}\left(\boldsymbol{d}-\boldsymbol{d}^{\mathrm{in}}\right)-\mathbf{L}\boldsymbol{\alpha}_0\dot{T}, \tag{7.55}$$

where the property was used that $\mathbf{L}\boldsymbol{\alpha}_0=\mathbf{L}^{(\mathrm{Je})}\boldsymbol{\alpha}_0$ (see (A7.39)).

The above relation requires specification of tensor of inelastic stretching rate $\boldsymbol{d}^{\mathrm{in}}$. Formal law of pseudoelastic flow defining tensor $\boldsymbol{d}^{\mathrm{in}}$ is hereby postulated in the form

$$\boldsymbol{d}^{\mathrm{in}}=\overset{\circ}{\overline{z\boldsymbol{\kappa}}}{}^{(\mathrm{Je})}=\dot{z}\boldsymbol{\kappa}+z\overset{\circ}{\boldsymbol{\kappa}}{}^{(\mathrm{Je})}. \tag{7.56}$$

Elastic Zaremba-Jaumann derivative of tensor $\boldsymbol{\kappa}$ can be expressed with the aid of elastic Zaremba-Jaumann derivative of Kirchoff stress as (see (7.53))

$$\overset{\circ}{\boldsymbol{\kappa}}{}^{(\mathrm{Je})}=\boldsymbol{\mathcal{P}}\overset{\circ}{\boldsymbol{\tau}}{}^{(\mathrm{Je})}. \tag{7.57}$$

Substituting in (7.55) relations (7.56) and (7.57), one obtains

$$\overset{\circ}{\boldsymbol{\tau}}{}^{(\mathrm{Je})}=J\mathbf{L}^{(\mathrm{Je})}\left[\boldsymbol{d}-\dot{z}\boldsymbol{\kappa}-\boldsymbol{\alpha}_0\dot{T}\right]-zJ\mathbf{L}^{(\mathrm{Je})}\boldsymbol{\mathcal{P}}\overset{\circ}{\boldsymbol{\tau}}{}^{(\mathrm{Je})}. \tag{7.58}$$

Grouping terms with corotational derivative of stress the following is obtained:

$$\overset{\circ}{\boldsymbol{\tau}}{}^{(\mathrm{Je})}=J\mathbf{L}^{\mathrm{S(Je)}}\left[\boldsymbol{d}-\dot{z}\boldsymbol{\kappa}-\boldsymbol{\alpha}_0\dot{T}\right]. \tag{7.59}$$

It proves convenient to introduce the following definitions:

$$\mathbf{L}^{\mathrm{S(Je)}} \equiv \left[\mathbf{I}^{(4)} + zJ\mathbf{L}^{\mathrm{(Je)}}\boldsymbol{\mathcal{P}}\right]^{-1}\mathbf{L}^{\mathrm{(Je)}}, \quad \mathbf{L}^{\mathrm{S}} \equiv \left[\mathbf{I}^{(4)} + zJ\mathbf{L}\boldsymbol{\mathcal{P}}\right]^{-1}\mathbf{L}. \tag{7.60}$$

Taking advantage of the mathematical formula $\left[\mathbf{A}^{-1}\mathbf{B}\right]^{-1} = \mathbf{B}^{-1}\mathbf{A}$ it can be shown that

$$\begin{aligned} &\mathbf{L}^{\mathrm{S(Je)}} = \left[\mathbf{M}^{\mathrm{(Je)}} + zJ\boldsymbol{\mathcal{P}}\right]^{-1}, \quad \mathbf{L}^{\mathrm{S}} = \left[\mathbf{M} + zJ\boldsymbol{\mathcal{P}}\right]^{-1}, \\ &\mathbf{M}^{\mathrm{(Je)}} = \left[\mathbf{L}^{\mathrm{(Je)}}\right]^{-1}, \quad \mathbf{M} = \mathbf{L}^{-1}. \end{aligned} \tag{7.61}$$

Using again properties (A7.39), $\mathbf{L}^{\mathrm{S(Je)}}\boldsymbol{\alpha}_0 = \mathbf{L}^{\mathrm{S}}\boldsymbol{\alpha}_0 = \mathbf{L}\boldsymbol{\alpha}_0$, and relations (7.41) the required rate form of mechanical equations of state is obtained

$$\overset{\circ}{\boldsymbol{\sigma}}{}^{(0)} = J^{-1}\overset{\circ}{\boldsymbol{\tau}}{}^{\mathrm{(J)}} = J^{-1}\overset{\circ}{\boldsymbol{\tau}}{}^{\mathrm{(Je)}} = \mathbf{L}^{\mathrm{S(Je)}}\left[\boldsymbol{d} - \dot{z}\boldsymbol{\kappa}\right] - \mathbf{L}\boldsymbol{\alpha}_0\dot{T}. \tag{7.62}$$

The rate equation of entropy change takes the following form, in the case of proposed hereby SMA model

$$\dot{s} = \left(c_p/T\right)\dot{T} + \left[-\Delta s_2^0 + s_{12}(1-2z)\right]\dot{z} + \boldsymbol{\alpha}_0\cdot\dot{\boldsymbol{\tau}}/\rho_0, \tag{7.63}$$

(see also $(7.33)_2$).

7.8.4 Thermal effects and mechanical work dissipation

The following relation is obtained upon contracting (7.56) with $\boldsymbol{\tau}$ and using property (7.52):

$$\boldsymbol{\tau}\cdot\boldsymbol{d}^{\mathrm{in}} = \dot{z}\boldsymbol{\tau}\cdot\boldsymbol{\kappa} + z\boldsymbol{\tau}\cdot\overset{\circ}{\boldsymbol{\kappa}}{}^{\mathrm{(Je)}} = \dot{z}\boldsymbol{\tau}\cdot\boldsymbol{\kappa} + \underbrace{\boldsymbol{\tau}\cdot\boldsymbol{\mathcal{P}}\overset{\circ}{\boldsymbol{\tau}}{}^{\mathrm{(Je)}}}_{=\mathbf{0}} = \dot{z}\boldsymbol{\tau}\cdot\boldsymbol{\kappa}. \tag{7.64}$$

The expression for mechanical work dissipation, taking place during any deformation process of SMA macroelement, can be obtained upon substituting in (7.38) expression (7.64) and using $(7.40)_3$

$$\begin{aligned} &\dot{w}_{\mathrm{Disp}} = \dot{w} - \dot{w}_{\mathrm{r}} = \pi^{1-2}\dot{z} \geq 0, \\ &\pi^{1-2} = \pi_{\mathrm{u}}^{1-2} + \boldsymbol{\kappa}\cdot\boldsymbol{\tau}/\rho_0 = \pi_0^{1-2}(T) - (1-2z)\phi_{\mathrm{it}}(T) + \overbrace{\boldsymbol{\tau}\cdot\boldsymbol{\kappa}/\rho^0}^{=g^{\mathrm{ult}}(\boldsymbol{\tau})}. \end{aligned} \tag{7.65}$$

Eliminating $\dot{u}$ from the relation resulting from definition of free energy $\dot{\phi} = \dot{u} - s\dot{T} - \dot{s}T$ and the equation expressing the first law of thermodynamics written in the form $\dot{u} = \dot{w} - \dot{q}$, next substituting such obtained expression for $\dot{\phi}$ into fundamental Gibbs equation of state $\dot{\phi} = -s\dot{T} + \dot{w}_r$ (see (7.23)), and using the expression for work dissipation $(7.65)_1$, the following relation is obtained:

$$\dot{q} = \pi^{1-2}\dot{z} - \dot{s}T. \tag{7.66}$$

After substituting $(7.65)_2$ and (7.63) in the above, one obtains

$$\dot{q} = -c_p\dot{T} + \left[\Delta u_2^0 - (1-2z)\phi_{12} + g^{\mathrm{ult}}(\boldsymbol{\tau})\right]\dot{z} - T\boldsymbol{\alpha}_0\cdot\dot{\boldsymbol{\tau}}/\rho_0. \tag{7.67}$$

Introducing denotation that $\dot{q}_{\mathrm{tr}}$ is specific power of heat sources connected with all thermal effects taking place in SMA macro element except for heat capacity, and using the formula $(7.42)_2$ for potential g^{ult}

$$\dot{q}_{\mathrm{tr}} \equiv c_p\dot{T} + \dot{q} = \left[\Delta u_2^0 - (1-2z)\phi_{12} + \gamma f(y)\tau_{\mathrm{ef}}/\rho_0\right]\dot{z} - T\boldsymbol{\alpha}_0\cdot\dot{\boldsymbol{\tau}}/\rho_0, \tag{7.68}$$

where,

$$\eta f(y)\tau' = \gamma f(y)\tau_{\mathrm{ef}}, \quad \tau_{\mathrm{ef}} = \sqrt{3/2}\tau', \quad \gamma = \sqrt{2/3}\eta, \quad \boldsymbol{\alpha}_0\cdot\dot{\boldsymbol{\tau}} = 3\alpha_0\dot{\tau}_{\mathrm{m}}. \tag{7.69}$$

The last relation of (7.69) results from assumed isotropy of thermal expansion tensor.

7.8.5 Kinetics of phase transformation ż

The general formal thermoelastic martensitic phase transformation (p.t.) kinetics relations fulfilling requirements imposed by the second law of thermodynamics were proposed in very comprehensive form by Raniecki and Lexcellent in papers (Raniecki et al., 1992), (Raniecki and Lexcellent, 1994); see also the text below formula (3.20). These relations, called here *R-L kinetics relations,* allow for the description of kinetics of various SMA alloys through proper identification of values of parameters $(m_\alpha, p_\alpha, r_\alpha)$ appearing in so-called specific p.t. kinetics rules defined with specific mathematical form of state functions $\lambda^\sigma_{(\alpha)}(\boldsymbol{\tau}, T, z)$. Specific functions (3.29) or alternatively (6.15) can be used.

The general R-L kinetics relations, being a counterpart of relations (3.27), written in objective form in Eulerian description are

$$\begin{aligned}
\dot{z} &= \lambda_1^\sigma(\boldsymbol{\tau}, T, z)\cdot\overset{\circ}{\overline{\tilde{\pi}_\tau^{1-2}}}^{(\mathrm{Je})} \quad \text{if} \left\{\pi^{1-2} \ge Y_1 \text{ and } \overset{\circ}{\overline{\tilde{\pi}_\tau^{1-2}}}^{(\mathrm{Je})} > 0\right\} (\mathcal{A} \to \mathcal{M} \text{ p.t.}),\\
\dot{z} &= \lambda_2^\sigma(\boldsymbol{\tau}, T, z)\cdot\overset{\circ}{\overline{\tilde{\pi}_\tau^{1-2}}}^{(\mathrm{Je})} \quad \text{if} \left\{\pi^{1-2} \le Y_2 \text{ and } \overset{\circ}{\overline{\tilde{\pi}_\tau^{1-2}}}^{(\mathrm{Je})} < 0\right\} (\mathcal{M} \to \mathcal{A} \text{ p.t.}),
\end{aligned} \tag{7.70}$$

$\dot{z} = 0$ in any other case, where π^{1-2}, the thermodynamic driving force of phase transformation, is defined with formula $(7.65)_2$. The lag threshold value functions

$0 \le Y_{(\alpha)}(\boldsymbol{\tau}, T, z)$, $(\alpha = 1,2)$, and $Y_{(1)}|_{z=0} = Y_{(2)}|_{z=1} = 0$ denote thresholds, which upon reaching, active forward (reverse) phase transition starts. These functions impose additional restrictions to those imposed by the second law of thermodynamics. The physical reason for such restrictions in phase transformation progress is not known at present; however, they are observed experimentally for some SMA alloys.

The power of thermodynamic driving force inducing process of active p.t. forward ($\mathcal{A} \to \mathcal{M}$) or reverse ($\mathcal{M} \to \mathcal{A}$) written in objective form is (see also (3.28))

$$\overset{\text{o}}{\overline{\tilde{\pi}_\tau^{1-2}}}{}^{(\text{Je})} \equiv \boldsymbol{\pi}_\tau^{1-2} . \overset{\text{o}}{\boldsymbol{\tau}}{}^{(\text{Je})} + \pi_T^{1-2} \dot{T},$$
$$\boldsymbol{\pi}_\tau^{1-2} \equiv \frac{\partial \pi^{1-2}}{\partial \boldsymbol{\tau}} = \boldsymbol{\kappa}/\rho^0, \quad \pi_T^{1-2} \equiv \frac{\partial \pi^{1-2}}{\partial T} = -\left[\Delta s_2^0 - s_{12}(1-2z)\right]. \tag{7.71}$$

The above form is inconvenient in numerical applications as it is expressed through $\dot{\boldsymbol{\tau}}$ ($\dot{\boldsymbol{\sigma}}$); that is, the quantity that is one of the output elements from the numerical material behavior procedure. In the general case, such a situation requires the application of some iteration scheme for checking whether the criterion of active phase transformation is fulfilled or not, which increases the cost of numerical computations. Hence, it is advisable to search such functional form of active p.t. criterion, which is expressed through input quantities to material behavior procedure (i.e., $\boldsymbol{d}$ and $\dot{T}$). In view of property $(7.46)_1$ it is

$$\boldsymbol{\kappa} \cdot \overset{\text{o}}{\boldsymbol{\tau}}{}^{(\text{Je})} = \boldsymbol{\kappa} \cdot \overset{\text{o}}{\boldsymbol{\tau}}{}^{(\text{e\_log})} = \boldsymbol{\kappa} \cdot \dot{\boldsymbol{\tau}}, \quad \overset{\text{o}}{\overline{\tilde{\pi}_\tau^{1-2}}}{}^{(\text{Je})} = \overset{\text{o}}{\overline{\tilde{\pi}_\tau^{1-2}}}{}^{(\text{e\_log})} = \dot{\tilde{\pi}}_\tau^{1-2}. \tag{7.72}$$

Let's calculate corotational elastic logarithmic derivative $(\text{A}7.34)_3$ of both sides of formula $(7.40)_1$ and use the property $(\text{A}7.34)_1$. Substituting next $(7.3)_1$ and (7.56), the rate constitutive relation of elasticity can be expressed in the form

$$\overset{\text{o}}{\boldsymbol{\tau}}{}^{(\text{e\_log})} = J\mathbf{L}\left[\underline{\mathbf{d}} - \dot{z}\underline{\boldsymbol{\kappa}} - z\overset{\text{o}}{\boldsymbol{\kappa}}{}^{(\text{Je})} - \boldsymbol{\alpha}_0 \dot{T}\right]. \tag{7.73}$$

In view of the property (7.53), it is

$$J^{-1}\overset{\text{o}}{\boldsymbol{\tau}}{}^{(\text{e\_log})} = \mathbf{L}\left[\boldsymbol{d} - \dot{z}\boldsymbol{\kappa} - z\boldsymbol{\mathcal{P}}\overset{\text{o}}{\boldsymbol{\tau}}{}^{(\text{e\_log})} - \boldsymbol{\alpha}_0 \dot{T}\right] - z\mathbf{L}\left[\boldsymbol{\kappa}\left(\boldsymbol{\omega}^{\text{e}} - \boldsymbol{\omega}^{\text{e\_log}}\right) - \left(\boldsymbol{\omega}^{\text{e}} - \boldsymbol{\omega}^{\text{e\_log}}\right)\boldsymbol{\kappa}\right],$$
$$\overset{\text{o}}{\boldsymbol{\tau}}{}^{(\text{e\_log})} = J\mathbf{L}^{\text{S}}\left[\boldsymbol{d} - \dot{z}\boldsymbol{\kappa} - \boldsymbol{\alpha}_0 \dot{T}\right] - z\left[\boldsymbol{\tau}^{\kappa}\left(\boldsymbol{\omega}^{\text{e}} - \boldsymbol{\omega}^{\text{e\_log}}\right) - \left(\boldsymbol{\omega}^{\text{e}} - \boldsymbol{\omega}^{\text{e\_log}}\right)\boldsymbol{\tau}^{\kappa}\right], \quad \boldsymbol{\tau}^{\kappa} \equiv J\mathbf{L}^{\text{S}}\boldsymbol{\kappa}. \tag{7.74}$$

Definition $(7.60)_2$ is used in the above. In view of isotropy of tensor $\mathbf{L}^{\text{S}}$ (see (7.61) and (7.51)), tensor $\boldsymbol{\tau}^{\kappa}$ is collinear with tensor $\boldsymbol{\kappa}$ ($\boldsymbol{\tau}^{\kappa}\boldsymbol{\kappa} = \boldsymbol{\kappa}\boldsymbol{\tau}^{\kappa}$) and $\mathbf{L}^{\text{S}}\left[\boldsymbol{\kappa}\boldsymbol{\omega}^{Q} - \boldsymbol{\omega}^{Q}\boldsymbol{\kappa}\right] = \boldsymbol{\tau}^{\kappa}\boldsymbol{\omega}^{Q} - \boldsymbol{\omega}^{Q}\boldsymbol{\tau}^{\kappa}$ (see (7.47)).

Contracting both sides of $(7.74)_3$ with $(\boldsymbol{\kappa}/\rho^0)$ it is

$$J^{-1}\boldsymbol{\kappa}\cdot\overset{\circ}{\boldsymbol{\tau}}^{(\mathrm{e\_log})}=\boldsymbol{\kappa}\cdot\mathbf{L}^{\mathrm{S}}(\boldsymbol{d}-\dot{z}\boldsymbol{\kappa}). \tag{7.75}$$

as $\boldsymbol{\kappa}\cdot\left[\boldsymbol{\tau}^{\kappa}\left(\boldsymbol{\omega}^{\mathrm{e}}-\boldsymbol{\omega}^{\mathrm{e\_log}}\right)-\left(\boldsymbol{\omega}^{\mathrm{e}}-\boldsymbol{\omega}^{\mathrm{e\_log}}\right)\boldsymbol{\tau}^{\kappa}\right]=0$ and $\boldsymbol{\kappa}\cdot\mathbf{L}^{S}\boldsymbol{\alpha}_0\dot{T}=0$.

Using $(7.72)_1$ and adding $\pi_T^{\mathrm{f}}\dot{T}$ on both sides of (7.75), it is obtained that

$$\left[\left(\boldsymbol{\kappa}/\rho^0\right)\cdot\overset{\circ}{\boldsymbol{\tau}}^{(\mathrm{Je})}+\pi_T^{1-2}\dot{T}\right]+\dot{z}(\boldsymbol{\kappa}/\rho)\cdot\mathbf{L}^{\mathrm{S}}\boldsymbol{\kappa}=(\boldsymbol{\kappa}/\rho)\cdot\mathbf{L}^{\mathrm{S}}\boldsymbol{d}+\pi_T^{1-2}\dot{T}. \tag{7.76}$$

When criteria for active p.t. are fulfilled, then $\dot{z}=\lambda_{(\alpha)}^{\sigma}\cdot\overset{\circ}{\widetilde{\pi}_{\tau}^{\mathrm{f}}}^{(\mathrm{Je})}$ (see (7.70)). Substituting this relation in the above equation, one obtains

$$\left[1+\lambda_{(\alpha)}^{\sigma}(\boldsymbol{\kappa}/\rho)\cdot\mathbf{L}^{\mathrm{S}}\boldsymbol{\kappa}\right]\overset{\circ}{\widetilde{\pi}_{\tau}^{1-2}}^{(\mathrm{Je})}=(\boldsymbol{\kappa}/\rho)\cdot\mathbf{L}^{\mathrm{S}}\boldsymbol{d}+\pi_T^{1-2}\dot{T}, \tag{7.77}$$

and finally,

$$\overset{\circ}{\widetilde{\pi}_{\tau}^{1-2}}^{(\mathrm{Je})}=\frac{(\boldsymbol{\kappa}/\rho)\cdot\mathbf{L}^{\mathrm{S}}\boldsymbol{d}+\pi_T^{\mathrm{f}}\dot{T}}{\left[1+\lambda_{(\alpha)}^{\sigma}(\boldsymbol{\kappa}/\rho)\cdot\mathbf{L}^{\mathrm{S}}\boldsymbol{\kappa}\right]}=f\left(\boldsymbol{d},\dot{T}\right). \tag{7.78}$$

7.8.6 Complete set of finite deformations, macroscopic, constitutive relations of elastically linear model of SMA pseudoelasticity

The rate constitutive equations of macroscopic elastically linear model of SMA pseudoelastic behavior undergoing finite deformations take the form specified below. The model input parameters are velocity gradient $\boldsymbol{L}$ and material derivative of temperature $\dot{T}$. The model predicts quantitatively the extent of mechanical, thermal, progress of thermoelastic martensitic phase transition effects (i.e., $\dot{\boldsymbol{\sigma}},\boldsymbol{d}^{\mathrm{in}},\dot{q}_{\mathrm{tr}},\dot{z}$). They are the output of the model. Mechanical equations of material behavior take the form (see (7.93) and (7.62))

$$\overset{(*)}{\boldsymbol{\Pi}}\left(\boldsymbol{L},\dot{T}\right)=\overset{\circ}{\boldsymbol{\sigma}}^{(0)}-\boldsymbol{d}\boldsymbol{\sigma}-\boldsymbol{\sigma}\boldsymbol{\omega},$$

$$\overset{\circ}{\boldsymbol{\sigma}}^{(0)}\left(\boldsymbol{d},\dot{T}\right)=\mathbf{L}^{\mathrm{S(Je)}}[\boldsymbol{d}-\dot{z}\boldsymbol{\kappa}]-\mathbf{L}\boldsymbol{\alpha}_0\dot{T}. \tag{7.79}$$

The first of the above equations describes in a combined way stress effects resulting from motion and deformation of a macroelement out of necessity expressed in non-objective form, adequate for weak formulation of finite element method (FEM) using updated Lagrangean technique. The second equation is given in the standard for constitutive description of material behavior objective form. In a majority of practical cases of problems involving behavior of shape memory alloys with perfect approximation, it can be adopted that (see also (7.34) and (7.35))

$$\mathbf{L}^{\mathrm{S(Je)}}\cong\mathbf{L}^{\mathrm{S}}=[\mathbf{M}+zJ\boldsymbol{\mathcal{P}}]^{-1},\quad \mathbf{M}=\mathbf{L}^{-1},\quad \left(J=\rho^0/\rho\right). \tag{7.80}$$

The formal thermoelastic martensitic phase transformation forward and reverse kinetic relations written in compact form are (see (7.70) and (7.78), (3.30), or (6.15))

$$\dot{z} = \lambda_1^{\sigma} \cdot \overline{H}(\pi^{1-2} - Y_1) \cdot \left\langle \overset{\mathrm{o}}{\overline{\tilde{\pi}_{\tau}^{1-2}}}^{(\mathrm{Je})} \right\rangle - \lambda_2^{\sigma} \cdot \overline{H}\left(-\pi^{1-2} - Y_2\right) \cdot \left\langle -\overset{\mathrm{o}}{\overline{\tilde{\pi}_{\tau}^{1-2}}}^{(\mathrm{Je})} \right\rangle,$$

$$\overset{\mathrm{o}}{\overline{\tilde{\pi}_{\tau}^{1-2}}}^{(\mathrm{Je})} = \overset{\mathrm{o}}{\overline{\tilde{\pi}_{\tau}^{1-2}}}^{(\mathrm{J})} = \dot{\tilde{\pi}}_{\tau}^{1-2} = \frac{(\boldsymbol{\kappa}/\rho) \cdot \mathbf{L}^{\mathrm{S}} \boldsymbol{d} + \pi_T^{1-2} \dot{T}}{\left(1 + \lambda_{(\alpha)}^{\sigma} (\boldsymbol{\kappa}/\rho) \cdot \mathbf{L}^{\mathrm{S}} \boldsymbol{\kappa}\right)} = f\left(\boldsymbol{d}, \dot{T}\right). \tag{7.81}$$

It is extremely advantageous from a numerical efficiency point of view that active phase transformation criterion is specified in the form dependent on input parameters to the material behavior subroutine (i.e., $\boldsymbol{d}, \dot{T}$). This favorable fact allows immediately finding out whether active phase transition will or will not take place in a particular time step. The extent of thermal effects is characterized by the power of specific heat source per unit of mass (see (7.68))

$$\begin{aligned}
&\dot{q}_{\mathrm{tr}} \equiv \left[\gamma f(y) \tau_{\mathrm{ef}}/\rho_0 + \Delta u_2^0 - (1-2z)\phi_{12}\right]\dot{z} - 3(T\alpha_0/\rho_0)\dot{\tau}_{\mathrm{m}}, \\
&c_p \partial T(\boldsymbol{x}, t)/\partial t + (1/\rho)\mathrm{div}(-\lambda_{\mathrm{c}} \nabla T(\boldsymbol{x}, t)) - \dot{q}_{\mathrm{tr}}(\boldsymbol{x}, t) = 0, \\
&T\boldsymbol{\alpha}_0 \cdot \dot{\boldsymbol{\tau}}/\rho_0 = T\boldsymbol{\alpha}_0 \cdot \dot{\boldsymbol{\sigma}}/\rho_0 = (3T\alpha_0/\rho_0)\dot{\tau}_{\mathrm{m}} = (3T\alpha_0/\rho)\dot{\sigma}_{\mathrm{m}} = (3KT\alpha_0/\rho)\dot{\varepsilon}_v,
\end{aligned} \tag{7.82}$$

The first term describes effects connected with latent heat of p.t. and mechanical work dissipation in a combined way; the second term describes piezocaloric effect. In a majority of cases regarding SMA alloys, heat power $\dot{q}_{\mathrm{tr}}$ (positive when removed to environment) is exchanged between SMA macro element and environment through the heat conduction. For isotropic SMA materials, this leads to the heat transport equation as in $(7.82)_2$, where λ_{c} denotes the value of isotropic heat conduction coefficient. Thermal effects resulting from the mechanical work dissipation alone are expressed by the formula $(7.65)_1$.

7.9 Field equations of finite deformations thermomechanics

Much freedom exists in mathematical specification of rate constitutive relations describing physical behavior of a continuum material. The limitation is the requirement of their objectivity; that is, material behavior response obtained as a result of using its constitutive relations must not depend on the observer describing this behavior. In mathematical terms this means a requirement for invariance of the constitutive relations with respect to transformation $\boldsymbol{x}' = \boldsymbol{Q}(t)\boldsymbol{x} + \boldsymbol{c}(t)$, where $\boldsymbol{Q}(t)$ is a rotation tensor dependent on time and $\boldsymbol{c}(t)$ is a shift vector. In other words, the requirement of objectivity means that regardless of whether specific physical process is described by the observer as motionless with respect to a fixed, global laboratory coordinate frame, or by another observer who rotates and moves in some manner in a global laboratory coordinate frame and additionally uses a different scale of time, the actual physical course of the process described by them is identical. Constitutive relations that are invariant with respect to transformation $\boldsymbol{x}' = \boldsymbol{Q}(t)\boldsymbol{x} + \boldsymbol{c}(t)$ are called objective. In general, it is difficult to say immediately whether a particular constitutive relation is objective or not. This requires detailed investigation. In order to simplify such assessment, it is common practice in formulating the rate constitutive relations to use so-called objective items (objects). The item is objective if it is invariant with respect to transformation $\boldsymbol{x}' = \boldsymbol{Q}(t)\boldsymbol{x} + \boldsymbol{c}(t)$); for example, corotational rates mentioned in the beginning of Section 7.6 in this chapter are objective items. Whenever all terms present in some specific material constitutive relation are expressed with objective items, then in view of the objectivity of each of such terms the whole constitutive relation is then also objective. Hence, it is not necessary to check this property in an additional procedure. Formulation of objective rate constitutive relations with the aid of some specific objective items can be more convenient than when they are expressed with some other objective items. The mathematical simplicity of the formulation or the possibility of applying some approximations is often a deciding factor. Here, elastic Zaremba-Jaumann corotational derivative is used to express objective rate constitutive relations of SMA materials in view of the possibility of applying simplifying approximations (see (7.35)).

While using objective derivatives of stress measures is convenient in formulating constitutive material behavior laws, their use in rate equations of mechanical balance is not handy. The nonsymmetric, nonobjective so-called rate of nominal stress tensor (7.86), plays a key role when formulating mechanical balance equations in rate form with the use of different reference configurations, as it is shown below and as was originally elucidated by Hill (1978). Mechanical balance equations in quasi-static case written in Eulerian description, in the weak form and local form, respectively, are

$$\oint_{\partial V^t} \boldsymbol{\sigma} \mathrm{d}\boldsymbol{a} + \oint_{V^t} \rho \boldsymbol{b} \mathrm{d}v = \boldsymbol{0}, \quad \mathrm{div}\boldsymbol{\sigma}^T + \rho \boldsymbol{b} = \boldsymbol{0}, \quad \left(\oint_{\partial V^t} \boldsymbol{\sigma} \mathrm{d}\boldsymbol{a} = \oint_{V^t} \nabla \cdot \boldsymbol{\sigma}^T \mathrm{d}v \right), \tag{7.83}$$

where $V^t, \partial V^t$ denote volume and surface area of actual configuration at time t, and $\boldsymbol{b}(\boldsymbol{x}, t)$ denotes intensity of volume forces per unit of mass expressed in Eulerian coordinates. The local form of balance equations is obtained from the global one after applying Gauss theorem (see $(7.83)_3$). The functions appearing under the integral must posses the regularity required by the assumptions of this theorem, which is assumed here. Using the known formula linking elemental area $\mathrm{d}\boldsymbol{a}$ in actual configuration with respective area $\mathrm{d}\boldsymbol{A}$ in initial configuration and performing change of coordinates ($\mathrm{d}\boldsymbol{a} = \boldsymbol{n}\mathrm{d}a = J\boldsymbol{F}^{-T}\boldsymbol{N}\mathrm{d}A = \mathrm{d}\boldsymbol{A}$), mechanical balance equations in initial configuration take the form

$$\begin{aligned} &\oint_{\partial V^0} \boldsymbol{N}^T \mathrm{d}\boldsymbol{A} + \oint_{V^0} \rho^0 \boldsymbol{b}_0 \mathrm{d}V = \boldsymbol{0}, \quad \mathrm{Div}\boldsymbol{N} + \rho^0 \boldsymbol{b}_0 = \boldsymbol{0}, \\ &\oint_{\partial V^0} \boldsymbol{N}^T \mathrm{d}\boldsymbol{A} = \oint_{V^0} \nabla \cdot \boldsymbol{N} \mathrm{d}V, \quad \boldsymbol{N} \equiv J\boldsymbol{F}^{-1}\boldsymbol{\sigma}, \quad \mathrm{d}\boldsymbol{a} = J\boldsymbol{F}^{-T}\mathrm{d}\boldsymbol{A}, \\ &\boldsymbol{b}_0(\boldsymbol{X}, t) = \boldsymbol{b}_0(\boldsymbol{X}(\boldsymbol{x}, t), t) = \boldsymbol{b}(\boldsymbol{x}(\boldsymbol{X}, t), t) = \boldsymbol{b}(\boldsymbol{x}, t), \end{aligned} \tag{7.84}$$

where $\boldsymbol{N}$ denotes nonsymmetric tensor of nominal stress, and $\boldsymbol{N}^T$ is known as first Piola-Kirchoff stress, in accordance with Ogden nomenclature (see Ogden, 1984), and $\boldsymbol{b}_0(\boldsymbol{X}, t)$ is intensity of volume forces per unit of mass expressed in total Lagrangean description.

It is allowed to enter time derivatives under the integrals in (7.85) below in view of the fact that the initial configuration does not change in time. If the transformation is next done to the actual configuration, the following is obtained:

$$\frac{\mathrm{D}}{\mathrm{D}t}\oint_{V^0}\nabla\cdot\boldsymbol{N}\mathrm{d}V=\oint_{V^0}\nabla\cdot\dot{\boldsymbol{N}}\mathrm{d}V=\oint_{\partial V^0}\dot{\boldsymbol{N}}^T\mathrm{d}\boldsymbol{A}=\oint_{\partial V^t}\overset{(*)}{\boldsymbol{\Pi}}{}^T\mathrm{d}\boldsymbol{a}=\oint_{V^t}\nabla\cdot\overset{(*)}{\boldsymbol{\Pi}}\mathrm{d}v,$$
$$\frac{\mathrm{D}}{\mathrm{D}t}\oint_{V^0}\rho^0\boldsymbol{b}_0\mathrm{d}v=\oint_{V^0}\rho^0\dot{\boldsymbol{b}}_0\mathrm{d}V=\oint_{V^t}\rho\overset{(*)}{\boldsymbol{b}}\,\mathrm{d}v. \tag{7.85}$$

In the formula (7.85), the rate of nominal stress tensor $\overset{(*)}{\boldsymbol{\Pi}}$ and the rate of intensity of volume forces vector defined below are used.

$$\overset{(*)}{\boldsymbol{\Pi}}\equiv J^{-1}\boldsymbol{F}\dot{\boldsymbol{N}},\;\;\dot{\boldsymbol{b}}_0=\dot{\boldsymbol{b}}_0(\boldsymbol{X},t)=\frac{\mathrm{D}}{\mathrm{D}t}\boldsymbol{b}(\boldsymbol{x}(\boldsymbol{X},t),t)=\frac{\partial\boldsymbol{b}}{\partial\boldsymbol{x}}\frac{\partial\boldsymbol{x}}{\partial t}+\frac{\partial\boldsymbol{b}}{\partial t}\equiv\overset{(*)}{\boldsymbol{b}}. \tag{7.86}$$

Calculating the material derivative of $(7.84)_1$ and using (7.85), the following forms of rate mechanical balance equations are obtained, written for time t

$$\oint_{V^t}\nabla\cdot\overset{(*)}{\boldsymbol{\Pi}}\mathrm{d}v+\oint_{V^t}\rho\overset{(*)}{\boldsymbol{b}}\mathrm{d}v=\boldsymbol{0},\;\;\mathrm{div}\overset{(*)}{\boldsymbol{\Pi}}+\rho\overset{(*)}{\boldsymbol{b}}=\boldsymbol{0}$$
$$\oint_{V^0}\nabla\cdot\dot{\boldsymbol{N}}\mathrm{d}V+\oint_{V^0}\rho^0\dot{\boldsymbol{b}}_0\mathrm{d}V=\boldsymbol{0},\;\;\mathrm{Div}\dot{\boldsymbol{N}}+\rho^0\dot{\boldsymbol{b}}_0=\boldsymbol{0} \tag{7.87}$$

In $(7.87)_{1,2}$, actual configuration $(t=t)$ is used as a reference configuration, and in $(7.87)_{3,4}$ initial configuration $(t=0)$ is used as reference configuration. Two techniques are most often used at present in solid-state mechanics for solving nonlinear problems computationally, the so-called *Total Lagrangean (TL) technique* and *Updated Lagrangean (UL) technique* (see Figure 7.2). In the case of Total Lagrangean technique, rate mechanical balance equations are written for consecutive time instants using the initial configuration $(t=0)$ as reference configuration at all times. In the case of Updated Lagrangean (UL) technique, rate mechanical balance equations are written for consecutive time instants using consecutive updated reference configurations, configurations from time $(t=t)$— the last known configuration on which state of mechanical balance has been reached. Methodologically, both techniques are identical. They start from a weak form of mechanical balance equations to obtain so-called rate virtual work principle. For that purpose, boundary conditions of the posed problem and mechanical balance equations in local form are used, that is, equation $(7.87)_4$ in TL technique and equation $(7.87)_2$ in UL technique to get

$$\oint_{V^t}\overset{(*)}{\boldsymbol{\Pi}}{}^T\cdot(\nabla\otimes\boldsymbol{\eta})\mathrm{d}v-\oint_{\partial V^t}\overset{(*)}{\boldsymbol{t}}\cdot\boldsymbol{\eta}\mathrm{d}\boldsymbol{a}-\oint_{V^t}\rho\overset{(*)}{\boldsymbol{b}}\cdot\boldsymbol{\eta}\mathrm{d}v=\boldsymbol{0},\;\;\overset{(*)}{\boldsymbol{\Pi}}\equiv J^{-1}\boldsymbol{F}\dot{\boldsymbol{N}},$$
$$\oint_{V^0}\dot{\boldsymbol{N}}^T\cdot(\nabla_0\otimes\boldsymbol{\eta}_0)\mathrm{d}V-\oint_{\partial V^0}\dot{\boldsymbol{t}}_0\cdot\boldsymbol{\eta}_0\mathrm{d}\boldsymbol{A}-\oint_{V^0}\rho^0\dot{\boldsymbol{b}}_0\cdot\boldsymbol{\eta}_0\mathrm{d}V=\boldsymbol{0},$$
$$\dot{\boldsymbol{N}}=\dot{\boldsymbol{N}}(\boldsymbol{\sigma},\boldsymbol{F},\boldsymbol{L})\;-\;\text{cf. (7.40)},\;\;\boldsymbol{t}\overset{(*)}{\boldsymbol{F}}\equiv J^{-1}\boldsymbol{F}\dot{\boldsymbol{t}}_0,\;\;\overset{(*)}{\boldsymbol{t}}=\overset{(*)}{\boldsymbol{\Pi}}\boldsymbol{n}^{\mathrm{out}},$$
$$J\overset{(*)}{\boldsymbol{\Pi}}{}^T=\dot{\boldsymbol{N}}^T\boldsymbol{F}^T,\;\;\nabla\otimes\boldsymbol{\eta}=(\nabla_0\otimes\boldsymbol{\eta}_0)\boldsymbol{F}^{-T},\;\;\dot{\boldsymbol{N}}=J\boldsymbol{F}^{-1}\overset{(*)}{\boldsymbol{\Pi}}, \tag{7.88}$$

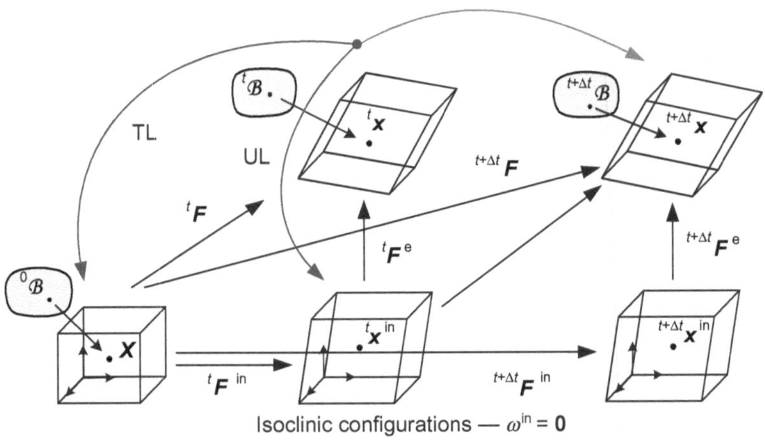

Figure 7.2 Schematic comparison of Total Lagrangean (TL) and Updated Lagrangean (UL) description.

where $\nabla_0 = (\partial/\partial X_i)\boldsymbol{E}_i$ is Lagrangean nabla operator, $\boldsymbol{\eta}_0$ denotes field of virtual velocity on initial configuration, $\dot{\boldsymbol{t}}_0$ denotes rate force boundary conditions on initial configuration, $\nabla = (\partial/\partial x_i)\boldsymbol{e}_i$ is nabla operator in Eulerian description, $\boldsymbol{\eta}$ denotes field of virtual velocity on actual configuration, $\overset{(*)}{\boldsymbol{t}}$ is rate force boundary conditions on actual configuration, and $\boldsymbol{n}^{out}$ denotes versor normal to RVE surface directed outward.

Rate virtual work principle equation $(7.88)_1$ or $(7.88)_2$, after its linearization around mechanical equilibrium state at time t, serves for the determination of instantaneous velocity field $\mathbf{v}(t)$. New equilibrium state and new equilibrium configuration at time $t + \Delta t$ is established using this determined instantaneous velocity field.

In theoretical terms both techniques discussed above are completely equivalent (as already mentioned) and lead to the same mathematical solution of the problem (see also Section 6.2.3 in Bathe (1996)) where the advantages and drawbacks of both techniques are discussed more extensively. The only difference between them is in the selection of different reference configurations.

The Total Lagrangean technique is advisable primarily in situations where the initial configuration has some important physical sense and it is justified to refer to it continuously. Such a situation arises when investigating problems of nonlinear elasticity, in which case unstressed initial configuration constitutes reference configuration for determination of elastic strains. Problems in which the initial configuration does not have such substantial physical relevance (e.g., in the case of the theory of elastoplasticity) or the material model discussed here, in which case elastic strains are determined based on the family of isoclinic configurations, the preferable computational technique seems to be the Updated Lagrangean one.

When the UL technique is used for solving a particular problem of nonlinear mechanics, the tensor of rate of nominal stress $(7.86)_1$ plays a key role, as can be noticed from formula (7.88). It is this tensor that brings in information about material behavior into mechanical balance equations. This requires specifying a linking relationship between nonobjective rate of nominal stress tensor and some objective corotational stress derivative $\overset{\circ}{\boldsymbol{\sigma}}^{(*)}$ appearing in rate material constitutive law $\overset{\circ}{\boldsymbol{\sigma}}^{(*)} = \mathbf{L}^{(*)}\boldsymbol{d}$. Such a relationship was given for the first time by Hill, who named it the "bridging equation." The bridging equation is linking information contained in the description of material properties of the body, expressed with the aid of rate constitutive equations using objective rates, with information on "properties" of motion of the body, expressed with the aid of mechanical balance equations. The bridging equation is derived below in a slightly different way than it was done originally by Hill. Calculating the material derivative of nonsymmetric nominal stress tensor, it is obtained (see $(7.84)_4$)

$$\dot{N} = \dot{J}\boldsymbol{F}^{-1}\boldsymbol{\sigma} + J\dot{\overline{\boldsymbol{F}^{-1}}}\boldsymbol{\sigma} + J\boldsymbol{F}^{-1}\dot{\boldsymbol{\sigma}}. \tag{7.89}$$

Care must be exercised when dealing with formulas (7.90)–(7.94) where symbol $\boldsymbol{L}$ (in italics) denotes velocity gradient tensor and symbol $\mathbf{L}$ (in roman) denotes elastic properties tensor. Multiplying the left and right hand side of (7.89) by $J^{-1}\underline{\mathbf{F}}$ and using formulas

$$\dot{J} = J\mathrm{tr}(\boldsymbol{d}), \left(\dot{\overline{\boldsymbol{F}^{-1}}} = -\boldsymbol{F}^{-1}\boldsymbol{L}\right) \Leftrightarrow \left(\mathrm{D}(\boldsymbol{F}\boldsymbol{F}^{-1})/\mathrm{D}t\right) = \mathrm{D}(\mathbf{I})/\mathrm{D}t = \mathbf{0}), \tag{7.90}$$

it is,

$$\overset{(*)}{\boldsymbol{\Pi}}(\boldsymbol{\sigma}, \boldsymbol{L}) = \dot{\boldsymbol{\sigma}} + \boldsymbol{\sigma}\mathrm{tr}(\boldsymbol{d}) - \boldsymbol{L}\boldsymbol{\sigma}, \quad \dot{N} = J\boldsymbol{F}^{-1}(\dot{\boldsymbol{\sigma}} + \boldsymbol{\sigma}\mathrm{tr}(\boldsymbol{d}) - \boldsymbol{L}\boldsymbol{\sigma}). \tag{7.91}$$

Using expression for Zaremba-Jaumann derivative of Cauchy stress the following is obtained:

$$\overset{(*)}{\boldsymbol{\Pi}} = \overset{\circ}{\boldsymbol{\sigma}}^{(J)} + \boldsymbol{\sigma}\mathrm{tr}(\boldsymbol{d}) - \boldsymbol{d}\boldsymbol{\sigma} - \boldsymbol{\sigma}\boldsymbol{\omega}. \tag{7.92}$$

Exactly such form of bridging equation was specified by Hill (see formula (1.91) in Hill (1978)). Naturally, any objective stress flux other than Zaremba-Jaumann can be used in the bridging equation, if this is expedient from the point of view of formulating material law. In the case discussed here, it is convenient to express $\overset{(*)}{\boldsymbol{\Pi}}$ by Zaremba-Jaumann derivative of Kirchoff stress referred to actual configuration $\overset{\circ}{\boldsymbol{\sigma}}^{(0)}$

$$\begin{gathered}\overset{(*)}{\boldsymbol{\Pi}}(\boldsymbol{\sigma}, \boldsymbol{L}) = \overset{\circ}{\boldsymbol{\sigma}}^{(0)} - \boldsymbol{d}\boldsymbol{\sigma} - \boldsymbol{\sigma}\boldsymbol{\omega} = \dot{\boldsymbol{\sigma}} - \boldsymbol{L}\boldsymbol{\sigma} + \boldsymbol{\sigma}\mathrm{tr}(\boldsymbol{d}), \\ \overset{\circ}{\boldsymbol{\sigma}}^{(0)} \equiv [\dot{\boldsymbol{\sigma}} + \boldsymbol{\sigma}\boldsymbol{\omega} - \boldsymbol{\omega}\boldsymbol{\sigma} + \boldsymbol{\sigma}\mathrm{tr}(\boldsymbol{d})] = J^{-1}\overset{\circ}{\boldsymbol{\tau}}^{(\mathrm{J})}, \quad J^{-1}\overset{\circ}{\boldsymbol{\tau}}^{(\mathrm{J})} = J^{-1}\overset{\circ}{\boldsymbol{\tau}}^{(\mathrm{Je})} + \left[\boldsymbol{\sigma}\boldsymbol{\omega}^{\mathrm{in}} - \boldsymbol{\omega}^{\mathrm{in}}\boldsymbol{\sigma}\right].\end{gathered} \tag{7.93}$$

Finally, taking advantage of the formulas (7.33), $(7.28)_3$, $(7.29)_1$, and $(7.3)_1$, the following constitutive material law is obtained expressed in the form suitable for solving the rate initial-boundary value problem with the use of the Updated Lagrangean technique

$$\begin{gathered}\overset{(*)}{\boldsymbol{\Pi}}(\boldsymbol{\sigma}, \boldsymbol{L}, \dot{T}) = \mathbf{L}^{(\mathrm{Je})}(\boldsymbol{d} - \boldsymbol{d}^{\mathrm{in}}) - \mathbf{L}\boldsymbol{\alpha}_0\dot{T} + \frac{\partial\boldsymbol{\sigma}}{\partial z}\dot{z} - \boldsymbol{d}\boldsymbol{\sigma} - \boldsymbol{\sigma}\boldsymbol{\omega} + \boldsymbol{\sigma}\boldsymbol{\omega}^{\mathrm{in}} - \boldsymbol{\omega}^{\mathrm{in}}\boldsymbol{\sigma}, \\ \overset{\circ}{\boldsymbol{\sigma}}^{(0)}(\boldsymbol{d}, \dot{T}) = \mathbf{L}^{(\mathrm{Je})}(\boldsymbol{d} - \boldsymbol{d}^{\mathrm{in}}) - \mathbf{L}\boldsymbol{\alpha}_0\dot{T} + (\partial\boldsymbol{\sigma}/\partial z)\dot{z} + \boldsymbol{\sigma}\boldsymbol{\omega}^{\mathrm{in}} - \boldsymbol{\omega}^{\mathrm{in}}\boldsymbol{\sigma}.\end{gathered} \tag{7.94}$$

It is worth comparing the advantages resulting from the application of mobile Lagrangean description (Mandel description), together with updated Lagrangean technique in relation to the application of the updated Total Lagrangean description used by Hill and Rice (1973)

$$\begin{aligned}
&\overset{\circ}{\boldsymbol{\sigma}}^{(0)} = \mathbf{L}^{(\mathrm{Je})}\left[\boldsymbol{d} - \boldsymbol{d}^{(0)\mathrm{in}}\right] - \mathbf{L}\boldsymbol{\alpha}_0\dot{T},\\
&\boldsymbol{d}^{(0)\mathrm{in}} = \boldsymbol{d}^{\mathrm{in}} - \overset{-1}{\mathbf{L}}{}^{(\mathrm{Je})}\left[(\partial\boldsymbol{\sigma}/\partial z)\dot{z} + \boldsymbol{\sigma}\boldsymbol{\omega}^{\mathrm{in}} - \boldsymbol{\omega}^{\mathrm{in}}\boldsymbol{\sigma}\right](\mathrm{Mandel}),\\
&\overset{\circ}{\boldsymbol{\sigma}}^{(0)} = \mathbf{L}^{(\mathrm{H-R})}\left[\boldsymbol{d} - \boldsymbol{d}^{(\mathrm{H-R})\mathrm{in}}\right] - \mathbf{L}\boldsymbol{\alpha}_0\dot{T}(\mathrm{Hill-Rice}).
\end{aligned} \tag{7.95}$$

In the case of using mobile Lagrangean description in a clear operational way, the structure of inelastic stretching rate tensor is revealed $\boldsymbol{d}^{(0)\mathrm{in}}$. This fact in an essential way facilitates posing conjectures regarding its explicit form.

7.10 Summary

In the chapter, the framework of nonequilibrium thermodynamics with internal state parameters has been expounded, combined with the concept of a collection of natural unloaded elastically, isoclinic configurations, and the notion of work conjugate stress and strain measures together delivering an adequate theoretical tool to describe SMA materials behavior in such situations where geometrical effects cannot be neglected. The tool is employed to formulate an explicit constitutive 3D, isotropic, finite deformations model of pseudoelastic behavior of SMA materials linear in terms of elastic logarithmic strain. The hyperelastic form of constitutive relations has been specified—equations of state, originating from adequate for SMA materials free energy function used as thermodynamic potential. A mathematically exact transition from hyperelastic relations to their corresponding rate (hypoelastic) form has been shown. In the process, it has been elucidated that the use of elastic logarithmic strain is favorable in a finite deformations constitutive model formulation. The reason is that when logarithmic strain measure is employed, then instantaneous elastic tangent present in nonrate form of constitutive relations, that is, elastic stiffness tensor $\mathbf{L}$ (see (7.28)) is practically the same as instantaneous, elastic Zaremba-Jaumann rate tangent present in corresponding rate form of constitutive relations $\left(\mathbf{L}^{(\mathrm{Je})} = \mathbf{L}\right)$ (see (7.33)–(7.35)). The use of the Newton-Raphson method is standard in computational mechanics of solid bodies for solving nonlinear problems. The quadratic rate of convergence, and in this way high efficiency of computational effort, is assured when exact rate tangent is used in the Newton-Raphson iterative procedure. Thus, when rate tangent is simply equal to elastic stiffness tensor and this is constant in linear elasticity, no costly numerical updates are necessary and the solution of the nonlinear problem can be attained robustly. The other advantage of using logarithmic strain measure is that it allows for relatively uncomplicated extension of the SMA model predictions capability to a range of impact loadings (high pressures) due to a known property of this measure of uncoupled description of volumetric and distortional effects. In order to achieve that, it is necessary to obtain reliable experimental data on pressure ↔ volumetric strain characteristic of SMA materials when submitted to impact (e.g., explosive loadings). As the model uses the formalism of nonequilibrium thermodynamics, its generalization to embrace the case when elastic modules depend on state of the material—taking into account different elastic modules of austenitic and martensitic phases also does not present significant difficulties—upon the condition that relevant experimental data are available. It is shown how information on deformation of SMA material embedded in SMA model rate constitutive relations combined with information on motion of the body through so-called bridging equation enters weak form of balance equations—a foundation for FEM. At the same time it is shown that the discussed SMA model constitutive relations are specified in the form most convenient for implementation into finite element code. The present finite deformation, SMA model belonging to the R_L models family upon its implementation into FEM code is very well-suited for performing engineering design computations of devices and structures containing elements made of shape memory alloy materials. This is true in analyzing situations where large rotations of structural elements appear during nominal operating conditions, or in analyzing emergency situations when also large strains may appear. Hypotheses on the basis of which geometrical effects are taken into account in the finite deformations SMA model discussed here are verified experimentally to rather little extent at present. A systematic experimental works are required to build a database of SMA materials behavior in response to multiaxial proportional and nonproportional loadings, isothermal and combined thermomechanical one, involving large deformations.

Annex 7.1 Relations between Lagrangean ω^{eL}, Eulerian ω^{eE}, and relative ω^{eR} elastic spins

The *Hill's method of principal axes* is used here to derive relations between Lagrangean, Eulerian, and relative elastic spins; see also Hill (1978) and Ogden (1984). Polar and spectral decomposition of elastic deformation gradient gives

$$\boldsymbol{F}^e = \boldsymbol{R}^e\boldsymbol{U}^e = \boldsymbol{V}^e\boldsymbol{R}^e, \quad \boldsymbol{U}^e = \sum \lambda_i^e \boldsymbol{N}_i^e \otimes \boldsymbol{N}_i^e, \quad \boldsymbol{V}^e = \sum \lambda_i^e \boldsymbol{n}_i^e \otimes \boldsymbol{n}_i^e. \tag{A7.1}$$

where $\boldsymbol{R}^e$ is proper orthogonal tensor, and $\boldsymbol{U}^e$ and $\boldsymbol{V}^e$ denote right and left symmetric elastic stretch tensors. The scalars λ_i^e denote principal values of elastic stretch. The $\boldsymbol{N}_i^e$ and $\boldsymbol{n}_i^e$; $i = 1,2,3$ denote Lagrangean and Eulerian principal directions of tensors $\boldsymbol{U}^e$ and $\boldsymbol{V}^e$—called Lagrangean and Eulerian elastic principal axes.

During elastic deformation of the body with respect to some fixed reference configuration, the triads $\boldsymbol{N}_i^e$ and $\boldsymbol{n}_i^e$ (Lagrangean and Eulerian) rotate in fixed global coordinates frame. This rotation is described by Lagrangean $\boldsymbol{\omega}^{eL}$ and Eulerian $\boldsymbol{\omega}^{eE}$ elastic spins, which in turn are generated by rotation tensors $\boldsymbol{R}^{eL}$ and $\boldsymbol{R}^{eE}$, respectively

$$\boldsymbol{\omega}^{eL} \equiv \dot{\boldsymbol{R}}^{eL} \boldsymbol{R}^{eL^T}, \quad \dot{\boldsymbol{N}}_i^e = \boldsymbol{\omega}^{eL}\boldsymbol{N}_i^e, \quad \boldsymbol{\omega}^{eE} \equiv \dot{\boldsymbol{R}}^{eE} \boldsymbol{R}^{eE^T}, \quad \dot{\boldsymbol{n}}_i^e = \boldsymbol{\omega}^{eE}\boldsymbol{n}_i^e. \tag{A7.2}$$

Relative elastic spin $\boldsymbol{\omega}^{\mathrm{eR}} \equiv \dot{\boldsymbol{R}}^{\mathrm{e}} \boldsymbol{R}^{\mathrm{e}T}$ defining rotation of Eulerian elastic principal axes with respect to Lagrangean elastic principal axes is generated by rotation tensor $\underline{\mathbf{R}}^{\mathrm{e}}$

$$\mathbf{R}^{\mathrm{eE}} = \mathbf{R}^{\mathrm{e}}\mathbf{R}^{\mathrm{eL}}, \quad \boldsymbol{\omega}^{\mathrm{eR}} \equiv \dot{\mathbf{R}}^{\mathrm{e}} \boldsymbol{R}^{\mathrm{e}T}, \quad \boldsymbol{n}_i^{\mathrm{e}} = \mathbf{R}^{\mathrm{e}}\boldsymbol{N}_i^{\mathrm{e}}, \quad \mathbf{R}^{\mathrm{e}} = \sum R_{ij}^{\mathrm{e}} \boldsymbol{n}_i^{\mathrm{e}} \otimes \boldsymbol{N}_j^{\mathrm{e}}. \tag{A7.3}$$

Differentiating (A7.3)$_1$ with respect to time—$\dot{\mathbf{R}}^{\mathrm{eE}} \equiv \boldsymbol{\omega}^{\mathrm{eE}}\mathbf{R}^{\mathrm{eE}}$, $\dot{\mathbf{R}}^{\mathrm{eE}} = \dot{\mathbf{R}}^{\mathrm{e}}\mathbf{R}^{\mathrm{eL}} + \mathbf{R}^{\mathrm{e}}\dot{\mathbf{R}}^{\mathrm{eL}} = \boldsymbol{\omega}^{\mathrm{eR}}\mathbf{R}^{\mathrm{e}}\mathbf{R}^{\mathrm{eL}} + \mathbf{R}^{\mathrm{e}}\boldsymbol{\omega}^{\mathrm{eL}}\mathbf{R}^{\mathrm{eL}}$ and using (A7.2), (A7.3) it is

$$\boldsymbol{\omega}^{\mathrm{eL}} = (\mathbf{R}^{\mathrm{e}})^{\mathrm{T}}\left[\boldsymbol{\omega}^{\mathrm{eE}} - \boldsymbol{\omega}^{\mathrm{eR}}\right]\mathbf{R}^{\mathrm{e}}. \tag{A7.4}$$

Upon material differentiation of (A7.1)$_1$ and using definitions of elastic stretching rate $\boldsymbol{d}^{\mathrm{e}}$ and elastic spin $\boldsymbol{\omega}^{\mathrm{e}}$

$$\boldsymbol{d}^{\mathrm{e}} = \frac{1}{2}\left[\boldsymbol{L}^{\mathrm{e}} + (\boldsymbol{L}^{\mathrm{e}})^{\mathrm{T}}\right], \quad \boldsymbol{\omega}^{\mathrm{e}} = \frac{1}{2}\left[\boldsymbol{L}^{\mathrm{e}} - (\boldsymbol{L}^{\mathrm{e}})^{\mathrm{T}}\right], \quad \dot{\boldsymbol{F}}^{\mathrm{e}} = \boldsymbol{L}^{\mathrm{e}}\boldsymbol{F}^{\mathrm{e}}, \tag{A7.5}$$

one obtains

$$\begin{aligned}
(\boldsymbol{R}^{\mathrm{e}})^{\mathrm{T}}\boldsymbol{d}^{\mathrm{e}}\boldsymbol{R}^{\mathrm{e}} &= \frac{1}{2}\left[\dot{\boldsymbol{U}}^{\mathrm{e}}(\boldsymbol{U}^{\mathrm{e}})^{-1} + (\boldsymbol{U}^{\mathrm{e}})^{-1}\dot{\boldsymbol{U}}^{\mathrm{e}}\right],\\
(\boldsymbol{R}^{\mathrm{e}})^{\mathrm{T}}(\boldsymbol{\omega}^{\mathrm{e}} - \boldsymbol{\omega}^{\mathrm{eR}})\boldsymbol{R}^{\mathrm{e}} &= \frac{1}{2}\left[\dot{\boldsymbol{U}}^{\mathrm{e}}(\boldsymbol{U}^{\mathrm{e}})^{-1} - (\boldsymbol{U}^{\mathrm{e}})^{-1}\dot{\boldsymbol{U}}^{\mathrm{e}}\right].
\end{aligned} \tag{A7.6}$$

Components of $(\boldsymbol{U}^{\mathrm{e}})^{-1}$, $\dot{\boldsymbol{U}}^{\mathrm{e}}$, and $\boldsymbol{\omega}^{\mathrm{eL}}$ (see (A7.1), (A7.2)$_1$) on Lagrangean elastic principal axes are denoted by $1/\lambda_i^{\mathrm{e}}$, $\overline{\dot{U}}_{ij}^{\mathrm{e}}$, $\omega_{ij}^{\mathrm{eL}}$

$$\dot{\boldsymbol{U}}^{\mathrm{e}} = \sum \overline{\dot{U}}_{ij}^{\mathrm{e}} \boldsymbol{N}_i^{\mathrm{e}} \otimes \boldsymbol{N}_j^{\mathrm{e}}, \quad (\boldsymbol{U}^{\mathrm{e}})^{-1} = \sum (1/\lambda_i^{\mathrm{e}}) \boldsymbol{N}_i^{\mathrm{e}} \otimes \boldsymbol{N}_i^{\mathrm{e}}, \boldsymbol{\omega}^{\mathrm{eL}} = \sum \omega_{ij}^{\mathrm{eL}} \boldsymbol{N}_i^{\mathrm{e}} \otimes \boldsymbol{N}_j^{\mathrm{e}}. \tag{A7.7}$$

The components of $\boldsymbol{d}^{\mathrm{e}}$, $\boldsymbol{\omega}^{\mathrm{e}}$, $\boldsymbol{\omega}^{\mathrm{eE}}$, $\boldsymbol{\omega}^{\mathrm{eR}}$ on Eulerian elastic principal axes are denoted by d_{ij}^{e}, ω_{ij}^{e} (see (A7.5)$_{1,2}$), $\omega_{ij}^{\mathrm{eE}}$ (see (A7.2)$_3$), and $\omega_{ij}^{\mathrm{eR}}$ (see (A7.3)$_2$)

$$\begin{aligned}
\boldsymbol{d}^{\mathrm{e}} &= \Sigma d_{ij}^{\mathrm{e}} \boldsymbol{n}_i^{\mathrm{e}} \otimes \boldsymbol{n}_j^{\mathrm{e}}, \quad \boldsymbol{\omega}^{\mathrm{e}} = \Sigma \omega_{ij}^{\mathrm{e}} \boldsymbol{n}_i^{\mathrm{e}} \otimes \boldsymbol{n}_j^{\mathrm{e}},\\
\boldsymbol{\omega}^{\mathrm{eE}} &= \Sigma \omega_{ij}^{\mathrm{eE}} \boldsymbol{n}_i^{\mathrm{e}} \otimes \boldsymbol{n}_j^{\mathrm{e}}, \quad \boldsymbol{\omega}^{\mathrm{eR}} = \Sigma \omega_{ij}^{\mathrm{eR}} \boldsymbol{n}_i^{\mathrm{e}} \otimes \boldsymbol{n}_j^{\mathrm{e}}.
\end{aligned} \tag{A7.8}$$

Then expressions appearing on the right-hand side of (A7.6) can be written on Lagrangean elastic principal axes as

$$\begin{aligned}
\frac{1}{2}\left[\dot{\boldsymbol{U}}^{\mathrm{e}}(\boldsymbol{U}^{\mathrm{e}})^{-1} + (\boldsymbol{U}^{\mathrm{e}})^{-1}\dot{\boldsymbol{U}}^{\mathrm{e}}\right] &= \sum \frac{1}{2}\left(\overline{\dot{U}}_{ij}^{\mathrm{e}}/\lambda_j^{\mathrm{e}} + \overline{\dot{U}}_{ij}^{\mathrm{e}}/\lambda_i^{\mathrm{e}}\right) \boldsymbol{N}_i^{\mathrm{e}} \otimes \boldsymbol{N}_j^{\mathrm{e}},\\
\frac{1}{2}\left[\dot{\boldsymbol{U}}^{\mathrm{e}}(\boldsymbol{U}^{\mathrm{e}})^{-1} - (\boldsymbol{U}^{\mathrm{e}})^{-1}\dot{\boldsymbol{U}}^{\mathrm{e}}\right] &= \sum \frac{1}{2}\left(\overline{\dot{U}}_{ij}^{\mathrm{e}}/\lambda_j^{\mathrm{e}} - \overline{\dot{U}}_{ij}^{\mathrm{e}}/\lambda_i^{\mathrm{e}}\right) \boldsymbol{N}_i^{\mathrm{e}} \otimes \boldsymbol{N}_j^{\mathrm{e}}.
\end{aligned} \tag{A7.9}$$

Using relations (A7.7)–(A7.9), the equations (A7.4) and (A7.6) reduce to the following relations:

$$\omega_{ij}^{\mathrm{eL}} = \omega_{ij}^{\mathrm{eE}} - \omega_{ij}^{\mathrm{eR}}, \quad d_{ij}^{\mathrm{e}} = \frac{\lambda_i^{\mathrm{e}} + \lambda_j^{\mathrm{e}}}{2\lambda_i^{\mathrm{e}}\lambda_j^{\mathrm{e}}} \overline{\dot{U}}_{ij}^{\mathrm{e}}, \quad \omega_{ij}^{\mathrm{e}} - \omega_{ij}^{\mathrm{eR}} = \frac{\lambda_i^{\mathrm{e}} - \lambda_j^{\mathrm{e}}}{2\lambda_i^{\mathrm{e}}\lambda_j^{\mathrm{e}}} \overline{\dot{U}}_{ij}^{\mathrm{e}}. \tag{A7.10}$$

Eliminating from the above relations $\overline{\dot{U}}_{ij}^{e}$ it is obtained that

$$\omega_{ij}^{e} - \omega_{ij}^{eR} = \frac{\lambda_i^e - \lambda_j^e}{\lambda_i^e + \lambda_j^e} d_{ij}^{e}. \tag{A7.11}$$

In the sequence, expressions for components $\overline{\dot{U}}_{ij}^{e}$ of tensor $\dot{\boldsymbol{U}}^{e}$ on Lagrangean elastic principal axes are needed.

$$\dot{\boldsymbol{U}}^{e} = \sum \overline{\dot{U}}_{ij}^{e} \boldsymbol{N}_i^e \otimes \boldsymbol{N}_j^e = \sum \dot{\lambda}_i^e \boldsymbol{N}_i^e \otimes \boldsymbol{N}_i^e + \sum \lambda_i^e \dot{\boldsymbol{N}}_i^e \otimes \boldsymbol{N}_i^e + \sum \lambda_i^e \boldsymbol{N}_i^e \otimes \dot{\boldsymbol{N}}_i^e. \tag{A7.12}$$

In order to obtain required formula corotational derivative of tensor $\boldsymbol{U}^e$ is calculated with the use of Lagrangean elastic spin $\boldsymbol{\omega}^{eL}$, which defines instantaneous rotation of Lagrangean elastic principal axes (see (A7.2))

$$\overset{o}{\boldsymbol{U}^{e}}(\mathrm{eL}) \equiv \boldsymbol{R}^{eL} \frac{\mathrm{D}}{\mathrm{D}t}\left(\boldsymbol{R}^{eL\,T} \boldsymbol{U}^e \boldsymbol{R}^{eL}\right) \boldsymbol{R}^{eL\,T} = \dot{\boldsymbol{U}}^e + \boldsymbol{U}^e \boldsymbol{\omega}^{eL} - \boldsymbol{\omega}^{eL} \boldsymbol{U}^e = \sum \dot{\lambda}_i^e \boldsymbol{N}_i^e \otimes \boldsymbol{N}_i^e. \tag{A7.13}$$

Using (A7.7) and performing simple manipulations, it is

$$\dot{\boldsymbol{U}}^{e} = \sum \overline{\dot{U}}_{ij}^{e} \boldsymbol{N}_i^e \otimes \boldsymbol{N}_j^e = \sum \left[\dot{\lambda}_i^e + \left(\lambda_j^e - \lambda_i^e\right) \omega_{ij}^{eL}\right] \boldsymbol{N}_i^e \otimes \boldsymbol{N}_j^e \tag{A7.14}$$

or, in scalar form,

$$\overline{\dot{U}}_{ij}^{e} = \dot{\lambda}_i^e, \ i = j, \ \overline{\dot{U}}_{ij}^{e} = \left(\lambda_j^e - \lambda_i^e\right) \omega_{ij}^{eL}, \ i \neq j. \tag{A7.15}$$

Substituting $(\mathrm{A7.15})_1$ to $(\mathrm{A7.10})_2$ for $i = j$, it is

$$d_{ii}^{e} = \dot{\lambda}_i^e / \lambda_i^e. \tag{A7.16}$$

The above relation is valid irrespective of the Lagrangean elastic principal axes rotation history. Substituting $(\mathrm{A7.15})_2$ to $(\mathrm{A7.10})_3$ for $i \neq j$, it is

$$\omega_{ij}^{eR} - \omega_{ij}^{e} = \frac{\left(\lambda_i^e - \lambda_j^e\right)^2}{2\lambda_i^e \lambda_j^e} \omega_{ij}^{eL}. \tag{A7.17}$$

Eliminating the ω_{ij}^{eR} from the above and $(\mathrm{A7.10})_1$, it is

$$\omega_{ij}^{eE} - \omega_{ij}^{e} = \frac{\left(\lambda_i^e\right)^2 + \left(\lambda_j^e\right)^2}{2\lambda_i^e \lambda_j^e} \omega_{ij}^{eL}. \tag{A7.18}$$

Finally, eliminating $\overline{\dot{U}}_{ij}^{e}$ from $(A7.15)_2$ and $(A7.10)_2$ gives

$$\omega_{ij}^{eL} = \frac{2\lambda_i^e \lambda_j^e}{\left(\lambda_j^e\right)^2 - \left(\lambda_i^e\right)^2} d_{ij}^e \quad \lambda_i^e \neq \lambda_j^e, \tag{A7.19}$$

and upon substituting the above to (A7.18), one obtains

$$\omega_{ij}^{eE} - \omega_{ij}^{e} = \frac{\left(\lambda_i^e\right)^2 + \left(\lambda_j^e\right)^2}{\left(\lambda_j^e\right)^2 - \left(\lambda_i^e\right)^2} d_{ij}^e \quad \lambda_i^e \neq \lambda_j^e. \tag{A7.20}$$

From the above, it results that instantaneous rotation of Lagrangean elastic principal axes depends only on the state of deformation $\boldsymbol{U}^e = \boldsymbol{U}^e\left(\lambda_i^e\right)$ and instantaneous elastic stretching rate $\boldsymbol{d}^e$.

Annex 7.2 Kinematical relations between $\dot{\boldsymbol{E}}^{e}(0) \leftrightarrow \boldsymbol{D}^{e}$, $\dot{\boldsymbol{e}}^{e}(0) \leftrightarrow \boldsymbol{d}^{e}$

Tensor of logarithmic elastic strains and its material derivative when expressed in the coordinate frame of Lagrangean elastic principal axes take the form

$$\begin{aligned} \boldsymbol{E}^e(0) &= \sum \ln\left(\lambda_i^e\right) \boldsymbol{N}_i^e \otimes \boldsymbol{N}_i^e, \ \dot{\boldsymbol{E}}^e(0) = \overline{\ln\left(\underline{\mathbf{U}}^e\right)}^{\boldsymbol{\cdot}} = \sum \overline{\dot{E}}_{ij}^{e} \boldsymbol{N}_i^e \otimes \boldsymbol{N}_j^e \\ &= \sum \left(\dot{\lambda}_i^e / \lambda_i^e\right) \boldsymbol{N}_i^e \otimes \boldsymbol{N}_i^e + \sum \ln\left(\lambda_i^e\right) \dot{\boldsymbol{N}}_i^e \otimes \boldsymbol{N}_i^e + \sum \ln\left(\lambda_i^e\right) \boldsymbol{N}_i^e \otimes \dot{\boldsymbol{N}}_i^e. \end{aligned} \tag{A7.21}$$

Using the formula for instantaneous rotation of Lagrangean elastic principal axes $\dot{\boldsymbol{N}}_i^e$ (see $(A7.2)_2$, $(A7.7)_3$), it is

$$\dot{\boldsymbol{E}}^e(0) = \sum \overline{\dot{E}}_{ij}^{e} \boldsymbol{N}_i^e \otimes \boldsymbol{N}_j^e = \sum \left[\dot{\lambda}_i^e / \lambda_i^e + \ln\left(\lambda_j^e / \lambda_i^e\right) \omega_{ij}^{eL}\right] \boldsymbol{N}_i^e \otimes \boldsymbol{N}_j^e, \tag{A7.22}$$

or, in scalar form,

$$\overline{\dot{E}}_{ij}^{e} = \dot{\lambda}_i^e / \lambda_i^e; \ i = j, \quad \overline{\dot{E}}_{ij}^{e} = \ln\left(\lambda_j^e / \lambda_i^e\right) \omega_{ij}^e \ i \neq j. \tag{A7.23}$$

Using formula (A7.16) for $i = j$ and formula (A7.19) for $i \neq j$, it is

$$\overline{\dot{E}}_{ij}^{e} = D_{ii}^e \ i = j, \quad \overline{\dot{E}}_{ij}^{e} = \frac{2\lambda_i^e \lambda_j^e \ln\left(\lambda_j^e / \lambda_i^e\right)}{\left(\lambda_j^e\right)^2 - \left(\lambda_i^e\right)^2} D_{ij}^e \quad i \neq j \text{ and } \lambda_i^e \neq \lambda_j^e. \tag{A7.24}$$

In formulas (A7.24) the property is used that components of rotated tensor of elastic stretching rate $\boldsymbol{D}^e = {}^T\boldsymbol{R}^e \boldsymbol{d}^e \boldsymbol{R}^e$ written on Lagrangean elastic principal axes take the same values as components of elastic stretching rate $\boldsymbol{d}^e$ written on Eulerian elastic principal axes $D_{ij}^e = d_{ij}^e$ (see $(7.17)_2$).

Hence, relations between components of $\dot{\boldsymbol{E}}^{\mathrm{e}}(0) \leftrightarrow \boldsymbol{D}^{\mathrm{e}}$ written on Lagrangean elastic principal axes take the form (A7.24). The same mathematical form (A7.24) have the relations between components of $\dot{\boldsymbol{e}}^{\mathrm{e}}(0) \leftrightarrow \boldsymbol{d}^{\mathrm{e}}$ written on Eulerian elastic principal axes, but then in place of $\overline{\dot{E}}{}^{\mathrm{e}}_{ij}$, D^{e}_{ij} there should be substituted $\overline{\dot{e}}{}^{\mathrm{e}}_{ij}$, d^{e}_{ij}.

The scalars λ^{e}_j denote all different principal values of tensor $\boldsymbol{U}^{\mathrm{e}}$ ($\boldsymbol{V}^{\mathrm{e}}$). When two principal values are the same $\lambda^{\mathrm{e}}_i = \lambda^{\mathrm{e}}_j$ then respective components of $\overline{\dot{E}}{}^{\mathrm{e}}_{ij}$ $i \neq j$ are obtained from (A7.24)$_2$ by performing limit operation $\lambda^{\mathrm{e}}_i \to \lambda^{\mathrm{e}}_j$.

Annex 7.3 Kinematical relations between $\overset{\circ}{\overline{\mathbf{e}^{\mathrm{e}}(0)}}{}^{(\mathrm{Je})} \leftrightarrow \boldsymbol{d}^{\mathrm{e}}$, tensor $\mathcal{E}^{\mathrm{de}}$ and its properties, elastic logarithmic corotational derivative $\overset{\circ}{\overline{\mathbf{e}^{\mathrm{e}}(0)}}{}^{(\mathrm{e\_log})}$

Zaremba-Jaumann elastic corotational derivative generated by spin $\boldsymbol{\omega}^{\mathrm{e}}$ (see (A7.5)$_2$), calculated for tensor $\boldsymbol{e}^{\mathrm{e}}(0)$, takes the form

$$\overset{\circ}{\overline{\boldsymbol{e}^{\mathrm{e}}(0)}}{}^{(\mathrm{Je})} = \dot{\boldsymbol{e}}^{\mathrm{e}}(0) + \boldsymbol{e}^{\mathrm{e}}(0)\boldsymbol{\omega}^{\mathrm{e}} - \boldsymbol{\omega}^{\mathrm{e}}\boldsymbol{e}^{\mathrm{e}}(0), \quad \overset{\circ}{\overline{\boldsymbol{e}^{\mathrm{e}}(0)}}{}^{(\mathrm{Je})} = \Sigma \overset{\mathrm{Je}}{e}{}^{\mathrm{e}}_{ij} \boldsymbol{n}^{\mathrm{e}}_i \otimes \boldsymbol{n}^{\mathrm{e}}_j. \tag{A7.25}$$

Similarly, derivative generated by elastic relative spin $\boldsymbol{\omega}^{\mathrm{eR}}$ (see (A7.3)$_2$), calculated for tensor $\boldsymbol{e}^{\mathrm{e}}(0)$, gives

$$\overset{\circ}{\overline{\boldsymbol{e}^{\mathrm{e}}(0)}}{}^{(\mathrm{Re})} \equiv \boldsymbol{R}^{\mathrm{e}} \frac{\mathrm{D}}{\mathrm{D}t}\left({}^T\boldsymbol{R}^{e}\boldsymbol{e}^{\mathrm{e}}(0)\boldsymbol{R}^{\mathrm{e}}\right)^T \boldsymbol{R}^{e} = \dot{\boldsymbol{e}}^{\mathrm{e}}(0) + \boldsymbol{e}^{\mathrm{e}}(0)\boldsymbol{\omega}^{\mathrm{Re}} - \boldsymbol{\omega}^{\mathrm{Re}}\boldsymbol{e}^{\mathrm{e}}(0). \tag{A7.26}$$

Taking advantage of the relation $\boldsymbol{E}^{\mathrm{e}}(0) = {}^T\boldsymbol{R}^{e}\boldsymbol{e}^{\mathrm{e}}(0)\boldsymbol{R}^{\mathrm{e}}$ between Lagrangean and Eulerian logarithmic elastic strains (see (A7.22)$_1$, (A7.3)$_3$), it is

$$\overset{\circ}{\overline{\boldsymbol{e}^{\mathrm{e}}(0)}}{}^{(\mathrm{Re})} = \Sigma \overset{\mathrm{Re}}{e}{}^{\mathrm{e}}_{ij} \boldsymbol{n}^{\mathrm{e}}_i \otimes \boldsymbol{n}^{\mathrm{e}}_j = \boldsymbol{R}^{\mathrm{e}}\dot{\boldsymbol{E}}^{\mathrm{e}}(0)\boldsymbol{R}^T\mathrm{e} = \Sigma \overline{\dot{E}}{}^{\mathrm{e}}_{ij} \boldsymbol{n}^{\mathrm{e}}_i \otimes \boldsymbol{n}^{\mathrm{e}}_j. \tag{A7.27}$$

Hence, components $\overset{\mathrm{Re}}{e}{}^{\mathrm{e}}_{ij}$ of corotational derivative $\overset{\circ}{\overline{\boldsymbol{e}^{\mathrm{e}}(0)}}{}^{(\mathrm{Re})}$ on Eulerian elastic principal axes are equal to components $\overline{\dot{E}}{}^{\mathrm{e}}_{ij}$ of material derivative $\dot{\boldsymbol{E}}^{\mathrm{e}}(0)$ on Lagrangean elastic principal axes

$$\overset{\mathrm{Re}}{e}{}^{\mathrm{e}}_{ij} = \overline{\dot{E}}{}^{\mathrm{e}}_{ij}. \tag{A7.28}$$

Eliminating $\dot{\mathbf{e}}^{\mathrm{e}}(0)$ from (A7.25) to (A7.26), the following relation is obtained:

$$\overset{\circ}{\overline{\boldsymbol{e}^{\mathrm{e}}(0)}}{}^{(\mathrm{Je})} = \overset{\circ}{\overline{\boldsymbol{e}^{\mathrm{e}}(0)}}{}^{(\mathrm{Re})} + \boldsymbol{e}^{\mathrm{e}}(0)\left(\boldsymbol{\omega}^{\mathrm{e}} - \boldsymbol{\omega}^{\mathrm{Re}}\right) - \left(\boldsymbol{\omega}^{\mathrm{e}} - \boldsymbol{\omega}^{\mathrm{Re}}\right)\boldsymbol{e}^{\mathrm{e}}(0). \tag{A7.29}$$

Writing the above formula on Eulerian elastic principal axes and using (A7.27), (A7.24), and (A7.11) one obtains formulas for components of elastic Zaremba-

Jaumann corotational derivative $\overset{\circ}{\overline{\boldsymbol{e}^{\mathrm{e}}(0)}}^{(\mathrm{Je})}$ of logarithmic elastic strain on Eulerian elastic principal axes expressed in terms of components of elastic stretching tensor $\boldsymbol{d}^{\mathrm{e}}$ expressed on the same axes

$$\begin{aligned} &\overset{\mathrm{Je}}{e^{\mathrm{e}}_{ij}} = d^{\mathrm{e}}_{ij} = \dot{\lambda}^{\mathrm{e}}_{i}/\lambda^{\mathrm{e}}_{i} \quad i=j, \\ &\overset{\mathrm{Je}}{e^{\mathrm{e}}_{ij}} = \frac{\left(\lambda^{\mathrm{e}}_{i}/\lambda^{\mathrm{e}}_{j}\right)^{2}+1}{\left(\lambda^{\mathrm{e}}_{i}/\lambda^{\mathrm{e}}_{j}\right)^{2}-1} \ln\left(\lambda^{\mathrm{e}}_{i}/\lambda^{\mathrm{e}}_{j}\right) d^{\mathrm{e}}_{ij} \quad \lambda^{\mathrm{e}}_{i} \neq \lambda^{\mathrm{e}}_{j} \quad \text{and} \quad i \neq j. \end{aligned} \tag{A7.30}$$

Diagonal components of tensor $\overset{\circ}{\overline{\boldsymbol{e}^{\mathrm{e}}(0)}}^{(\mathrm{Je})}$ on Eulerian elastic principal axes are equal to respective material derivatives of principal values of logarithmic elastic strain $\overset{\bullet}{\overline{\ln\left(\lambda^{\mathrm{e}}_{i}\right)}}$, regardless of history of elastic deformation gradient.

Formulas (A7.30) are already specified in Raniecki and Nguyen (1984). In tensorial form, these relations can be expressed as

$$\overset{\circ}{\overline{\boldsymbol{e}^{\mathrm{e}}(0)}}^{(\mathrm{Je})} = \boldsymbol{d}^{\mathrm{e}} - \boldsymbol{e}^{\mathrm{e}}(0)\boldsymbol{x}^{\mathrm{e}} + \boldsymbol{x}^{\mathrm{e}}\boldsymbol{e}^{\mathrm{e}}(0), \quad \boldsymbol{x}^{\mathrm{e}} = -\boldsymbol{x}^{T}\mathrm{e}, \quad \boldsymbol{x}^{\mathrm{e}} = \Sigma x^{\mathrm{e}}_{ij}\boldsymbol{n}^{\mathrm{e}}_{i}\otimes\boldsymbol{n}^{\mathrm{e}}_{j}, \tag{A7.31}$$

where $\boldsymbol{x}^{\mathrm{e}}$ is a spin tensor. Its components can be calculated by writing (A7.31) on Eulerian elastic principal axes. After using (A7.30) and simple manipulations, it is

$$\begin{aligned} &x^{\mathrm{e}}_{ij} = 0 \quad i=j \\ &x^{\mathrm{e}}_{ij} = \left(\frac{1}{\ln\left(\lambda^{\mathrm{e}}_{i}/\lambda^{\mathrm{e}}_{j}\right)} + \frac{1+\left(\lambda^{\mathrm{e}}_{i}/\lambda^{\mathrm{e}}_{j}\right)^{2}}{1-\left(\lambda^{\mathrm{e}}_{i}/\lambda^{\mathrm{e}}_{j}\right)^{2}}\right) d^{\mathrm{e}}_{ij} \quad \lambda^{\mathrm{e}}_{i} \neq \lambda^{\mathrm{e}}_{j} \text{ and } i \neq j. \end{aligned} \tag{A7.32}$$

Whatever deformation in the form of dilatation does not change the value of tensor $\boldsymbol{x}^{\mathrm{e}}$, while for purely dilatational deformation $\lambda^{\mathrm{e}}_{1} = \lambda^{\mathrm{e}}_{2} = \lambda^{\mathrm{e}}_{3}$ it is uniquely equal to zero $\boldsymbol{x}^{\mathrm{e}} \equiv \boldsymbol{0}$.

Xiao et al. (1997), searching for corotational derivative of logarithmic total strain tensor, which would be equal to the total stretching rate tensor $\boldsymbol{d}$, introduced the concept of *logarithmic spin* $\boldsymbol{\omega}^{\log}$, and next the concept of corotational logarithmic derivative of any second-order tensor $\boldsymbol{T}$, as specified below:

$$\begin{aligned} &\overset{\circ}{\overline{\boldsymbol{e}(0)}}^{(\log)} \equiv \dot{\boldsymbol{e}}(0) + \boldsymbol{e}(0)\boldsymbol{\omega}^{\log} - \boldsymbol{\omega}^{\log}\boldsymbol{e}(0) = \boldsymbol{d}, \\ &\overset{\circ}{\boldsymbol{T}}^{(\log)} \equiv \dot{\boldsymbol{T}} + \boldsymbol{T}\boldsymbol{\omega}^{\log} - \boldsymbol{\omega}^{\log}\boldsymbol{T}, \quad \boldsymbol{\omega}^{\log} = \boldsymbol{\omega} + \boldsymbol{x}. \end{aligned} \tag{A7.33}$$

These authors also derived explicit formulas for components of spin $\boldsymbol{\omega}^{\log}$ as functions of components of $\boldsymbol{d}$, both expressed on Eulerian principal axes. The expressions for these components can be obtained using formula $(\mathrm{A}7.33)_3$. The components of spin tensor $\boldsymbol{x}$ appearing in this formula, expressed on Eulerian principal axes, can be obtained by substituting formulas (A7.32) in place of components d^{e}_{ij} and λ^{e}_{i} the components d_{ij} and λ_i of total deformation velocity gradient $\boldsymbol{d}$ and total principal stretches $(\boldsymbol{U} \rightarrow \lambda_i)$, expressed on Eulerian principal axes.

In analogy, the concept of elastic logarithmic corotational derivative and elastic logarithmic spin is introduced as follows (see (A7.31), $(7.4)_2$)

$$\overset{\mathrm{o}}{\boldsymbol{T}}^{(\mathrm{e\_log})} \equiv \dot{\boldsymbol{T}} + \boldsymbol{T}\boldsymbol{\omega}^{\mathrm{e\_log}} - \boldsymbol{\omega}^{\mathrm{e\_log}}\boldsymbol{T}, \quad \boldsymbol{\omega}^{\mathrm{e\_log}} = \boldsymbol{\omega}^{\mathrm{e}} + \boldsymbol{x}^{\mathrm{e}},$$
$$\overset{\mathrm{o}}{\overline{\boldsymbol{e}(0)}}^{(\mathrm{e\_log})} \equiv \dot{\boldsymbol{e}}(0) + \boldsymbol{e}(0)\boldsymbol{\omega}^{\mathrm{e\_log}} - \boldsymbol{\omega}^{\mathrm{e\_log}}\boldsymbol{e}(0) = \boldsymbol{d}^{\mathrm{e}}. \tag{A7.34}$$

It is also useful to introduce the following definition of fourth order tensor $\boldsymbol{\mathcal{E}}^{\mathrm{de}}$, being a function of current elastic deformation only (λ_i^{e}), and describing the difference between Zaremba-Jaumann elastic corotational derivative of logarithmic elastic strain and elastic stretching rate

$$\boldsymbol{\mathcal{E}}^{\mathrm{de}}\boldsymbol{d}^{\mathrm{e}} \equiv \overset{\mathrm{o}}{\overline{\boldsymbol{e}^{\mathrm{e}}(0)}}^{(\mathrm{Je})} - \boldsymbol{d}^{\mathrm{e}}, \quad \overset{\mathrm{Je}}{e^{\mathrm{e}}_{ij}} - d^{\mathrm{e}}_{ij} = \mathcal{E}^{\mathrm{de}}_{ijkl}d^{\mathrm{e}}_{kl}. \tag{A7.35}$$

Writing down the above relation on Eulerian elastic principal axes (see (A7.30)), it is

$$\begin{bmatrix} e_{11}^{\mathrm{e(Je)}} - d_{11} \\ e_{22}^{\mathrm{e(Je)}} - d_{22} \\ e_{33}^{\mathrm{e(Je)}} - d_{33} \\ e_{23}^{\mathrm{e(Je)}} - d_{23} \\ e_{13}^{\mathrm{e(Je)}} - d_{13} \\ e_{12}^{\mathrm{e(Je)}} - d_{12} \end{bmatrix} = \begin{bmatrix} 0&0&0&0&0&0 \\ 0&0&0&0&0&0 \\ 0&0&0&0&0&0 \\ 0&0&0&A_{23}^{\mathrm{e}}&0&0 \\ 0&0&0&0&A_{13}^{\mathrm{e}}&0 \\ 0&0&0&0&0&A_{12}^{\mathrm{e}} \end{bmatrix} \begin{bmatrix} d_{11} \\ d_{22} \\ d_{33} \\ d_{23} \\ d_{13} \\ d_{12} \end{bmatrix}, \quad \overset{\mathrm{o}}{\overline{\boldsymbol{e}^{\mathrm{e}}(0)}}^{(\mathrm{Je})} - \boldsymbol{d}^{\mathrm{e}} = \boldsymbol{\mathcal{E}}^{\mathrm{de}}\boldsymbol{d}^{\mathrm{e}}$$

$$A_{ij}^{\mathrm{e}} = \left[\frac{\left(\lambda_i^{\mathrm{e}}/\lambda_j^{\mathrm{e}}\right)^2 + 1}{\left(\lambda_i^{\mathrm{e}}/\lambda_j^{\mathrm{e}}\right)^2 - 1} \ln\left(\lambda_i^{\mathrm{e}}/\lambda_j^{\mathrm{e}}\right) - 1 \right]; \quad ij{=}23,32,\ ij{=}13,31,\ ij{=}12,21,$$

$$\mathcal{E}^{\mathrm{de}}_{ijkl}d^{\mathrm{e}}_{kl} = \overset{\mathrm{Je}}{e^{\mathrm{e}}_{ij}} - d^{\mathrm{e}}_{ij} = \begin{cases} 0, & i=j \\ \left[\dfrac{\left(\lambda_i^{\mathrm{e}}/\lambda_j^{\mathrm{e}}\right)^2 + 1}{\left(\lambda_i^{\mathrm{e}}/\lambda_j^{\mathrm{e}}\right)^2 - 1} \ln\left(\lambda_i^{\mathrm{e}}/\lambda_j^{\mathrm{e}}\right) - 1 \right] d^{\mathrm{e}}_{ij}; & i \neq j. \end{cases} \tag{A7.36}$$

The above indicates that $\mathcal{E}^{\mathrm{de}}_{ijkl}d^{\mathrm{e}}_{kl} = 0$ when $\lambda_i^{\mathrm{e}} \to \lambda_j^{\mathrm{e}}$.

Components of tensor $\boldsymbol{\mathcal{E}}^{\mathrm{de}}$ depend on elastic state of the material only. It has interesting properties in the case of materials exhibiting isotropy of elastic properties. It can be shown by direct calculation that

$$\mathbf{L}\boldsymbol{\mathcal{E}}^{\mathrm{de}} = \boldsymbol{\mathcal{E}}^{\mathrm{de}}\mathbf{L},$$
$$\mathbf{L}\boldsymbol{\mathcal{E}}^{\mathrm{de}} = \begin{bmatrix} 0&0&0&0&0&0 \\ 0&0&0&0&0&0 \\ 0&0&0&0&0&0 \\ 0&0&0&\mu A_{23}^{\mathrm{e}}&0&0 \\ 0&0&0&0&\mu A_{13}^{\mathrm{e}}&0 \\ 0&0&0&0&0&\mu A_{12}^{\mathrm{e}} \end{bmatrix}, \quad \mathbf{L} = \begin{bmatrix} \lambda+2\mu&\lambda&\lambda&0&0&0 \\ \lambda&\lambda+2\mu&\lambda&0&0&0 \\ \lambda&\lambda&\lambda+2\mu&0&0&0 \\ 0&0&0&\mu&0&0 \\ 0&0&0&0&\mu&0 \\ 0&0&0&0&0&\mu \end{bmatrix}. \tag{A7.37}$$

where $\mathbf{L}$ is classical isotropic elastic properties tensor.

The following property holds for any second order tensor $\boldsymbol{s}$, which is colinear with tensor $\boldsymbol{V}^{\mathrm{e}}$

$$\boldsymbol{\mathcal{E}}^{\mathrm{de}}\boldsymbol{s}=0. \tag{A7.38}$$

Hence, in the case of elastic isotropy, it is, for example, $\boldsymbol{\mathcal{E}}^{\mathrm{de}}\boldsymbol{\tau}=\boldsymbol{\mathcal{E}}^{\mathrm{de}}\boldsymbol{\sigma}=\boldsymbol{\mathcal{E}}^{\mathrm{de}}\boldsymbol{e}^{\mathrm{e}}(0)=\mathbf{0}$, as then all Eulerian elastic strain tensors and elastic work conjugated with them stress tensors and also Cauchy and Kirchoff stress tensors are colinear with $\boldsymbol{V}^{\mathrm{e}}$.

For isotropic thermal expansion coefficients tensor $\boldsymbol{\alpha}_0=\alpha_0\mathbf{I}$, it is

$$\mathbf{L}^{\mathrm{S(Je)}}\boldsymbol{\alpha}_0=\mathbf{L}^{\mathrm{S}}\boldsymbol{\alpha}_0=\mathbf{L}^{\mathrm{(Je)}}\boldsymbol{\alpha}_0=\mathbf{L}\boldsymbol{\alpha}_0,$$
$$\mathbf{L}^{\mathrm{(Je)}}\boldsymbol{\alpha}_0=\mathbf{L}\boldsymbol{\alpha}_0=\begin{bmatrix}\alpha_0(3\lambda+2\mu)\\ \alpha_0(3\lambda+2\mu)\\ \alpha_0(3\lambda+2\mu)\\ 0\\ 0\\ 0\end{bmatrix}=\begin{bmatrix}\lambda+2\mu & \lambda & \lambda & 0 & 0 & 0\\ \lambda & \lambda+2\mu & \lambda & 0 & 0 & 0\\ \lambda & \lambda & \lambda+2\mu & 0 & 0 & 0\\ 0 & 0 & 0 & \mu & 0 & 0\\ 0 & 0 & 0 & 0 & \mu & 0\\ 0 & 0 & 0 & 0 & 0 & \mu\end{bmatrix}\begin{bmatrix}\alpha_0\\ \alpha_0\\ \alpha_0\\ 0\\ 0\\ 0\end{bmatrix}. \tag{A7.39}$$

Annex 7.4 Family of isoclinic natural configurations—$\boldsymbol{\omega}^{\mathrm{in}}=\mathbf{0}$

For the purpose of multiplicative decomposition of total deformation gradient into elastic and inelastic part, the concept of unloaded elastically, instantaneous, reference configurations is used, also called "natural" configurations ${}^{t}\mathcal{B}^{\mathrm{in}}$ (see also Section 7.2). The requirement for zeroing the value of macroscopic stress tensor in these configurations does not constitute a sufficient condition for their unique determination. Additional information is necessary to fix their orientation with respect to global laboratory coordinates frame. From a mathematical point of view, this means necessity for specification of three geometrical constraints relations. For that purpose, Mandel (1974) introduced the concept of "director triad" and families of *isoclinic natural configurations*. Director triad is a set of three mutually orthogonal versors $\{\boldsymbol{m}_i\}$, $i=1,2,3$. It can be associated with mesostructural elements of macroelement (RVE), if premises of physical nature for that exist. Then director triad allows, within the scope of the macroscopic constitutive model of material, the description in a certain averaged (macroscopic) way of evolution distinguished with the triad mesostructural elements of the material (e.g., evolution of texture with respect to material fibers). Director triad can also be selected in such a way as to align instantaneously with some material fibers in unloaded elastically configurations. Then, it allows for the description of instantaneous relative rotation of some material fibers due to inelastic deformation, but not their rigid body rotation in unloaded configuration.

Director triad is not a material object and the evolution of its orientation, that is, rotation with respect to fixed laboratory coordinated frame, can only take place as a result of nonzero at particular time instant change of inelastic deformation—$\dot{\boldsymbol{F}}^{\mathrm{in}}(t)\neq\mathbf{0}$.

The family of *isoclinic natural configurations* $\mathcal{B}_{\text{iso}}$ is such a set of natural configurations ${}^t\mathcal{B}^{\text{in}}$ (configurations unloaded elastically), for which certain prefixed director triad $\{\boldsymbol{m}_i\}$ does not change its orientation (does not rotate), at any instant of inelastic deformation advancement described with tensor $\boldsymbol{F}^{\text{in}}$, (i.e., $\dot{\boldsymbol{m}}_i(t) \equiv \boldsymbol{0}$).

From a mathematical point of view, the preselection of director triad $\{\boldsymbol{m}_i\}$ and the requirement of its isoclinicity $\{\dot{\boldsymbol{m}}_i = \boldsymbol{0}\}$ is equivalent to imposing at each time instant three kinematic constraint conditions leading to a unique orientation of family of configurations ${}^t\mathcal{B}^{\text{in}}$ and, hence, a unique determination of gradient $\boldsymbol{F}^{\text{in}}(t)$.

In accordance with the Mandel concept, the director triad can be operationally defined (in a physically justified manner) in accordance with the following procedure. Let us assume that in consecutive, instantaneous, unloaded configurations there can be distinguished at each material point certain material plane with unit normal $\boldsymbol{m}_2(t)$ and lying in this plane material direction with versor $\boldsymbol{m}_1(t)$. A set composed of versors $\boldsymbol{m}_1$, $\boldsymbol{m}_2, \boldsymbol{m}_3 = \boldsymbol{m}_1 x \boldsymbol{m}_2$ is a director triad, where symbol "x" denotes vector product. A spectacular example of director triad at a particular material neighborhood, fixed in accordance with the above procedure, is the triad associated with crystallographic lattice of polycrystalline material; a concept frequently used in crystallographic theory of plasticity. Then, plane with normal $\boldsymbol{m}_2$ is slip plane, and material direction with versor $\boldsymbol{m}_1$ is plastic slip direction. Employment of the family of isoclinic natural configurations for description of plastic deformation process results in the invariability of plastic slip planes and versors; that is, they do not change their orientation in time and, thus, it makes the description simpler.

In the generic case of any family of natural configurations (nonisoclinic in particular), orientation of the director triad $\{\boldsymbol{m}_i\}$ evolves as a result of inelastic deformation process leading to consecutive unloaded elastically configurations ${}^t\mathcal{B}^{\text{in}}$. This evolution can be described with the aid of spin $\boldsymbol{\Omega}^m(t)$ defined by the formula (see also Raniecki and Samanta, 1989)

$$\boldsymbol{\Omega}^m(t) \equiv \tfrac{1}{2} \sum_{i=1,3} [\dot{\boldsymbol{m}}_i \otimes \boldsymbol{m}_i - \boldsymbol{m}_i \otimes \dot{\boldsymbol{m}}_i], \quad \boldsymbol{\Omega}^m = \Sigma \omega_{ij}^m \boldsymbol{m}_i \otimes \boldsymbol{m}_j, \; -\overset{T}{\boldsymbol{\Omega}}{}^m = \boldsymbol{\Omega}^m,$$
$$\boldsymbol{\Omega}^m(t) = \dot{\boldsymbol{m}}_1 \otimes \boldsymbol{m}_1 + \dot{\boldsymbol{m}}_2 \otimes \boldsymbol{m}_2 + \dot{\boldsymbol{m}}_3 \otimes \boldsymbol{m}_3 = -[\boldsymbol{m}_1 \otimes \dot{\boldsymbol{m}}_1 + \boldsymbol{m}_2 \otimes \dot{\boldsymbol{m}}_2 + \boldsymbol{m}_3 \otimes \dot{\boldsymbol{m}}_3]. \tag{A7.40}$$

Upper index "m" indicates that this spin describes evolution of freely selected, but then fixed director triad $\{\boldsymbol{m}_i\}$. It is worth noting that for the same inelastic deformation process described with tensor $\boldsymbol{F}^{\text{in}}(t)$, it takes different values for different, fixed director triads $\{\boldsymbol{m}_i\}$. It can be shown by direct calculation upon the use of definition (A7.40) that

$$\dot{\boldsymbol{m}}_1 = \boldsymbol{\Omega}^m \boldsymbol{m}_1 = -\omega_{12}^m \boldsymbol{m}_2 - \omega_{13}^m \boldsymbol{m}_3, \quad \dot{\boldsymbol{m}}_2 = \boldsymbol{\Omega}^m \boldsymbol{m}_2 = \omega_{12}^m \boldsymbol{m}_1 - \omega_{23}^m \boldsymbol{m}_3,$$
$$\dot{\boldsymbol{m}}_3 = \boldsymbol{\Omega}^m \boldsymbol{m}_3 = \omega_{13}^m \boldsymbol{m}_1 + \omega_{23}^m \boldsymbol{m}_2, \quad \dot{\boldsymbol{m}}_i \equiv \boldsymbol{\Omega}^m \boldsymbol{m}_i, \tag{A7.41}$$

and

$$(\boldsymbol{m}_2 \cdot \dot{\boldsymbol{m}}_1) = -\omega_{12}^m = -(\boldsymbol{m}_1 \cdot \dot{\boldsymbol{m}}_2), (\boldsymbol{m}_3 \cdot \dot{\boldsymbol{m}}_1) = -\omega_{13}^m = -(\boldsymbol{m}_1 \cdot \dot{\boldsymbol{m}}_3),$$
$$(\boldsymbol{m}_3 \cdot \dot{\boldsymbol{m}}_2) = -\omega_{23}^m = -(\boldsymbol{m}_2 \cdot \dot{\boldsymbol{m}}_3). \tag{A7.42}$$

It is convenient to introduce the notation

$$A_i = \dot{\boldsymbol{m}}_i \otimes \boldsymbol{m}_i - \boldsymbol{m}_i \otimes \dot{\boldsymbol{m}}_i. \tag{A7.43}$$

Upon elimination from (A7.41)$_3$, the terms $\omega_{13}^m \boldsymbol{m}_3$ and $\omega_{23}^m \boldsymbol{m}_3$ with the use of (A7.41)$_{1,2}$, the term $A_3 = \dot{\boldsymbol{m}}_3 \otimes \boldsymbol{m}_3 - \boldsymbol{m}_3 \otimes \dot{\boldsymbol{m}}_3$ can be expressed as

$$A_3 = \omega_{13}^m \boldsymbol{m}_1 \otimes \boldsymbol{m}_3 - \boldsymbol{m}_3 \otimes \omega_{13}^m \boldsymbol{m}_1 + \omega_{23}^m \boldsymbol{m}_2 \otimes \boldsymbol{m}_3 - \boldsymbol{m}_3 \otimes \omega_{23}^m \boldsymbol{m}_2, \tag{A7.44}$$

next using (A7.42)$_1$, it can be obtained that

$$\begin{aligned} A_3 &= A_1 + A_2 - 2\omega_{12}^m \left[\boldsymbol{m}_1 \otimes \boldsymbol{m}_2 - \boldsymbol{m}_2 \otimes \boldsymbol{m}_1\right], \\ A_3 &= A_1 + A_2 + \left[(\boldsymbol{m}_2 \cdot \dot{\boldsymbol{m}}_1) - (\boldsymbol{m}_1 \cdot \dot{\boldsymbol{m}}_2)\right] \left[\boldsymbol{m}_1 \otimes \boldsymbol{m}_2 - \boldsymbol{m}_2 \otimes \boldsymbol{m}_1\right]. \end{aligned} \tag{A7.45}$$

Similarly, formulas for A_1 are obtained as

$$\begin{aligned} A_1 &= A_2 + A_3 - 2\omega_{23}^m \left[\boldsymbol{m}_2 \otimes \boldsymbol{m}_3 - \boldsymbol{m}_3 \otimes \boldsymbol{m}_2\right], \\ A_1 &= A_2 + A_3 + \left[(\boldsymbol{m}_3 \cdot \dot{\boldsymbol{m}}_2) - (\boldsymbol{m}_2 \cdot \dot{\boldsymbol{m}}_3)\right] \left[\boldsymbol{m}_2 \otimes \boldsymbol{m}_3 - \boldsymbol{m}_3 \otimes \boldsymbol{m}_2\right], \end{aligned} \tag{A7.46}$$

and, finally for A_2,

$$\begin{aligned} A_2 &= A_1 + A_3 - 2\omega_{13}^m \left[\boldsymbol{m}_1 \otimes \boldsymbol{m}_3 - \boldsymbol{m}_3 \otimes \boldsymbol{m}_1\right], \\ A_2 &= A_1 + A_3 + \left[(\boldsymbol{m}_3 \cdot \dot{\boldsymbol{m}}_1) - (\boldsymbol{m}_1 \cdot \dot{\boldsymbol{m}}_3)\right] \left[\boldsymbol{m}_1 \otimes \boldsymbol{m}_3 - \boldsymbol{m}_3 \otimes \boldsymbol{m}_1\right]. \end{aligned} \tag{A7.47}$$

The spin $\boldsymbol{\Omega}^m$ can be expressed as $\boldsymbol{\Omega}^m = (A_1 + A_2 + A_3)/2$ in accordance with formula (A7.40)$_1$. Summing up right and left sides of formulas (A7.45)$_1$, (A7.46)$_1$, and (A7.47)$_1$ it is obtained that

$$\begin{aligned} \boldsymbol{\Omega}^m &= \omega_{12}^m \left[\boldsymbol{m}_1 \otimes \boldsymbol{m}_2 - \boldsymbol{m}_2 \otimes \boldsymbol{m}_1\right] + \omega_{13}^m \left[\boldsymbol{m}_1 \otimes \boldsymbol{m}_3 - \boldsymbol{m}_3 \otimes \boldsymbol{m}_1\right] \\ &\quad + \omega_{23}^m \left[\boldsymbol{m}_2 \otimes \boldsymbol{m}_3 - \boldsymbol{m}_3 \otimes \boldsymbol{m}_2\right], \\ \boldsymbol{\Omega}^m &= -(\boldsymbol{m}_2 \cdot \dot{\boldsymbol{m}}_1)\left[\boldsymbol{m}_1 \otimes \boldsymbol{m}_2 - \boldsymbol{m}_2 \otimes \boldsymbol{m}_1\right] - (\boldsymbol{m}_3 \cdot \dot{\boldsymbol{m}}_1)\left[\boldsymbol{m}_1 \otimes \boldsymbol{m}_3 - \boldsymbol{m}_3 \otimes \boldsymbol{m}_1\right] \\ &\quad - (\boldsymbol{m}_3 \cdot \dot{\boldsymbol{m}}_2)\left[\boldsymbol{m}_2 \otimes \boldsymbol{m}_3 - \boldsymbol{m}_3 \otimes \boldsymbol{m}_2\right]. \end{aligned} \tag{A7.48}$$

In obtaining the last formula, relation (A7.42) above is used.

In order to find an explicit relation expressing spin $\boldsymbol{\Omega}^m$ as a function of $\boldsymbol{L}^*$, ($\boldsymbol{L}^* = \dot{\boldsymbol{F}}^{\mathrm{in}} \boldsymbol{F}^{-\mathrm{in}}$) (see (7.2)$_4$) and $\{\boldsymbol{m}_i\}$ material derivatives of director triad versors must be calculated. Let us take incremental vector $\mathrm{d}\boldsymbol{x}$ with direction of versor $\boldsymbol{m}_1$ lying in plane $f(\boldsymbol{x}, t) = 0$ with normal $\boldsymbol{m}_2$. This vector deforms in accordance with the formula $\mathrm{d}\boldsymbol{x} = \boldsymbol{F}^{\mathrm{in}} \mathrm{d}\boldsymbol{X}$. Hence, it is $\mathrm{d}\dot{\boldsymbol{x}} = \boldsymbol{L}^* \mathrm{d}\boldsymbol{x}$. Material derivative of versor $\boldsymbol{m}_1$ can be determined by calculation of the derivative of the expression $\boldsymbol{m}_1 = \mathrm{d}\boldsymbol{x}/|\mathrm{d}\boldsymbol{x}|$ to get

$$\dot{\boldsymbol{m}}_1 = \boldsymbol{L}^* \boldsymbol{m}_1 - (\boldsymbol{m}_1 \cdot \boldsymbol{L}^* \boldsymbol{m}_1)\boldsymbol{m}_1 \tag{A7.49}$$

As vector $\mathbf{dx}$ is in plane $f(\mathbf{x}, t) = 0$; hence, at any fixed time instant, the constraint relation $(\nabla f)^{\mathrm{T}} \cdot \boldsymbol{E}\mathbf{x} = 0$ is valid. Wherefrom

$$\left((\dot{\nabla f})^{\mathrm{T}} \cdot \mathrm{d}\boldsymbol{x} + (\nabla f)^{\mathrm{T}} \cdot (\boldsymbol{L}^* \mathrm{d}\boldsymbol{x}) = 0\right) \Leftrightarrow \left(\dot{\nabla f} = -(\boldsymbol{L}^*)^{\mathrm{T}} \nabla f\right) \tag{A7.50}$$

The material derivative of versor $\boldsymbol{m}_2$ can be determined by calculation of the derivative of the expression $\boldsymbol{m}_2 = \nabla f / |\nabla f|$ to obtain

$$\dot{\boldsymbol{m}}_2 = -(\boldsymbol{L}^*)^{\mathrm{T}}\boldsymbol{m}_2 + (\boldsymbol{m}_2 \cdot \boldsymbol{L}^* \boldsymbol{m}_2)\boldsymbol{m}_2. \tag{A7.51}$$

Using formulas (A7.49) and (A7.51), the following scalar products can be computed:

$$\begin{aligned}
(\boldsymbol{m}_2 \cdot \dot{\boldsymbol{m}}_1) &= (\boldsymbol{m}_2 \cdot \boldsymbol{L}^* \boldsymbol{m}_1) = (\boldsymbol{m}_2 \cdot \boldsymbol{\Omega}^* \boldsymbol{m}_1) + (\boldsymbol{m}_2 \cdot \boldsymbol{D}^* \boldsymbol{m}_1) = -\omega_{12}^* + (\boldsymbol{m}_2 \cdot \boldsymbol{D}^* \boldsymbol{m}_1),\\
(\boldsymbol{m}_3 \cdot \dot{\boldsymbol{m}}_1) &= (\boldsymbol{m}_3 \cdot \boldsymbol{L}^* \boldsymbol{m}_1) = (\boldsymbol{m}_3 \cdot \boldsymbol{\Omega}^* \boldsymbol{m}_1) + (\boldsymbol{m}_3 \cdot \boldsymbol{D}^* \boldsymbol{m}_1) = -\omega_{13}^* + (\boldsymbol{m}_3 \cdot \boldsymbol{D}^* \boldsymbol{m}_1),\\
(\boldsymbol{m}_3 \cdot \dot{\boldsymbol{m}}_2) &= -\left(\boldsymbol{m}_3 \cdot (\boldsymbol{L}^*)^T \boldsymbol{m}_2\right) = (\boldsymbol{m}_3 \cdot \boldsymbol{\Omega}^* \boldsymbol{m}_2) - (\boldsymbol{m}_3 \cdot \boldsymbol{D}^* \boldsymbol{m}_2) = -\omega_{23}^* - (\boldsymbol{m}_3 \cdot \boldsymbol{D}^* \boldsymbol{m}_2).
\end{aligned} \tag{A7.52}$$

Substituting the above to $(\text{A7.48})_2$, it is obtained that

$$\begin{aligned}
\boldsymbol{\Omega}^m = &\left(\omega_{12}^* - (\boldsymbol{m}_2 \cdot \boldsymbol{D}^* \boldsymbol{m}_1)\right)[\boldsymbol{m}_1 \otimes \boldsymbol{m}_2 - \boldsymbol{m}_2 \otimes \boldsymbol{m}_1]\\
&+ \left(\omega_{23}^* + (\boldsymbol{m}_3 \cdot \boldsymbol{D}^* \boldsymbol{m}_2)\right)[\boldsymbol{m}_2 \otimes \boldsymbol{m}_3 - \boldsymbol{m}_3 \otimes \boldsymbol{m}_2]\\
&+ \left(\omega_{13}^* - (\boldsymbol{m}_3 \cdot \boldsymbol{D}^* \boldsymbol{m}_1)\right)[\boldsymbol{m}_1 \otimes \boldsymbol{m}_3 - \boldsymbol{m}_3 \otimes \boldsymbol{m}_1],
\end{aligned}$$

and finally,

$$\begin{aligned}
\boldsymbol{\Omega}^m(\boldsymbol{L}^*, \boldsymbol{m}_i) = &\boldsymbol{\Omega}^* - (\boldsymbol{m}_2 \cdot \boldsymbol{D}^* \boldsymbol{m}_1)[\boldsymbol{m}_1 \otimes \boldsymbol{m}_2 - \boldsymbol{m}_2 \otimes \boldsymbol{m}_1]\\
&- (\boldsymbol{m}_3 \cdot \boldsymbol{D}^* \boldsymbol{m}_1)[\boldsymbol{m}_1 \otimes \boldsymbol{m}_3 - \boldsymbol{m}_3 \otimes \boldsymbol{m}_1]\\
&- (\boldsymbol{m}_3 \cdot \boldsymbol{D}^* \boldsymbol{m}_2)[\boldsymbol{m}_3 \otimes \boldsymbol{m}_2 - \boldsymbol{m}_2 \otimes \boldsymbol{m}_3].
\end{aligned} \tag{A7.53}$$

Alternatively, spin $\boldsymbol{\Omega}^m$ can be expressed in a form dependent on only two versors of director triad $\boldsymbol{m}_1$ and $\boldsymbol{m}_2$ with the formula

$$\begin{aligned}
\boldsymbol{\Omega}^m = &\boldsymbol{\Omega}^* + (\boldsymbol{m}_2 \cdot \boldsymbol{D}^* \boldsymbol{m}_1)[\boldsymbol{m}_1 \otimes \boldsymbol{m}_2 - \boldsymbol{m}_2 \otimes \boldsymbol{m}_1]\\
&+ (\boldsymbol{D}^* \boldsymbol{m}_1) \otimes \boldsymbol{m}_1 - \boldsymbol{m}_1 \otimes (\boldsymbol{D}^* \boldsymbol{m}_1) + \boldsymbol{m}_2 \otimes (\boldsymbol{D}^* \boldsymbol{m}_2) - (\boldsymbol{D}^* \boldsymbol{m}_2) \otimes \boldsymbol{m}_2.
\end{aligned} \tag{A7.54}$$

In order to obtain (A7.54), (A7.44) should be substituted to (A7.40) and advantage should be taken of (A7.49) and (A7.51).

As was already stated, if the orientation of a family of unloaded elastically configurations ${}^t\mathcal{B}^{\text{in}}$ is selected in such a way that certain director triad $\{\boldsymbol{m}_i\}$ does not change its orientation as a result of inelastic deformation ($\dot{\boldsymbol{m}}_i = \boldsymbol{0}$), then such a family is isoclinic $\mathcal{B}_{\text{iso}}$, more strictly $\boldsymbol{m}_i$—isoclinic. In accordance with formula (A7.40), this means that in such a case, spin $\boldsymbol{\Omega}^m(t) \equiv \boldsymbol{0}$. Substituting the condition $\boldsymbol{\Omega}^m = 0$ in relation (A7.53) leads to three constraint relations

$$\begin{aligned}
\boldsymbol{\Omega}^* = \boldsymbol{\Omega}^*(\boldsymbol{D}^*) = \hat{\boldsymbol{\Omega}}(\boldsymbol{D}^*; \boldsymbol{m}_i), \hat{\boldsymbol{\Omega}} \equiv \widehat{\omega}_{ij}\left[\boldsymbol{m}_i \otimes \boldsymbol{m}_j - \boldsymbol{m}_j \otimes \boldsymbol{m}_i\right]/2, \widehat{\omega}_{12} = (\boldsymbol{m}_2 \cdot \boldsymbol{D}^* \boldsymbol{m}_1)\\
= -\widehat{\omega}_{21}, \widehat{\omega}_{13} = (\boldsymbol{m}_3 \cdot \boldsymbol{D}^* \boldsymbol{m}_1) = -\widehat{\omega}_{31}, \widehat{\omega}_{32} = (\boldsymbol{m}_3 \cdot \boldsymbol{D}^* \boldsymbol{m}_2) = -\widehat{\omega}_{23}.
\end{aligned} \tag{A7.55}$$

The versors $\{\boldsymbol{m}_i\}$ fulfill here a role of fixed parameter. Hence, the property of isoclinicity of the family of natural configurations means the requirement for the existence of the relation between spin tensor $\boldsymbol{\Omega}^*$ and inelastic stretching tensor $\boldsymbol{D}^*$. While, in view of (A7.55), this relation must be such that

$$\boldsymbol{\Omega}^*(\boldsymbol{D}^*=\mathbf{0})=\mathbf{0}, \quad (\boldsymbol{D}^*=\mathbf{0}) \Rightarrow (\boldsymbol{\Omega}^*=\mathbf{0}). \tag{A7.56}$$

On the other hand, the existence of relation $\boldsymbol{\Omega}^*(\boldsymbol{D}^*)$ with property (A7.56) is a necessary and sufficient condition for the family of natural configurations defined by tensor $\boldsymbol{F}^{\text{in}}$ to be isoclinic.

As can be easily noticed from formula (A7.55), the same inelastic deformation — $\boldsymbol{F}^{\text{in}}(t)$—leads to different families of isoclinic natural configurations at different selection of director triads $\{\boldsymbol{m}_i\}$. Hence, an infinite number of such families can be distinguished for particular inelastic deformation.

In the case of the development of a macroscopic model of isotropic material, in particular of isotropic shape memory alloys, no physical premises exist for selection of orientation of family of unloaded elastically configurations. In view of that, this orientation is usually fixed by the condition that, at any time instant of inelastic deformation process, Eulerian inelastic spin is identically equal to zero $\boldsymbol{\omega}^{\text{in}}(t)\equiv\mathbf{0}$. This particular condition leads to considerable simplification of mathematical formulas of constitutive model. Using kinematic condition $\boldsymbol{\omega}^{\text{in}}(t)=\mathbf{0}$ in $(7.2)_4$, $(7.3)_4$, and $(7.4)_4$, the following relations are obtained:

$$\begin{aligned} &\boldsymbol{d}^{\text{in}}=\boldsymbol{L}^{\text{in}}=\overset{T}{\boldsymbol{L}^{\text{in}}}, \quad \boldsymbol{L}^*=\overset{-1}{\boldsymbol{F}^{\text{e}}}\boldsymbol{d}^{\text{in}}\boldsymbol{F}^{\text{e}}, \\ &\boldsymbol{D}^*=\frac{1}{2}\left[\overset{-1}{\boldsymbol{F}^{\text{e}}}\boldsymbol{d}^{\text{in}}\boldsymbol{F}^{\text{e}}+\overset{T}{\boldsymbol{F}^{\text{e}}}\boldsymbol{d}^{\text{in}}\overset{-T}{\boldsymbol{F}^{\text{e}}}\right], \quad \boldsymbol{\Omega}^*=\frac{1}{2}\left[\overset{-1}{\boldsymbol{F}^{\text{e}}}\boldsymbol{d}^{\text{in}}\boldsymbol{F}^{\text{e}}-\overset{T}{\boldsymbol{F}^{\text{e}}}\boldsymbol{d}^{\text{in}}\overset{-T}{\boldsymbol{F}^{\text{e}}}\right]. \end{aligned} \tag{A7.57}$$

In an implicit way, the above formulas define relation $\boldsymbol{\Omega}^*=\boldsymbol{\Omega}^*(\boldsymbol{D}^*)$ (see (A7.55)). This relation has the property (A7.56). Indeed, if $(\boldsymbol{d}^{\text{in}}\to\mathbf{0})$ then $(\boldsymbol{D}^*\to\mathbf{0})$ and $(\boldsymbol{\Omega}^*\to\mathbf{0})$. Thus, the family of natural configurations $\mathcal{B}_{\omega^{\text{in}}=0}$ defined with condition $\boldsymbol{\omega}^{\text{in}}=\mathbf{0}$ is isoclinic.

It is worth it to discuss how the family of isoclinic configurations $\mathcal{B}_{\omega^{\text{in}}=0}$ is related to the family of isoclinic configurations $\mathcal{B}_{\Omega^*=0}$, defined with kinematical condition $\boldsymbol{\Omega}^*(t)\equiv\mathbf{0}$; often used in classical theory of isotropic plasticity. In classical theory of isotropic plasticity, the property of isotropy is assumed for two elements of the model

- First, it is assumed isotropy of elastic properties of the material resulting in the property that Kirchoff, Cauchy stress tensors and Eulerian strain tensors are all collinear with tensor $\boldsymbol{V}^{\text{e}}$ and, hence, they mutually commute with each other and with tensor $\boldsymbol{V}^{\text{e}}$ (e.g., $\boldsymbol{\sigma}\boldsymbol{V}^{\text{e}}=\boldsymbol{V}^{\text{e}}\boldsymbol{\sigma}$). On natural configurations, this means colinearity of Lagrangean stress tensors elastic work conjugate with respective Lagrangean strain tensors with tensor $\boldsymbol{U}^{\text{e}}$, and in particular $\boldsymbol{T}^{\text{Re}}\boldsymbol{U}^{\text{e}}=\boldsymbol{U}^{\text{e}}\boldsymbol{T}^{\text{Re}}=\boldsymbol{T}^{\text{e}}(0)\boldsymbol{U}^{\text{e}}=\boldsymbol{U}^{\text{e}}\boldsymbol{T}^{\text{e}}(0)$.
- Second, it is assumed isotropy of plastic flow rule in the sense that it is postulated on actual configuration with the aid of tensor function isotropic with respect to $\boldsymbol{V}^{\text{e}}$. The plastic flow rule is very often postulated in the form $\boldsymbol{d}^{\text{in}}(\boldsymbol{V}^{\text{e}})=\dot{\lambda}(\overline{\boldsymbol{\sigma}}/|\overline{\boldsymbol{\sigma}}|)$, which results in colinearity of tensors $\boldsymbol{d}^{\text{in}}$ and $\boldsymbol{V}^{\text{e}}$—$\boldsymbol{d}^{\text{in}}\boldsymbol{V}^{\text{e}}=\boldsymbol{V}^{\text{e}}\boldsymbol{d}^{\text{in}}$. This results in colinearity of $\boldsymbol{D}^*$ and $\boldsymbol{U}^{\text{e}}$ on natural configurations—$\boldsymbol{D}^*\boldsymbol{U}^{\text{e}}=\boldsymbol{U}^{\text{e}}\boldsymbol{D}^*$.

Using formulas (7.2)–(7.4), and $\boldsymbol{F}^{\mathrm{e}} = \boldsymbol{R}^{\mathrm{e}}\boldsymbol{U}^{\mathrm{e}}$, the following general formulas are obtained:

$$
\begin{aligned}
\boldsymbol{d}^{\mathrm{in}} &= \boldsymbol{R}^{\mathrm{e}}0.5\left[\boldsymbol{U}^{\mathrm{e}}\boldsymbol{D}^{*}\overset{-1}{\boldsymbol{U}}{}^{\mathrm{e}} + \overset{-1}{\boldsymbol{U}}{}^{\mathrm{e}}\boldsymbol{D}^{*}\boldsymbol{U}^{\mathrm{e}}\right]\overset{T}{\boldsymbol{R}}{}^{\mathrm{e}} + \boldsymbol{R}^{\mathrm{e}}0.5\left[\boldsymbol{U}^{\mathrm{e}}\boldsymbol{\Omega}^{*}\overset{-1}{\boldsymbol{U}}{}^{\mathrm{e}} + \overset{-1}{\boldsymbol{U}}{}^{\mathrm{e}}\overset{T}{\boldsymbol{\Omega}}{}^{*}\boldsymbol{U}^{\mathrm{e}}\right]\overset{T}{\boldsymbol{R}}{}^{\mathrm{e}},\\
\boldsymbol{\omega}^{\mathrm{in}} &= \boldsymbol{R}^{\mathrm{e}}0.5\left[\boldsymbol{U}^{\mathrm{e}}\boldsymbol{D}^{*}\overset{-1}{\boldsymbol{U}}{}^{\mathrm{e}} - \overset{-1}{\boldsymbol{U}}{}^{\mathrm{e}}\boldsymbol{D}^{*}\boldsymbol{U}^{\mathrm{e}}\right]\overset{T}{\boldsymbol{R}}{}^{\mathrm{e}} + \boldsymbol{R}^{\mathrm{e}}0.5\left[\boldsymbol{U}^{\mathrm{e}}\boldsymbol{\Omega}^{*}\overset{-1}{\boldsymbol{U}}{}^{\mathrm{e}} - \overset{-1}{\boldsymbol{U}}{}^{\mathrm{e}}\overset{T}{\boldsymbol{\Omega}}{}^{*}\boldsymbol{U}^{\mathrm{e}}\right]\overset{T}{\boldsymbol{R}}{}^{\mathrm{e}}.
\end{aligned}
\tag{A7.58}
$$

Using the assumed property of isotropy of plastic flow rule in Eulerian description and, hence, colinearity of tensors $\boldsymbol{D}^*$ and $\boldsymbol{U}^{\mathrm{e}}$, $\boldsymbol{D}^*\boldsymbol{U}^{\mathrm{e}} = \boldsymbol{U}^{\mathrm{e}}\boldsymbol{D}^*$, it is

$$
\begin{aligned}
\boldsymbol{d}^{\mathrm{in}} &= \boldsymbol{R}^{\mathrm{e}}\boldsymbol{D}^{*}\overset{T}{\boldsymbol{R}}{}^{\mathrm{e}} + \boldsymbol{R}^{\mathrm{e}}0.5\left[\boldsymbol{U}^{\mathrm{e}}\boldsymbol{\Omega}^{*}\overset{-1}{\boldsymbol{U}}{}^{\mathrm{e}} + \overset{-1}{\boldsymbol{U}}{}^{\mathrm{e}}\overset{T}{\boldsymbol{\Omega}}{}^{*}\boldsymbol{U}^{\mathrm{e}}\right]\overset{T}{\boldsymbol{R}}{}^{\mathrm{e}},\\
\boldsymbol{\omega}^{\mathrm{in}} &= \boldsymbol{R}^{\mathrm{e}}0.5\left[\boldsymbol{U}^{\mathrm{e}}\boldsymbol{\Omega}^{*}\overset{-1}{\boldsymbol{U}}{}^{\mathrm{e}} - \overset{-1}{\boldsymbol{U}}{}^{\mathrm{e}}\overset{T}{\boldsymbol{\Omega}}{}^{*}\boldsymbol{U}^{\mathrm{e}}\right]\overset{T}{\boldsymbol{R}}{}^{\mathrm{e}}.
\end{aligned}
\tag{A7.59}
$$

Using the condition defining the family of isoclinic configurations $\mathcal{B}_{\Omega^*=0}$ ($\boldsymbol{\Omega}^* = \mathbf{0}$), the following relations are obtained:

$$
\boldsymbol{d}^{\mathrm{in}} = \boldsymbol{R}^{\mathrm{e}}\boldsymbol{D}^{*}\overset{T}{\boldsymbol{R}}{}^{\mathrm{e}}, \quad \boldsymbol{\omega}^{\mathrm{in}} = \mathbf{0}. \tag{A7.60}
$$

The above relations mean that in the case of classical theory of plasticity, isotropic elastically, and inelastically, the kinematic conditions $\boldsymbol{\Omega}^* = \mathbf{0}$ and $\boldsymbol{\omega}^{\mathrm{in}} = \mathbf{0}$ imply selection of the same family of isoclinic configurations. The director triad of this family is the set of three different principal directions of tensor $\boldsymbol{D}^*$.

In the case of the SMA material model discussed here, tensor $\boldsymbol{d}^{\mathrm{in}}$ is not expressed by tensor valued function isotropic with respect to stress tensor $\boldsymbol{\tau}$ (see (7.56)). Hence, in a general case tensors $\boldsymbol{d}^{\mathrm{in}}$ and $\boldsymbol{V}^{\mathrm{e}}$ do not commute and, as a result, conditions $\boldsymbol{\omega}^{\mathrm{in}} = \mathbf{0}$ and $\boldsymbol{\Omega}^* = \mathbf{0}$ define different families of isoclinic configurations.

Summary and future trends

8

8.1 Summary and future trends

In the previous chapters of this book, state-of-the-art information is delivered on the pseudoelastic behavior of shape memory alloys materials. An effort has been made to divide the attention of the reader in a rather balanced way between fundamental research and engineering applications issues. The main emphasis is put on macroscopic constitutive models of SMA materials elaborated in a form convenient for implementation into finite element codes, a powerful design and study tool for engineering structures and devices containing elements made of SMA materials. Information is harmoniously divided between different scales of observation. First, experimental evidence on SMA materials behavior and underlying physical phenomena originating from materials science research is reported, followed by macroscopic thermomechanical experimental results and theoretical developments of macroscopic modeling description. The reason for the limitation to the pseudoelastic range of SMA materials behavior is that, in this range, consensus slowly emerges between researchers in the field on the common theoretical basis on which macroscopic constitutive models of SMA materials should rely; that is, the framework of nonequilibrium thermodynamics with internal state variables. This generally accepted theoretical modeling base provides an opportunity for reaching agreement on universally understood and reliable tools and methods for characterization of SMA materials operating in the pseudoelastic range of behavior.

The interest in shape memory alloy materials originates in scientific curiosity from economic and societal pressures to find or elaborate materials with better properties superior to those already attainable, more useful for a specific purpose, or exhibiting a combination of quite contradictory qualities. For example, typical intermetallics are brittle, breaking at strains not exceeding several percent. In comparison, NiTi intermetallic with its fracture strains reaching 25%, decently can be called ductile, yet with the possibility of reaching a strength of 1 GPa. Great opportunity for achieving new special features has been spotted in materials sensitive to several physical fields; for example, thermal and mechanical or magnetic and mechanical. Materials of this kind have been qualified to a group called active or *adaptive materials*. Shape memory alloys belong to this group, along with magnetostrictive, ferroelectric, or birefringence materials. In principle, special (unattainable with "conventional" materials) or additional functionality is searched for in order to facilitate manufacturing of engineering devices and structures satisfying human needs in a simpler, faster, and/or cheaper fashion. The large initial investment required, in terms of time and money, for development work on active materials usually originates from one or the other

Pseudoelasticity of Shape Memory Alloys. http://dx.doi.org/10.1016/B978-0-12-801697-8.00008-9

sources of military spending, an activity with inherent high risks and with the emphasis put on performance to gain advantage even if high initial costs are necessary. Only later will technological achievements spill out to civil applications. This is a quite standard pattern. For example, progress in efficiency of jet aircraft engines results primarily in superalloys development and their coatings technology, allowing for higher and higher entry temperatures. In the very first place, new superalloys developments are used in turbine blades of military aircraft engines, to be later used in civil jet engines when positively verified. All that is usually financed by government subsidies. The other application area with "natural" high risks and acceptable initial high costs is medicine. Indeed, military application in F-14 pipe joints, was the first ever commercial application of SMA materials and the medical industry is by far a prime driving force for present commercial use and developments of tools and devices made from or containing elements made of SMA materials.

As frequently happens, technological efforts targeted at the development of SMA materials with required functional features such as pseudoelasticity or one-way memory effect, for example, with a preferred range of characteristic temperatures of phase transition—so-called high temperature shape memory alloys—or favorite size of hysteresis loop resulted in the discovery of quite unexpected other advantages of SMA materials. For example, it has been found that NiTi alloys, besides active (adaptive) features connected with shape memory effects, possess a number of utility features such as biocompatibility, resistance to corrosion, resistance to wear, and kinking resistance; properties of prime importance in medical applications. The other interesting discovery is connected with so-called Nitinol 60 alloy ($Ni_{60wt\%}$-Ti). It has been discovered that this alloy has excellent tribological properties in conditions of boundary regime lubrication; speaking more colloquially, when there is a very small amount of lubricant substance. Tribological properties of Nitinol 60, in terms of friction coefficient and rate of wear, proved to be better than standard at present bearing material (i.e., steel 440C). These advantages, together with dimensional stability, high hardness (on the order of 62 HRC), wear resistance, nongalling (which is quite unusual for Ti containing alloys), and high brittleness resistance at ultra low temperatures, indicates Nitinol 60 is a promising candidate for numerous (low temperatures) aerospace applications. Coupling of functional and utility features of SMA materials makes them even more interesting, and delivers additional momentum to devote time and effort for further research work on this class of materials.

While SMA materials deliver high promise for intellectual and other satisfactions, the problems arise in identifying existing research and development trends, the need for their possible correction, and/or formulating of new tasks of high priority. The reason for this is that although more than 50 years have passed since the spectacular growth of interest in SMA materials aroused by Buhler in 1963, the interesting and challenging problems are accumulating instead of being cleared out. There can be formulated so many interesting future trends in research and development devoted to SMA materials that no attempt has been undertaken here to make an exhaustive list. Rather, some problems have been noted that have the highest gravity for the author of this book, and the addressing of which can possibly bring about considerable progress in the field of SMA materials.

A discussion on SMA materials research trends is worth to commence with formulating a question belonging to fundamental research problems: What are the key "ingredients" necessary for shape memory effects to appear? Or, said another way: What are the key properties that specific metallic material must have or conditions that it must fulfill in order that, for example, nice and stabile pseudoelastic behavior occurs? It seems that after many years of research devoted to martensitic transformations, the answer to that question is still to a large extent absent. It is neither well known, nor fully understood, why NiTi-based alloys exhibit excellent shape memory properties (martensitic transformation taking place in them is thermoelastic), while the dominant majority of Fe-based alloys do not exhibit nice shape memory effects (martensitic transformation taking place in them is not thermoelastic). Recently, an interesting contribution on this issue was published by Otsuka et al. (2011). The authors do not limit their attention to strictly crystallographic issues, which has been customary for many authors for many years, but try to take a broader perspective. They found that fulfillment of crystallographic symmetry conditions between parent and product phase crystallographic lattices (i.e., the presence of so-called G-subG symmetry relations) introduced by Bhattacharya et al. (2004) and regarded to be a prerequisite for martensitic transformation reversibility is not a sufficient condition for good SME. Otsuka et al. (2011) state that the other necessary conditions for good SME are the existence of sufficient "driving force" and the existence of some reversible kinematic mechanism, which they identify to be "reversible twinning modes" with reasonably small twinning shear. The wording is pretty ambiguous, which suggests the complexity of the discussed problem. Anyway, crystallographic conditions alone certainly are not the key factor deciding on the existence of good SME. It seems that the landscape of important elements should be enlarged even more. A hint to discover the key ingredients of SME may come upon studying other materials also exhibiting pseudoelastic behavior, but not necessarily metallic ones. It is probably worth it to direct careful attention to "biological materials". Taking into account that life on earth has a much longer history (it had to survive harsh and demanding conditions during millions of years) in comparison to the several hundred years of humankind technological civilization, it is interesting to analyze what solutions nature worked out to deal with the impediments encountered.

A very interesting survey work on biological materials exhibiting very remarkable properties was published recently by Chen et al. (2012). Information can be found there that whelk egg capsules exhibit excellent pseudoelastic behavior (see Figure 8.1a). A striking difference exists between SMA materials and whelk egg pseudoelasticity. In the case of SMA materials, the pseudoelastic loop shifts toward higher stresses with growing temperature, and in the case of whelk egg material it shifts toward lower stresses with growing temperature. The other significant difference is that the size of the hysteresis loop changes considerably in the case of whelk egg material. This suggests care in drawing too far-reaching conclusions originating from straightforward comparisons. In Figure 8.1b the mechanism of whelk egg material pseudoelasticity is elucidated. Let us analyze this mechanism of pseudoelasticity in which clearly no crystallography is involved, at least in the classical metallurgical sense. Initially, at no mechanical load, whelk egg capsules material has mesostructure

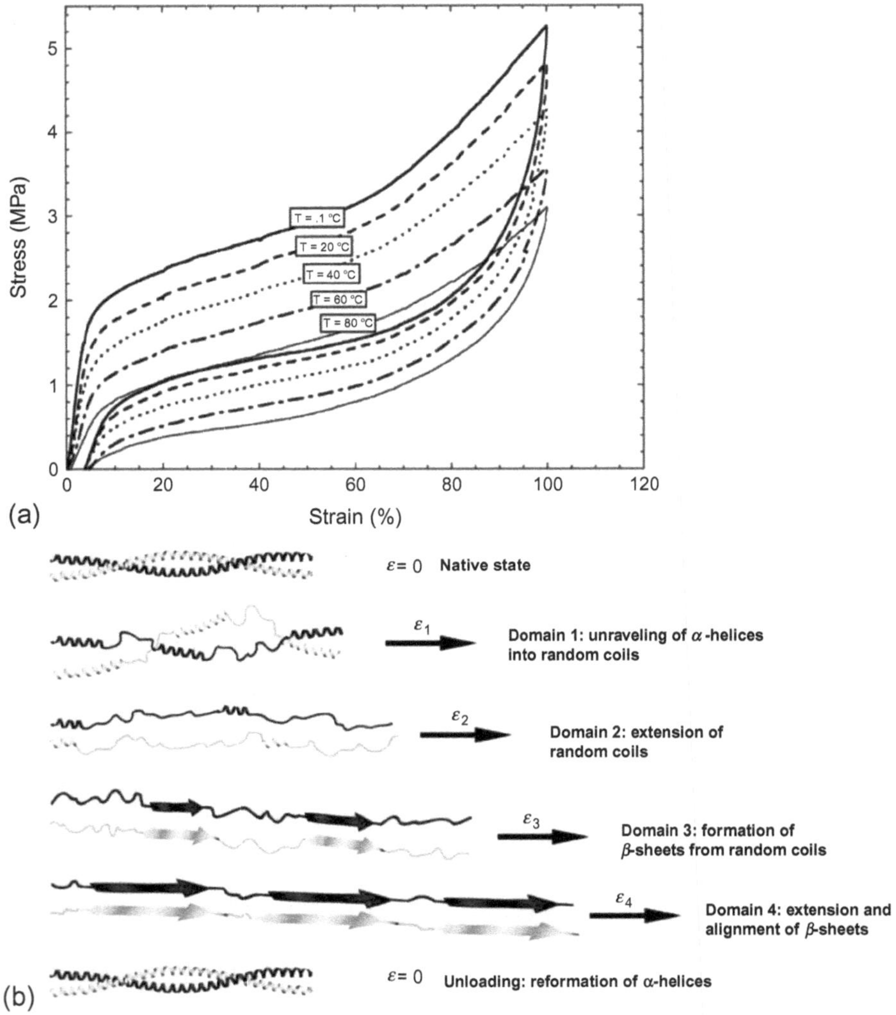

Figure 8.1 (a) Pseudoelastic behavior of whelk egg capsules submitted to large deformation (maximum engineering strain up to ~100%) uniaxial tension, loading-unloading cycle.
(b) Schematic representation of geometrically reversible structural transformation of whelk egg capsule material. After Figure 120 in Chen et al. (2012).
Originally from Fratzl, P., Misof, K., Zizak, I., Rapp, G., Amenitsch, H., Bernstorff, S., 1997. Fibrillar structure and mechanical properties of collagen. J. Struct. Biol. 122, 119–122.

of coiled-coil of α-helices. These helices unravel under an increase in stress, due to overcoming forces of van der Waals and hydrogen bonds, which results in a pseudoelastic strain plateau. At higher stresses, stretched chains undergo further ordering into β-sheets. Upon stress, unloading opposite processes take place in reverse order and, in a geometrical sense, the process can be qualified as reversible. Searching for an analogy with thermoelastic martensitic transformation, the unraveling of coiled α-helices can be matched to the reordering of martensitic variants and β-sheets formations can be treated as a process analogous to HPV variants detwinning. As in biological material, no crystallographic symmetry problems obscure the picture so we can more clearly understand what is actually necessary in order that shape memory effect appears. Namely, it is worth it to clearly formulate that some stable mesostructural objects must be present that make not a kind of smooth "clay" continuum but a kind of "crunchy" one, with individual mesostructural elements not being able to match smoothly to one another (geometrically incompatible objects). These mesostructural objects can be identified to be α-helices in biological material and HPV martensitic variants in SMA materials. Naturally, absolute size and specific shape of these generic (crunchy) objects can have an influence on exhibited SME effects; for example, size of the pseudoelastic hysteresis loop. After having the basic ingredients of α-helices or HPV variants in place, the issue of a source for a driving force arises to induce return to original shape, as pointed out by Otsuka et al. (2011). In the case of biological material, this driving force for forming coiled-coil of α-helices originates from van der Waals and hydrogen bonds, and in the case of SMA materials it originates from incompatibility of martensitic eigenstrain field, which translates into nonzero macroscopic coherency energy, as has been demonstrated in Chapter 4 of this book. Now, the question arises concerning how incompatibility of martensitic phase eigenstrain field can be controlled and, thus, the "driving force" of the SME effect. The second necessary ingredient mentioned by Otsuka et al. (2011) for good SME of SMA materials (i.e., the existence of reversible kinematic mechanism), does not pose a problem in the case of biological material because fibers can bend and twist, having many degrees of freedom, but in the case of martensitic transformation kinematics of (reversible) martensitic variants, evolution mechanism requires careful investigation. Revealing details of this mechanism may have a connection and may be helpful in solving one of the fundamental mysteries regarding nearly equiatomic NiTi alloys—its exceptional ductility, as indicated by Duerig (2006). Problems of crystallographic symmetry are, to a large extent, not discussed here, but this does not mean that they are neglected. They are extensively discussed in the literature devoted to SMA materials and calling attention to this issue does not require further promotion.

An interesting problem belonging to fundamental research, the solution of which can considerably facilitate obtaining SMA materials with required properties, is the question of what factors influence characteristic temperatures of thermoelastic martensitic transformation. For example, why does A_s temperature of $Ni_{50at\%}$-Ti amount to 32 °C and not 140°? Taking the thermodynamic R_L model of pseudoelastic behavior, the answer is seemingly simple: It is the internal energy and entropy of the formation of martensitic phase Δu^0, Δs^0 ($T^{eq} = \Delta u^0 / \Delta s^0$) that decides on the specific value of thermodynamic equilibrium temperature of parent and product phase in the neighborhood where characteristic temperatures of A_s and M_s are located. But this answer, while correct, does not deliver much insight into the posed problem. Similarly, in the case of NiTi alloys it is well known that Hf or Zr increase p.t. temperatures of NiTi alloy and Fe, Co alloying additives shift characteristic temperatures of p.t. to lower temperatures. The problem is what these addictives actually change in material structure that causes characteristic temperatures of p.t. to shift? The theoretical developments connected with the family of thermodynamic R_L model already give some insight into better understanding of SMA material properties originating from material mesostructural characteristics. For example, it helped to identify that the key reason for the split of parent and product phase equilibrium temperature into martensite start and austenite start temperatures observed experimentally is the nonzero value of coherence energy. Mesostructural analyses show, in turn, that the source of nonzero coherency energy originates from martensitic phase eigenstrains incompatibility (see (3.16)).

New challenges and research efforts worth undertaking in the area of SMA materials are abundant: experimental and theoretical, fundamental and application, technological and numerical. Among them are development work on a new high temperature, low temperature, Ni-free (for medical applications), Fe-based (for civil structures applications) SMA materials and the development of new manufacturing and design methods of SMA components. Large research areas constitute surface treatment and joining methods of SMA materials, fatigue, and wear problems. Qualitatively new and more challenging research and development tasks are connected with the exploitation of SMA materials as active materials. For example, notice that a cyclist when riding toward the finishing line, best when in the leading position, does not use his legs only (simply pedaling), but uses literally his whole body. Similarly, now is the time for a transfer from the traditional concept of engineering structures composed of rather stiff elements that move in some kinematic chains to a new paradigm of engineering structures with elements of actively controlled flexibility in a working cycle. The control should allow changing the stiffness of the engineering element to the required level and at the required time. In fact, adaptive flexibility is an ordinary operating scheme for practically all living creatures, and is tried to be mimicked in modern structures such as drones.

In the remainder of this chapter, attention on new trends and tasks regarding SMA materials is focused on constitutive modeling development and related necessary experimental works. The constitutive modeling task of highest priority is the development of a unified thermodynamic model of pseudoelasticity and one-way memory effect. The earlier chapters of the book signaled that distinction of at least two martensitic phases seems to be a reasonable step toward a unified theory and, at the same time, a minor complication cost that must be necessarily borne. While distinguishing self-accommodating and oriented martensite "phases" is not possible with standard crystallographic lattice diffraction methods, as both martensitic phases have the same crystallographic lattices, it is possible to do that on macroscopic grounds with the DSC method, as has been already successfully demonstrated (see text to Figure 3.5). The extended Gibbs free energy function, valid for such three-phase SMA macroelement, has to be elaborated and proposed in explicit form. Actually, a step in this direction has already been taken by Raniecki and Ziółkowski (2015), where the methodology used for the development of composite materials models has been extended to include the macroelement of a multiphase mixture composed in general of n-phase linear thermoelastic materials with eigenstrains. The development work of key importance here is connected with the extension of rule for evolution of macroscopic ultimate phase eigenstrain tensor $\boldsymbol{\kappa}$ into the low temperatures range of one-way memory effect.

Macroscopic kinetics rules of thermoelastic martensitic transformation have high priority on the SMA materials research tasks list. While microscopic mechanisms of martensitic transformation in steel have been extensively researched, the thermoelastic variant of this transformation taking place in SMA materials received much less attention, and reliable, macroscopic, theoretical description of thermoelastic martensitic transformation is in its initial stage. Probably one of the reasons for that is that it can't be done in an uncoordinated manner between theory and experiment. Relatively involved, combined theoretical and experimental works are required in this case, as not only broad program of thermomechanical tests involving complex stress states should be executed, preferably with direct measurement of the volume fraction of martensitic phase simultaneously with the measurement of strain effects assisting the transformation, but also an attempt should be undertaken to find links between the formal mathematical form of equations of transformation kinetics and the underlying physical mechanisms.

A return to closer mutual interaction between theoretical and experimental research in a classical way is required. The validity of theoretical assumptions should be subject to experimental inquiry without excessive delay, and the results of experimental scrutiny should, in turn, deliver guidelines for improved theoretical conjectures. However, experimental efforts are time-consuming, costly, and undervalued by administrative bodies responsible for financing scientific research and, regretably, also by a sizable part of academia. Due to that, there is simply too small a pool of experimentalist researchers who could and are eager to cooperate to check ingenious and abundant theoretical concepts. Unfortunately, the contemporary system of financing science and its evaluation based not on real but tabular achievements with bias placed on booms and busts and not on systematic research efforts adds to the problem. An example of this inefficiency is the lack of even a single comprehensive experimental program, involving multiaxial loadings, on a single SMA material that will cover a set of experimental tests enabling reasonable physical validation of theoretical assumptions in the full range of temperatures embracing one-way memory and pseudoelasticity range of SMA material behavior. Such an experimental program enabling the validation of a 3D constitutive model necessarily must embrace a series of isothermal tests executed every 10 °C apart, with several multiaxial, proportional loading paths at each test temperature. Next, a subprogram of isostatic (stress-assisted) tests must be executed with temperature cycling in the range embracing characteristic temperatures of phase transitions. Several multiaxial stress states must be applied and several levels of values of equivalent assisting stress (for example, every 25 MPa in the range from 0 to 400 MPa). This very basic fragment of the experimental program should be supplemented by nonproportional isothermal tests, and combined thermomechanical tests with simultaneously changing of stress and temperature. Industrial company laboratories will not execute the SMA material experimental program described above because no customer will pay for it. Customers are interested in already good SMA product and not in the effort on how to make them. Governmental and/or university laboratories also have considerable difficulties executing such a program because it requires multiunit coordination and a long-term financing program. The present standard 3-years' time perspective for research projects is too short a time to realize this rather complex project. It seems that strategies of financing science should be somehow changed or enriched, as the present system seems to reach certain limits. For example, a trend can be noticed in SMA materials literature that every year very similar experimental results on NiTi wires are published, and every year a "new" 1D constitutive model of SMA materials is proposed.

Last, but not least, the effort is worth it to regularly implement improved thermodynamic models of SMA material behavior into FEM codes to create strong design tools for structures and devices made of SMA materials. These same tools can serve for research purposes enabling checking of the admissibility of some theoretical concepts numerically and, in this way, help in better directing or limiting the scope of actual experimental efforts.

Complexity of SMA materials behavior and broad application opportunities originating form their special properties indicate that persons who decide to engage in this topic can expect many intellectual and quite practical pleasures and rewards.

References

Aerofit web page, http://www.aerofit.com/sma.html.

Alam, M.S., Youssef, M.A., Nehdi, M., 2007. Utilizing shape memory alloys to enhance the performance and safety of civil infrastructure: a review. Can. J. Civ. Eng. 34, 1075–1086.

Annand, L., Gurtin, M., 2003. Thermal effects in the superelasticity of crystalline shape-memory materials. J. Mech. Phys. Solids 51, 1015–1058.

Asanovic, V., Delijic, K., 1999. The mechanical behavior and shape memory recovery of Cu-Zn-Al alloys. Metalurgija—J. Metall. 13, 59–64.

Ashby, M., 2011. Materials Selection in Mechanical Design. Butterworth-Heinemann is an imprint of Elsevier, Oxford, UK.

ASTM web page, http://www.astm.org.

Auricchio, F., Taylor, R., 1997. Shape memory alloys: modeling and numerical simulations of the finite-strain superelastic behavior. Comput. Methods Appl. Mech. Eng. 143, 175–194.

Ball, J., James, R., 1987. Fine phase mixtures as minimizers of energy. Arch. Ration. Mech. Anal. 100, 13–52.

Basinski, Z., Christian, J., 1954. Experiments on the martensitic transformation in single crystals of indium-thallium alloys. Acta Metall. 2, 148–166.

Bathe, K.J., 1996. Finite Element Procedures. Prentice Hall, Upper Saddle River, NJ.

Bernardini, D., 2001. On the macroscopic free energy functions for shape memory alloys. J. Mech. Phys. Solids 49, 813–837.

Bhattacharya, K., 2003. Microstructure of Martensite. Oxford University Press, Oxford, UK.

Bhattacharya, K., Conti, S., Zanzotto, G., Zimmer, J., 2004. Crystal symmetry and the reversibility of martensitic transformation. Nature 428, 55–59.

Biscarini, A., Mazzolai, G., Tuissi, A., 2008. Enhanced nitinol properties for biomedical applications. Recent Pat. Biomed. Eng. 1, 180–196.

Bouvet, C., Calloch, S., Lexcellent, C., 2002. Mechanical behavior of a Cu-Al-Be shape memory alloy under multiaxial proportional and non-proportional loadings. Trans. ASME 124, 112–124.

Bouvet, C., Calloch, S., Lexcellent, C., 2004. A phenomenological model for pseudoelasticity of shape memory alloys under multiaxial proportional and non-proportional loadings. Eur. J. Mech. A 23, 37–61.

Bowles, J., MacKenzie, J., 1954. The crystallography of martensitic transformations I and II. Acta Metall. 2, 129–137, 138–147.

Braz Fernandez, F.M. (Ed.), 2013. Shape Memory Alloys – Processing, Characterization and Applications. InTech, Rijeka, Croatia.

Brinson, L.C., 1993. One dimensional constitutive behavior of shape memory alloys: thermomechanical derivation with non-constant functions and redefined martensite internal variable. J. Intell. Mater. Syst. Struct. 4, 229–242.

Brinson, C., Schmidt, I., Lammering, R., 2004. Stress-induced transformation behavior of a polycrystalline NiTi shape memory alloy: micro and macromechanical investigations via in situ optical microscopy. J. Mech. Phys. Solids 52, 1549–1571.

Buehler, W., Gilfrich, J., Weiley, R., 1963. Effect of low-temperature phase changes on mechanical properties of alloys near composition TiNi. J. Appl. Phys. 34, 1475–1477.

Burkart, M., Read, T., 1953. Diffusionless phase change in the indium-thallium system. Trans. AIME 197, 1516–1524.

Chang, L., Read, T., 1951. Plastic deformation and diffusionless phase changes in metals: the gold-cadmium beta phase. Trans. AIME 191, 47–52.

Chemisky, Y., Duval, A., Patoor, E., Ben, Zineb T., 2011. Constitutive model for shape memory alloys including phase transformation, martensitic reorientation and twins accommodation. J. Mech. Mater. 44 (6), 953–980.

Chen, J., Li, Z., Zhao, Y.Y., 2009. A high-working-temperature CuAlMnZr shape memory alloy. J. Alloys Compd. 480, 481–484.

Chen, P.-Y., McKittrick, J., Meyers, M., 2012. Biological materials: functional adaptations and bioinspired designs. Prog. Mater. Sci. 57, 1492–1704.

Cherkaoui, M., Berveiller, M., Sabar, H., 1998. Micromechanical modeling of martensitic transformation induced plasticity (TRIP) in austenitic single crystals. Int. J. Plast. 14 (7), 597–626.

Cherkaoui, M., Berveiller, M., Lemoine, X., 2000. Couplings between plasticity and martensitic phase transformation: overall behavior of polycrystalline TRIP steels. Int. J. Plast. 16 (10–11), 1215–1241.

Christian, J., 1982. Deformation by moving interfaces. Metall. Trans. 13A, 509–538.

Christian, J.W., 2002. The Theory of Transformations in Metals and Alloys, (Part I, II), third ed. Pergamon Press, Oxford.

Computational Materials Science Network, http://www.phys.washington.edu/users/cmsn/links/ssi/VIC.doc.

Delaey, L., 1991. Diffusionless transformations. In: Cahn, R.W., Haasen, P., Kramer, E.J. (Eds.), Materials Science and Technology, Phase Transformations in Materials. VCH, Weinheim, pp. 341–404.

Delaey, L., Krishnan, R., Tas, H., Warlimont, H., 1974. Thermoelasticity, pseudoelasticity and the memory effects associated with martensitic transformations, part 1, 2 and 3. J. Mater. Sci. 9, 1521–1555.

Della Corte, C., 2010. Nickel-titanium: corrosion "proof" alloys for space bearing, components and mechanism applications. NASA/TM-2010-216334, Glenn Research Center, Cleveland, Ohio, pp. 1–8.

Della Corte, C., Pepper, S., Noebe, R., Hull, D., Glennon G., 2009. Intermetallic nickel-titanium alloys for oil-lubricated bearing applications. NASA/TM-2009-215646, Glenn Research Center, Cleveland, Ohio, pp. 1–26.

DesRoches, R., Delemont, M., 2003. Seismic retrofit of simply supported bridges using shape memory alloys. Eng. Struct. 24, 325–332.

Duerig, T.W., 2006. Some unsolved aspects of nitinol. Mater. Sci. Eng. A 438–440, 69–74.

Duerig, T., Pelton, A., Trepanier, C., 2011. Alloying and composition. chapter 9, In: Nitinol, Part I: Mechanisms and Behavior. ASM International, pp. 1–30. SMST e-Elastic newsletter January 2011, www.asminternational.org.

Dvorak, G.J., Benveniste, Y., 1992. On transformation strains and uniform fields in multiphase elastic media. Proc. R. Soc. Lond. A 437, 291–310.

Elahinia, M., Hashemi, M., Tabesh, M., Bhaduri, S., 2012. Manufacturing and processing of NiTi implants: a review. Prog. Mater. Sci. 57, 911–946.

ESOMAT, 2012. In: European Symposia on Martensitic Transformations and Shape Memory Alloys, September 9–16, 2012, Sankt Peterburg, Russia. http://www.esomat-2012.ru/.

Falk, F., 1983. One dimensional model of shape memory alloys. Arch. Mech. 35, 63–84.

Fan, G., Chen, W., Yang, S., Zhu, J., Ren, X., Otsuka, K., 2004. Origin of abnormal multi-stage martensitic transformation behavior in aged Ni-rich Ti–Ni shape memory alloys. Acta Mater. 52, 4351–4362.

Follansbee, P., 2014. On the definition of state variables for an internal state variable constitutive model describing metal deformation. Mater. Sci. Appl. 5, 603–609.

Frenzel, J., Pfetzing, J., Neuking, K., Eggeler, G., 2008. On the influence of thermomechanical treatments on the microstructure and phase transformation behavior of Ni–Ti–Fe shape memory alloys. Mater. Sci. Eng. A 481–482, 635–638.

Grabe, C., Bruhns, O., 2008. Tension/torsion tests of pseudoelastic, polycrystalline NiTi shape memory alloys under temperature control. Mater. Sci. Eng. A 481–482, 109–113.

Grabe, C., Bruhns, O., 2009. Path dependence and multiaxial behavior of a polycrystalline NiTi alloy within the pseudoelastic and pseudoplastic temperature regimes. Int. J. Plast. 25, 513–545.

Greninger, A., Mooradian, V., 1938. Strain transformation in metastable beta copper-zinc and beta copper-tin alloy. Trans. AIME 128, 337–368.

Helm, D., Haupt, P., 2003. Shape memory behavior: modeling within continuum thermomechanics. Int. J. Solids Struct. 40, 827–849.

Hill, R., 1963. Elastic properties of reinforced solids: some theoretical principles. J. Mech. Phys. Solids 11, 357–372.

Hill, R., 1978. Aspects of invariance in solid mechanics. Adv. Appl. Mech. 18, 1–75.

Hill, R., Rice, J., 1973. Elastic potentials and the structure of inelastic constitutive laws. SIAM J. Appl. Math. 25, 448–461.

Hornbogen, E., Wassermann, G., 1956. Zeit. Met. 47, 427–433, after information in Christian, J.W., 2002. The Theory of Transformations in Metals and Alloys, (Part I, II), third ed. Pergamon Press, Oxford.

Horstenmeyer, M., Bammann, D., 2010. History historical review of internal state variable theory for inelasticity. Int. J. Plast. 26, 1310–1334.

Hu, J., 2007. Shape Memory Polymers and Textiles. Woodhead Publishing Limited, Cambridge, England.

Huang, W., 2002. On the selection of shape memory alloys for actuators. Mater. Des. 23, 11–19.

Humbeeck, J., Stalmans, R., 1999. Characteristics of shape memory alloys. In: Otsuka, K., Wayman, C.M. (Eds.), Shape Memory Materials. Cambridge University Press, Cambridge, pp. 149–183.

Humbeeck, J., Stalmans, R., 2002. Shape memory alloys, types and functionalities. In: Schwartz, M. (Ed.), The Encyclopedia of Smart Materials. John Wiley & Sons, Inc, Chichester, pp. 941–963.

Huo, Y., Müller, I., 1993. Non-equilibrium thermodynamics of pseudoelasticity. Contin. Mech. Thermodyn. 5 (3), 163–204.

Ichinose, S., Funatsu, Y., Otsuka, K., 1985. Type II deformation twinning in γ'_1 martensite in a Cu-Al-Ni alloy. Acta Metall. 33, 1613–1620.

ICOMAT, 2014. In: International Conference on Martensitic Transformations 2014, July 6–11, Bilbao, Spain. http://www.icomat2014.com/.

Inamura, T., Nishiura, T., Kawano, H., Hosoda, H., Nishida, M., 2012. Self-accommodation of B19′ martensite in Ti–Ni shape memory alloys – part III. Analysis of habit plane variant clusters by the geometrically nonlinear theory. Philos. Mag. 92, 2247–2263.

Intrinsic Devices web page, http://www.intrinsicdevices.com/index.html.

Ishida, K., Kainuma, R., Ohnuma, I., Omori, T., Ando, K., 2012. Fe-based shape memory alloy and its production method. US Patent Nr 0199253.

Jani, J.M., Leary, M., Subic, A., Gibson, M.A., 2014. A review of shape memory alloy research, applications and opportunities. Mater. Des. 56, 1078–1113.

Janke, L., Czaderski, C., Motavalli, M., Ruth, J., 2005. Applications of shape memory alloys in civil engineering structures—overview, limits and new ideas. Mater. Struct. 38, 578–592.

Kaieda, Y., 2003. Fabrication of composition-controlled TiNi shape memory wire using combustion synthesis process and the influence of Ni content on phase transformation behavior. Sci. Technol. Adv. Mater. 4, 239–246.

Kohl, M., 2004. Shape Memory Microactuators. Springer, Berlin.

Kohl, M., Dittmann, D., Quandt, E., Winzek, B., 2000. Thin film shape memory microvalves with adjustable operation temperature. Sensors Actuators A 83, 214–219.

Koistinen, D., Marburger, R., 1959. A general equation prescribing the extent of the austenite martensite transformation in pure iron-carbon alloys and plain carbon steels. Acta Metall. 7, 59–69.

Kurdjumov, G., Khandros, L., 1949. Dokl. Akad. Nauk SSSR 66, 211–214, after information in Christian, J.W., 2002. The Theory of Transformations in Metals and Alloys, (Part I, II), third ed. Pergamon Press, Oxford.

Lagoudas, D.C. (Ed.), 2008. Shape Memory Alloys Modeling and Engineering Applications. Springer, New York.

Lagoudas, D., Entchev, P., Popov, P., Patoor, E., Brinson, C., Gao, X., 2006. Shape memory alloys, part II: modeling of polycrystals. Mech. Mater. 38, 430–462.

Leclercq, S., Lexcellent, C., 1996. A general macroscopic description of the thermomechanical behavior of shape memory alloys. J. Mech. Phys. Solids 44 (6), 953–980.

Lexcellent, C., 2013. Shape-memory Alloys Handbook. ISTE Ltd and John Wiley & Sons, London.

Liang, C., Rogers, C.A., 1990. One-dimensional thermomechanical constitutive relations for shape memory materials. J. Intell. Mater. Syst. Struct. 1, 207–234.

Lieberman, D., Wechsler, M., Read, T., 1955. Cubic to othorombic phase change—experimental and theoretical studies of AuCd. J. Appl. Phys. 26 (4), 473–484.

Liu, Y., Xie, Z., 2007. Detwinning in shape memory alloy. In: Reece, P. (Ed.), Progress in Smart Materials and Structures. Nova Science Publishers, New York, pp. 29–65 (Chapter 3).

Liu, Y., Xie, Z., Humbeeck, J., Delaey, L., Liu, Y., 2000. On the deformation of the twinned domain in NiT shape memory alloys. Philos. Mag. A 8, 1935–1953.

Mabe, J.H., Calkins, F., Butler, G., 2006. Boeing's variable geometry chevron, morphing aerostructure for jet noise reduction. In: Proceedings of the 47th AIAA/ASME/ASCE/AHS/ASC Structures, Structural Dynamics and Materials Conference, Newport, Rhode Island, 2006–2142, pp. 1–19.

Magee, C., 1968. The nucleation of martensite. In: Aaronson, H. (Ed.), Phase Transformations. ASM International, Menlo Park, pp. 115–156.

Maki, T., 1999. Ferrous shape memory alloys. In: Wayman, C., Otsuka, K. (Eds.), Shape Memory Materials. Cambridge University Press, Cambridge, pp. 117–132.

Maki, T., Kobayashi, K., Minato, M., Tamura, I., 1984. Thermoelastic martensite in an ausaged Fe-Ni-Ti-Co alloy. Scr. Metall. 18, 1105–1109.

Mandel, J., 1974. Thermodynamics and plasticity. In: Delgado, J., Nina, N. (Eds.), Foundations of Continuum Thermodynamics, Macmillan, New York, pp. 283–304.

Maruyama, T., Kubo, H., 2011. Ferrous (Fe-based) shape memory alloys (SMAs): properties, processing and applications. In: Yamauchi, K., Ohkata, I., Tsuchiya, K., Miyazaki, S. (Eds.), Shape Memory and Superelastic Alloys, Technologies and Applications. Woodhead Publishing Limited, Cambridge, pp. 141–162.

Melton, K., 1999. General applications of SMA and smart materials. In: Otsuka, K., Wayman, C. (Eds.), Shape Memory Materials. Cambridge University Press, Cambridge, pp. 220–239.

Melzer, A., Stoeckel, D., 2010. Function and performance of nitinol vascular implants. Open Med. Dev. J. 2, 32–41.

Memory Metalle web page, http://memry.com/.

Memry Corporation, 2004. http://www.specialchem4coatings.com/news-trends/displaynews.aspx?id=1691.

Mertmann, 2000. Processing and quality control of binary NiTi shape memory alloys. In: Yahia, L. (Ed.), Shape Memory Implants. Springer, New York.

Mertmann, 2008. Design and application of shape memory actuators. Eur. Phys. J.-Spec. Top. 158, 221–230.

Miyazaki, S., 1999. Medical and dental applications of shape memory alloys. In: Otsuka, K., Wayman, C. (Eds.), Shape Memory Materials. Cambridge University Press, Cambridge, pp. 267–282.

Miyazaki, S., Ishida, A., 1999. Martensitic transformation and shape memory behavior in sputter-deposited TiNi-base thin films. Mater. Sci. Eng. A 273–275, 106–133.

Miyazaki, S., Kim, H.Y., Hosoda, H., 2006. Development and characterization of Ni-free Ti-base shape memory and superelastic alloys. Mater. Sci. Eng. A 438–440, 18–24.

Miyazaki, S., Fu, Y., Huang, W. (Eds.), 2009. Thin Film Shape Memory, Fundamentals and Device Applications. Cambridge University Press, Cambridge.

Morgan, N.B., 2004. Medical shape memory alloy applications—the market and its products. Mater. Sci. Eng. A 378, 16–23.

Müller, I., Xu, H., 1991. On the pseudoelastic hysteresis. Acta Metall. Mater. 39, 263–271.

Musolff, A., Müller, I., Santopuoli, N., Seccia, L., 2004. Shape memory alloys for the conservation of cultural heritage: new applications also for real reversibility in the restoration operation. DFG Preprint 104, http://www.mathematic.uni-stuttgart.de/~mehrskalen/.

Nakahata, T., 2011. Industrial processing of titanium – nickel (TiNi) shape memory alloys (SMAs) to achieve key properties. In: Yamauchi, K., Ohkata, I., Tsuchiya, K., Miyazaki, S. (Eds.), Shape Memory and Superelastic Alloys, Technologies and Applications. Woodhead Publishing Limited, Oxford, Cambridge, Philadelphia, New Delhi, pp. 53–62.

Nemat, N.S., Hori, M., 1993. Micromechanics: Overall Properties of Heterogeneous Materials. North Holland, London.

Nespoli, A., Besseghini, S., 2011. A complete thermo-mechanical study of a NiTiCu shape memory alloy wire. J. Therm. Anal. Calorim. 103, 821–826.

Nishida, M., Kawano, H., Nishiura, T., Inamura, T., 2012. Self-accommodation of B19′ martensite in Ti–Ni shape memory alloys – part I. Morphological and crystallographic studies of the variant selection rule. Philos. Mag. 92, 2215–2233.

Ogden, R., 1984. Nonlinear Elastic Deformations. Ellis Horwood, Chichester.

Ogden, R.W., 2001. Pseudo-elasticity and stress softening. In: Fu, Y.B., Ogden, R.W. (Eds.), Nonlinear Elasticity: Theory and Applications. In: London Mathematical Society Lecture Note Series, vol. 283. Cambridge University Press, Cambridge, UK (Chapter 13).

Okhata, I., 2011. Medical applications of superelastic nickel–titanium (Ni-Ti) alloys. In: Yamauchi, K., Ohkata, I., Tsuchiya, K., Miyazaki, S. (Eds.), Shape Memory and Superelastic Alloys, Technologies and Applications. Woodhead Publishing Limited, Oxford, Cambridge, Philadelphia, New Delhi, pp. 176–196.

Okhata, I., Suzuki, Y., 1999. The design of shape memory alloy actuators and their applications. In: Otsuka, K., Wayman, C. (Eds.), Shape Memory Materials. Cambridge University Press, Cambridge, pp. 240–266.

Olander, A., 1932. An electrochemical investigation of solid cadmium-gold alloys. J. Am. Chem. Soc. 54, 3819–3833.

Olson, G.B., Owen, W.S. (Eds.), 1992. Martensite: A Tribute to Morris Cohen. ASM International, Ohio, USA.

Omori, T., Ando, K., Okano, M., Xu, X., Tanaka, Y., Ohnuma, I., Kainuma, R., Ishida, K., 2011. Superelastic effect in polycrystalline ferrous alloys. Science 333, 68–70.

Orgeas, L., Favier, D., 1996. Non symmetric tension compression behavior of NiTi alloy. J. Phys. IV C8, 605–610.

Otsuka, K., Ren, X., 2005. Physical metallurgy of TiNi based shape memory alloys. Prog. Mater. Sci. 50, 511–678.

Otsuka, K., Wayman, C. (Eds.), 1999. Shape Memory Materials. Cambridge University Press, Cambridge.

Otsuka, K., Saxena, A., Dengac, J., Ren, X., 2011. Mechanism of the shape memory effect in martensitic alloys: an assessment. Philos. Mag. 91 (36), 4514–4535.

Panico, M., Brinson, L., 2007. A three-dimensional phenomenological model for martensite reorientation in shape memory alloys. J. Mech. Phys. Solids 55, 2491–2511.

Pascal, Yu., Monasevich, L., 1981. Hysteresis features of the martensitic transformation of titanium nickelide. Phys. Met. Metallogr. 52, 95–99.

Patoor, E., Eberhardt, A., Berveiller, M., 1988. Thermomechanical behaviour of shape memory alloys. Arch. Mech. 40, 775–794.

Patoor, E., Lagoudas, D., Entchev, P., Brinson, C., Gao, X., 2006. Shape memory alloys, part I: general properties and modeling of single crystals. Mech. Mater. 38, 391–429.

Petrini, L., Migliavacca, F., 2011. Biomedical applications of shape memory alloys. J. Metall. 1–15. http://dx.doi.org/10.1155/2011/501483. Vol 2011. Art Id 501483.

Popov, S., Lagoudas, D., 2007. A 3-D constitutive model for shape memory alloys incorporating pseudoelasticity and detwinning of self-accommodated martensite. Int. J. Plast. 23, 1679–1720.

Porter, D., Easterling, K., 1992. Phase Transformations in Metals and Alloys. Chapman & Hall, London.

Rachinger, W., 1958. Brit. J. Appl. Phys. 9, 250–252, after information in Christian, J.W., 2002. The Theory of Transformations in Metals and Alloys, (Part I, II), third ed. Pergamon Press, Oxford.

Raghavan, V., 1992. Kinetics of martensitic transformation. In: Owen, W., Olson, G. (Eds.), Martensite A: Tribute to Morris Cohen. ASM International, Materials Park, OH.

Raniecki, B., Lexcellent, C., 1994. R_L-models of pseudoelasticity and their specifications for some shape memory alloys. Eur. J. Mech. A. Solids 13 (1), 21–50.

Raniecki, B., Lexcellent, C., 1998. Thermodynamics of isotropic pseudoelasticity in shape memory alloys. Eur. J. Mech. A. Solids 17, 185–205.

Raniecki, B., Mróz, Z., 2008. Yield or martensitic phase transformation conditions and dissipation functions for isotropic, pressure-insensitive alloys exhibiting SD effect. Acta Mech. 195, 81–102.

Raniecki, B., Nguyen, H., 1984. Isotropic elastic–plastic solids at finite strain and arbitrary pressure. Arch. Mech. 36 (5–6), 687–704.

Raniecki, B., Samanta, S., 1989. The thermodynamic model of rigid-plastic solid with kinematic hardening, plastic spin and orientation variables. Arch. Mech. 41 (5), 747–758.

Raniecki, B., Ziółkowski, A., 2015. On the macroscopic coherence energy function for SMA and instabilities of phase equilibriums during formation of self-accommodating martensite, (to be submitted in Mechanics in Materials).

Raniecki, B., Lexcellent, C., Tanaka, K., 1992. Thermodynamic models of pseudoelastic behaviour of shape memory alloys. Arch. Mech. 44 (3), 261–284.

Raniecki, B., Dietrich, L., Kowalewski, Z., Socha, G., Miyazaki, S., Tanaka, K., Ziółkowski, A., 1999a. Experimental methodology for NiTi shape memory alloy testing under complex stress state. Arch. Mech. 51, 726–744.

Raniecki, B., Miyazaki, S., Tanaka, K., Dietrich, L., Lexcellent, C., 1999b. Deformation behaviour of NiTi shape memory alloy undergoing R-phase reorientation in torsion-tension (compression) tests. Arch. Mech. 51, 745–784.

Raniecki, B., Tanaka, K., Ziółkowski, A., 2001. Testing and modeling of NiTi SMA at complex stress state—selected results of Polish-Japanese research cooperation. Mater. Sci. Res. Int. 2, 327–334.

Rice, J.R., 1971. Inelastic constitutive relations for solids: an internal variable theory and its application to metal plasticity. J. Mech. Phys. Solids 19, 433–455.

Rogueda, C., Lexcellent, C., Bocher, R., 1996. Experimental study of pseudoelastic behavior of CuZnAl polycrystalline shape memory alloy under tension-torsion proportional and non-proportional loading tests. Arch. Mech. 48 (6), 1025–1045.

Saburi, T., 1999. TiNi shape memory alloys. In: Otsuka, K., Wayman, C.M. (Eds.), Shape Memory Materials. Cambridge University Press, Cambridge, pp. 49–96.

Saburi, T., Wayman, C., Takata, K., Nenno, S., 1980. The shape memory mechanism in 18R martensitic alloys. Acta Metall. 28, 15–32.

Sadd, M.H., 2005. Elasticity, Theory Applications and Numerics. Elsevier Butterworth-Heinemann, Oxford.

Sakamoto, S., 2002. Distinction between thermal and stress-induced martensitic transformations and inhomogeneity in internal stress. Mater. Trans. 43 (9), 2249–2255.

Saleeb, A., Padula II, S., Kumar, A., 2010. A multi-axial, multimechanism based constitutive model for the comprehensive representation of the evolutionary response of SMAs under general thermomechanical loading conditions. Int. J. Plast. 27, 655–687. http://dx.doi.org/10.1016/j.ijplas.2010.08.012.

Sato, A., Chischima, E., Soma, K., Mori, T., 1982. Shape memory effect in $\gamma \leftrightarrow \varepsilon$ transformation in Fe-30Mn-1Si alloy single crystals. Acta Metall. 30, 1177–1183.

Scheil, E., 1932. Uber die Umwandlung des Austenits in Martensit in Eisen-Nickel-Legierungen unter Belastung. Z. Anorg. Allg. Chem. 207, 21–40.

Schwartz, M. (Ed.), 2002. Encyclopedia of Smart Materials. Wiley and Sons, New York.

Shaw, J., Churchill, C., Iadicola, M., 2008. Tips and tricks for characterizing shape memory alloy wire: part 1—differential scanning calorimetry and basic phenomena. Exp. Tech. 32, 55–62.

Sittner, P., Hara, Y., Tokuda, M., 1995. Experimental study on the thermoelastic martensitic transformation in shape memory alloy polycrystal induced by combined external forces. Metall. Mater. Trans. A 26a, 2923–2935.

Sittner, P., Heller, L., Pilch, J., Curfs, C., Alonso, T., Favier, D., 2014. Young's modulus of austenite and martensite phases in superelastic NiTi wires. J. Mater. Eng. Perform. 3, 1–12. http://dx.doi.org/10.1007/s11665-014-0976-x.

SMST, 2014. Shape Memory and Superelastic Technologies Conferences, Pacific Grove, California, USA, May 12–16. http://www.asminternational.org/content/Events/smst/index.jsp.

Socha, G., Raniecki, B., Dietrich, L., Ziółkowski, A., Kowalewski, Z.L., 2000. Deformation behaviour of Ti-51% at Ni undergoing martensitic phase transformation under complex stress state. Wyd. IPPT, PAN, Warszawa, pp. 1–151.

Song, C., 2010. History and current situation of shape memory alloys devices for minimally invasive surgery. Open Med. Dev. J. 2, 24–31.

Stoeckel, D., Pelton, A., Duerig, T., 2009. Self-expanding nitinol stents for the treatment of vascular disease. In: Yoneyama, T., Miyazaki, S. (Eds.), Shape Memory Alloys for Biomedical Applications. CRC, Woodhead Publishing, Boca Raton, pp. 237–256.

Sutou, Y., Omori, T., Kainuma, R., Ishida, K., 2008. Ductile Cu–Al–Mn based shape memory alloys: general properties and applications. Mater. Sci. Technol. 4 (8), 896–901.

Suzuki, A., 2011. Applications of shape memory alloys (SMAs) in hot water supplies. In: Yamauchi, K., Ohkata, I., Tsuchiya, K., Miyazaki, S. (Eds.), Shape Memory and Superelastic Alloys, Technologies and Applications. Woodhead Publishing Limited, Oxford, Cambridge, Philadelphia, New Delhi, pp. 100–109.

Tanaka, K., Kobayashi, S., Sato, Y., 1986. Thermomechanics of transformation pseudoelasticity and shape memory effect in alloys. Int. J. Plast. 2, 59–72.

Tanaka, K., Nishimura, F., Tobushi, H., 1995. Transformation start lines in TiNi and Fe-based shape memory alloys after incomplete transformations induced by mechanical and/or thermal loads. Mech. Mater. 19, 271–280.

Tanaka, K., Kitamura, K., Miyazaki, S., 1999. Shape memory alloy preparation for multiaxial tests and identification of fundamental alloy performance. Arch. Mech. 51 (6), 785–803.

Tanaka, Y., Himuro, Y., Kainuma, R., Sutou, Y., Omori, T., Ishida, K., 2010. Ferrous polycrystalline shape-memory alloy showing huge superelasticity. Science 327, 1488–1490.

Treppmann, D., Hornbogen, E., 1997. On the influence of thermomechanical treatments on shape memory alloys. J. Phys. IV 7 (C5), 211–220.

Tuissi, A., Bassani, P., Mangioni, A., Toia, L., Butera, F., 2004. Fabrication Process and Characterization NiTi Wires for Actuators. SAES Getters, Lainate, MI, pp. 1–8.

Uehara, T., Asai, C., Ohno, N., 2009. Molecular dynamics simulations of the shape memory behaviour using a multi-grain model. Model. Simul. Mater. Sci. Eng. 17, 035011.

Vacher, P., Lexcellent, C., 1991. Study of pseudoelastic behavior of polycrystalline shape memory alloys by resistivity measurements and acoustic emission. Proc. ICM VI, Kyoto 3, 231–236.

Warlimont, H., Delaey, L., 1974. Martensitic transformations in copper-silver- and gold-based alloys. Pergamon Press, New York.

Wayman, C., 1971. On memory effects related to martensitic transformations and observations in β-brass and Fe_3Pt. Scr. Metall. 5, 489–492.

Wayman, C., 1992. The phenomenological theory of martensite crystallography: Interrelations. Metall. Mater. Trans. 25A, 1787–1795.

Wayman, C.M., Duerig, T.W., 1990. An introduction to martensite and shape memory. In: Duerig, T., Melton, K., Stockel, D., Wayman, C. (Eds.), Engineering Aspects of Shape Memory Alloys. Butterworth-Heinemann, London, pp. 3–20.

Wechsler, M., Lieberman, D., Read, T., 1953. On the theory of the formation of martensite. Trans. AIME, J. Metals 197, 1503–1515.

Wu, M., 1990. Cu-based shape memory alloys. In: Duerig, T., Melton, K., Stockel, D., Wayman, C. (Eds.), Engineering Aspects of shape memory alloys. Butterworth-Heinemann, London.

Xiao, H., Bruhns, O.T., Meyers, A., 1997. Logarithmic strain, logarithmic spin and logarithmic rate. Acta Mech. 24, 89–105.

Yahia, L. (Ed.), 2000. Shape Memory Implants. Springer, Berlin.

Yamauchi, K., Ohkata, I., Tsuchiya, K., Miyazaki, S. (Eds.), 2011. Shape Memory and Superelastic Alloys Technologies and Applications. Woodhead Publishing Limited, Oxford, Cambridge, Philadelphia, New Delhi.

Yoneyama, T., Miyazaki, S. (Eds.), 2009. Shape Memory Alloys for Biomedical Applications. CRC, Woodhead, Boca Raton.

Zhang, X., Rogers, C., Liang, C., 1992. Modeling of the two-way shape memory effect. Philos. Mag. A 65, 1199–1215.

Ziółkowski, A., 2006. Pseudosprężystość stopów z pamięcią kształtu badania doświadczalne i opis teoretyczny. IPPT, PAN, Warsaw, Poland (in Polish).

Ziółkowski, A., 2007. Three-dimensional phenomenological thermodynamic model of pseudoelasticity of shape memory alloys at finite strains. Contin. Mech. Thermodyn. 19, 379–398.

Ziółkowski, A., 2012. On analysis of DSC curves for characterization of intrinsic properties of NiTi shape memory alloys. Acta Phys. Pol., A 122, 601–605.

Ziółkowski, A., Raniecki, B., 1999. On the free energy function for shape memory alloys treated as a two-phase continuum. Arch. Mech. 51 (6), 785–911.

Ziółkowski, A., Raniecki, B., Miyazaki, S., 2004. Stress induced martensitic transformation kinetics of polycrystalline NiTi shape memory alloy. Mater. Sci. Eng. A 378, 86–91.

Index

H

I

J

K

L

M

N

O

U

W

Kolibri 360 GmbH
Ferdinand-Jühlke-Straße 7
99095 Erfurt, Deutschland
produktsicherheit@kolibri360.de